Barnard Smith, Archibald McMurchy

Advanced Arithmetic for Canadian Schools

Barnard Smith, Archibald McMurchy

Advanced Arithmetic for Canadian Schools

ISBN/EAN: 9783337190828

Printed in Europe, USA, Canada, Australia, Japan

Cover: Foto ©Paul-Georg Meister /pixelio.de

More available books at **www.hansebooks.com**

Canadian Series of School Books.

ADVANCED ARITHMETIC

FOR

CANADIAN SCHOOLS.

BY

BARNARD SMITH, M.A.,

St. Peter's College, Cambridge.

AND

ARCHIBALD McMURCHY, M.A.,

University College, Toronto.

Authorized by the Council of Public Instruction of Ontario.

TORONTO:
ADAM MILLER & CO., 11 WELLINGTON STREET WEST.

The favourable reception given to the "Elementary Arithmetic" has induced the Authors to prepare this Treatise for the use of the more advanced pupils in the Schools of the Dominion, with a view to completing the course of instruction in the subject.

TORONTO, July, 1870.

CONTENTS.

SECTION I.

	PAGE
Definitions, Notation, and Numeration	9
Simple Addition	13
" Subtraction	17
Roman Notation	20
Simple Multiplication	21
" Division	31
Miscellaneous Questions, &c.	38

SECTION II.

Concrete Numbers (Tables). Money	42
Measure of Weight	44
" Length	46
" Surface	47
" Solidity	48
" Capacity	49
" Time	51
Reduction	54
Compound Addition	59
" Subtraction	62
" Multiplication	65
" Division	67
New Decimal Coinage	73
Miscellaneous Questions	77

SECTION III.

Greatest Common Measure	82
Least Common Multiple	85

SECTION IV.

Fractions	89
Vulgar Fractions	90
Addition of Vulgar Fractions	100

CONTENTS.

	PAGE
Subtraction of Vulgar Fractions	101
Multiplication of Vulgar Fractions	102
Division of Vulgar Fractions	104
Reduction of Vulgar Fractions	106
Miscellaneous Questions	113
Decimal Fractions	117
Addition of Decimal Fractions	120
Subtraction " "	122
Multiplication " "	123
Division " "	124
Circulating Decimals	129
Reduction of Decimals	134
Miscellaneous Questions, &c	141
Practice	145
Miscellaneous Questions, &c	157

SECTION V.

Ratio and Proportion	166
Rule of Three	169
Double Rule of Three	182
Simple Interest	191
Compound Interest	197
Present Worth and Discount	199
Stocks	203
Applications of the Term Per Cent	210
Division into Proportional Parts	221
Simple Fellowship	223
Compound Fellowship	224
Equation of Payments	225
Exchange	227
Value of Foreign Coins	232

SECTION VI.

Square Root	236
Cube Root	242
Scales of Notation	248
Application of Arithmetic to Geometry	255
Examination Questions	268

ARITHMETIC.

SECTION I.

ARTICLE 1. By a UNIT is meant a single object or thing, considered as one and undivided.

2. NUMBER is the name by which we signify how many objects or things are considered, whether *one or more*. When, for instance, we speak of one horse, two apples, three yards, or four hours, the number of the things referred to will be one, two, three, or four, according to the case; and so one, two, three, four, and the rest, are called numbers.

3. NUMBERS are considered either as ABSTRACT or CONCRETE.

Abstract numbers are those which have no reference to any particular kind of unit; thus, five, as an abstract number, signifies five units only, without any regard to particular objects.

Concrete numbers are those which have reference to some particular kind of unit; thus, when we speak of five hours, six yards, seven horses, the numbers five, six, seven, are said to be concrete numbers, having reference to the particular units one hour, one yard, one horse, respectively.

4. ARITHMETIC is the science of Numbers.

5. All numbers in common Arithmetic are expressed by means of the figure 0, commonly called zero or a cypher, which has no value in itself, and nine significant figures, 1, 2, 3, 4, 5, 6, 7, 8, 9, which denote respectively the numbers one, two, three, four, five, six, seven, eight, nine. These ten figures are sometimes called DIGITS.

The number one, which is represented by the figure 1, is called UNITY.

6. When any of these figures stands by itself, it expresses its simple or intrinsic value; thus, 9 expresses nine abstract units, or nine particular things: but when it is followed by another figure, it then expresses ten times its simple value; thus, 94 expresses ten times nine units, together with four units more: when it is followed by two figures, it then expresses one hundred times its simple value; thus, 943 expresses one hundred times nine units, together with ten times four units, and also three units more; and so on by a tenfold increase for each additional figure that follows it.

The value, which thus belongs to a figure in consequence of its position or place, is called its LOCAL VALUE.

Therefore all numbers have a simple or intrinsic value, and also a local value.

7. It appears then, that in common Arithmetic we proceed towards the left from units to tens of units; from tens of units to tens of tens of units, or hundreds of units; from hundreds of units to tens of hundreds of units, or thousands of units; from thousands of units to tens of thousands of units; from tens of thousands of units to tens of tens of thousands of units, that is, to hundreds of thousands of units; thence to tens of hundreds of thousands of units, or millions of units; thence to tens of millions of units, hundreds of millions of units, &c., till we come to billions, trillions, quadrillions, &c.

Thus, 10 represents one ten of units, together with no units; or, as it is briefly read, ten. 11 represents one ten of units, together with one unit; or as it is briefly read eleven. Similarly 12, 13, 14, 15, 16, 17, 18, 19, respectively represent one ten of units together with two, three, four, five, six, seven, eight, nine units; they are respectively read twelve, thirteen, fourteen, fifteen, sixteen, seventeen, eighteen, nineteen.

The next ten numbers are expressed by 20, 21, 22, 23, 24, 25, 26, 27, 28, 29, which respectively represent two tens of units together with no, one, two, three, four, five, six, seven, eight, nine units; they are briefly read twenty, twenty-one, twenty-two, twenty-three, twenty-four, twenty-five, twenty-six, twenty-seven, twenty-eight, twenty-nine.

The next ten numbers are expressed by 30, 31, 32, 33, 34, 35, 36, 37, 38, 39, which are respectively read thirty, thirty-one, thirty-two, thirty-three, thirty-four, thirty-five, thirty-six, thirty-seven, thirty-eight, thirty-nine; wo.thus arrive at 40 (forty), 50 (fifty), 60 (sixty), 70 (seventy), 80 (eighty), 90 (ninety).

99 is the largest number which can be expressed by two figures, since it represents nine tens of units together with nine units; the next number to this is 100, which represents ten tens of units, or one hundred of units, together with no tens of units, together with no units; or, as it is briefly read, one hundred.

By pursuing the same system in higher numbers, the figure occupying the fourth place from the right hand will represent so many tens of hundreds of units, or thousands of units; the figure in the fifth place will represent so many tens of thousands of units; and so on.

205 represents two hundreds of units, together with no tens of units, together with five units; or, as it is briefly read, two hundred and five.

5473 represents five thousands of units, together with four hundreds of units, together with seven tens of units, together with three units; or, as it is briefly read, five thousand, four hundred and seventy-three.

7040730 represents seven millions of units, together with no hundreds of thousands of units, together with four tens of thousands of units, together with no thousands of units, together with seven hundreds of units, together with three tens of units, together with no units; or, as it is briefly read, seven millions, forty thousand, seven hundred and thirty.

8. NOTATION is the art of expressing any number by figures or letters which is already given in words. There are two methods of Notation: 1st ARABIC; 2nd ROMAN.

9. NUMERATION is the converse of Notation, being the art of expressing any number in words which is already given in figures.

10. The method above explained of denoting numbers by means of the symbols 0, 1, 2, 3, 4, 5, 6, 7, 8, 9, and combinations of them, was brought into Europe by the Arabs, and it therefore is often called the ARABIC NOTATION. It was derived by the Arabs from the Hindoos. This method of notation is now in common use, not only in the British Empire, but throughout Europe.

Ex. I.

Exercises in Notation and Numeration.

Express the following numbers in figures:

(1) Sixty-three ; eighty-one ; ninety-nine ; forty ; thirteen.

(2) Two hundred ; three hundred and three ; seven hundred and sixty-four: eight hundred and eighty-eight.

(3) Four thousand; one thousand, four hundred and seventy-one; six thousand, nine hundred and thirty ; nine thousand and nine.

(4) Twenty-seven thousand, five hundred and four : thirty-three thousand; nine thousand and sixteen.

(5) One hundred thousand ; six hundred and seventy-six thousand and fifty ; two hundred and two thousand, five hundred and ninety-three.

(6) Seven millions, three thousand ; eleven millions, one hundred and eight thousand, one hundred and six ; fifty-four millions, fifty-four thousand and eighty-eight ; six hundred and thirteen millions, twenty thousand, three hundred and three.

(7) Two billions; nine billions ; three hundred thousand and twenty-one ; ninety-four billions, ninety millions,, ninety-four thousand, nine hundred and four.

Write down in words at full length the following numbers:

(1) 43 ; 60 ; 88 ; 97 ; 59 ; 12 ; 21 ; 19.
(2) 256 ; 401 ; 500 ; 999 ; 365 ; 578 ; 837.
(3) 2000 ; 1724 ; 3003 ; 7584 ; 1075 ; 4541.
(4) 37003 ; 47049 ; 63090 ; 80008 ; 341323.
(5) 6850406 ; 8080808 ; 7849630 ; 418254.
(6) 10000001 ; 20220022 ; 92568987 ; 30180070.
(7) 2560530200 ; 800309560 ; 9738413208.
(8) 7070000423 ; 987654321 ; 5707068080.
(9) 100198700010090 ; 487268706341032264.

ADDITION.

11. ADDITION is the method of finding a number, which is equal to two or **more** numbers taken together.

ADDITION. 13

The numbers to be added together are called ADDENDS.

The number found by adding two or more numbers together is called the SUM or AMOUNT of the several numbers so added.

12. There are two kinds of Addition, SIMPLE and COMPOUND.

It is Simple Addition, when the numbers to be taken together are all abstract numbers; or when they are all concrete numbers of the same denomination, as *all* pence, *all* days, *all* pints.

It is Compound Addition, when the numbers to be taken together are concrete numbers of the same kind, but of different denominations of that kind; as pounds, shillings, and pence; or years, months, and days; or gallons, quarts, and pints.

13. The sign $+$, PLUS, placed between two or more numbers, signifies that the numbers are to be added together: thus $2+5+7$ signifies that 2, 5, and 7 are to be added together, and denotes their sum.

The sign $=$, EQUAL, placed between two numbers, signifies that the numbers are equal to one another.

The ‾, VINCULUM, placed over numbers, and the sign ( ) or { }, called a BRACKET, enclosing numbers within it, are used to denote that all numbers under the vinculum, or within the bracket, are equally affected by all numbers not under the vinculum or within the bracket: thus $\overline{2+3}$ or $(2+3)$ or $\{2+3\}$, each signify, that whatsoever is outside the vinculum or bracket which affects 2 in any way, must also affect 3 in the same way, and conversely.

The sign $\therefore$ signifies 'therefore.'

SIMPLE ADDITION.

14. RULE. **Write** down the given **numbers under each other,** so that **units may come** under units, **tens under** tens, **hundreds under hundreds, and so on**; then draw a straight line under the lowest line.

Find the sum of the column of units; if it be under ten, write it down under the column of units, below the line just drawn; if it exceed ten, then write down the last figure of the sum under the column of units, **and carry to the next column** the remaining figure or figures; treat each succeeding column in the same way, and write down the full sum of the extreme left-hand column. The entire sum so marked **down will be the sum or amount of the separate numbers.**

ARITHMETIC.

Ex. Add together 5469, 743, and 27.
Proceeding by the Rule given above, we obtain

$$\begin{array}{r} 5469 \\ 743 \\ 27 \\ \hline 6239 \end{array}$$

The reason for the Rule will appear from the following considerations.

When we take the sum of 7 units and 3 units and 9 units, we get 19 units; we therefore place the 9 units under the column of units, and carry on the 1 ten units to the next column, viz., the column of tens.

Now the sum of 1 ten, 2 tens, 4 tens, and 6 tens, is 13 tens; we therefore place the 3 tens under the column of tens, and carry on the 1 hundred units to the next column, viz., the column of hundreds.

Again, the sum of one hundred, 7 hundreds, and four hundreds, is 12 hundreds; we therefore place the 2 hundreds under the column of hundreds, and carry on the 1 thousand units to the next column, viz., the column of thousands.

Again, the sum of 1 thousand and 5 thousand, is six thousands; we therefore place the 6 under the column of thousands, and the entire sum is 6239.

The above example might have been worked thus, putting down at full length the local value of all the figures.

Thus $5469 = 5000 + 400 + 60 + 9$
$+743 = 700 + 40 + 3$
$+27 = 20 + 7$

Now adding the columns, we get the sum
$= 5000 + 1100 + 120 + 19$
$= 5000 + \overline{1000 + 100} + \overline{100 + 20} + \overline{10 + 9}$,
(since $1100 = 1000 + 100$, $120 = 100 + 20$, and $19 = 10 + 9$)
$= 6000 + 200 + 30 + 9$.
(collecting the thousands together, the hundreds together, and so on)
$= 6239$.

NOTE. The truth of all results in Addition may be proved by adding the columns first upwards as in the above example, and then adding them downwards; if the results be the same, the operation in each case will in all probability have been performed correctly.

SIMPLE ADDITION.

Ex. II.

Examples in Simple Addition.

(1) 256783	(2) 627432	(3) 892764	(4) 1807353	(5) 117064
21003	543201	93687	298743	92973
5734	678641	9482	5987	827569
40036	548200	100	760003	351
21	868759	152346	247	777777
100001	345678	11	50705	65656

(6) 792085579	(7) 92837465321	(8) 186457902
909672916	39765689079	920749365
495987650	42900	467458729
868253926	59678009678	814382257
150470279	87064571493	376914936
918065933	45902768537	743790051
672409308	94555672990	210706090
542984945	63749568	631721987
709365197	73594693845	598765432
994883772	56789987654	119647896
185938976	49596878397	952081638
847090439	84309724606	749327965
706050896	29684371989	574981672
685675	11234562	324680357
842378	23678521	891076813

(9) 17649+35+3910400708+8566000+9687+7150030090074+498
   +80700769+14083+709+9083809400094.

(10) 459800071+605040+907+30056049+50005+00087078
   +9200299200+1002003.

(11) 4921863+294698+9035075+59507+8407268+4390+5960507
   +3075089+862+7485996+6739448+19+5479074.

ARITHMETIC.

(12) Add together 7384, 326, 6780, and 57; also 6740, 9745, 5769, 8031, 6543, 2002, and 9999; also 89, 4500, 423, 2024, 5408, 60546, and 9401.

(13) Add together 83746, 2478, 692577, 456, and 7; also 935473, 262, 13897, 598453, 25, 3734, 724008, and 649763.

(14) Find the sum of 4788685, 237869513, 148794343978, 865, 4647, and 250; also of 68539582, 786020145, 370489000, 7055591234, 276, 9123456789, and 5000; also of 888929944, 73600, 27978462, 333, 587539600, 4827532, 486684836, 80032148379, 12345, 1112858673, and 53800000835.

(15) Add together one thousand, four hundred and eighty-three; seven hundred and ninety-six; thirty-nine; forty thousand, seven hundred and forty-four; five thousand, eight hundred and sixty; fifty thousand and seven.

(16) Add together the following numbers: fifteen thousand, seven hundred and ninety-six; four hundred and nine; two hundred and thirty-four thousand and fifty; four millions, three thousand and seventy-six; forty thousand and thirty-six; ten thousand, nine hundred and one.

(17) Add together the following figures: twenty-two millions, six hundred thousand, five hundred and three; five hundred and sixty-three millions, seventy-six thousand and thirty-four; one hundred and eleven millions, six hundred and fifty thousand and fifty; three hundred and twenty-six millions, seven thousand, nine hundred and ninety-one; one thousand seven hundred and ten millions, one thousand seven hundred and ten; one quadrillion, three hundred thousand and five.

(18) The cost of building the Welland Canal was $7000000; the Rideau, $4380000; the St Lawrence, $8550000; the Ottawa, $1500000; the Chambly and St. Ours' lock, $550000; Burlington and Desjardins, $560000. Find the total cost of the above Canadian canals.

(19) The tolls collected for the year ending 30th June, 1866, from the above named canals were as follows: Welland Canal, $174603; St. Lawrence, $63052; Chambly and St. Ours' lock, $28324; Burlington Bay Canal, $14917; St. Ann's lock, $6024; Ottawa and Rideau Canals, $13369; to which add the amount on tolls on vessels, $39797. Find the whole sum collected.

SUBTRACTION. 17

(20) **The** exports of Canada for the years 1855, 1856, 1857, were respectively as follows: Products of Agriculture, $14625580, $17536332, $10990064; Products of the Forest, $7947920, $10019880, $11730064; other products, $2249944, $2251916, $2729968. Find the total amount of the exports of the country for those three years.

(21) **The** value of Exports of Canada for the years 1865, 1866, 1867, were respectively as follows, viz.: Products of Agriculture, $14283207, $16651074, $16765981; Products of the Forest, $10451509, $13846986, $13948648; other products, $13176012, $17361201, $9761473. Find the total value of the exports for those years.

SUBTRACTION.

15. Subtraction is the method of finding what number remains when a smaller number is taken from a greater number.

The number to be substracted is called the SUBTRAHEND; the number subtracted from, the MINUEND.

The number found by subtracting the smaller of two numbers from the greater is called the Remainder.

16. There are two kinds of Subtraction, SIMPLE and COMPOUND, which differ from each other in precisely the same way, in which Simple and Compound Addition differ from each other.

17. The sign —, minus, placed between two numbers, signifies that the second number is to be subtracted from the first number.

SIMPLE SUBTRACTION.

18. RULE. Place the less number under the greater number, so that units may come under units, tens under tens, hundreds under hundreds, and so on; then draw a straight line under the lower line.

Take, if possible, the number of units in each figure of the lower line from the number of units in each figure of the upper line which stands immediately over it, and put the remainder below the line just drawn, units under units, tens under tens, and so on; but if the units in any figure in the lower line exceed the number of units in the figure above it, add ten to the upper figure, and then take the number of units in the lower figure from the number in the upper figure thus increased; put the remainder down as before, and then carry one to the next figure of the lower line. The entire difference or remainder, so marked down, will be the difference or remainder of the given numbers.

Ex. Subtract 4938 from 5123.

Proceeding by the Rule given above, we obtain

$$\begin{array}{r} 5123 \\ 4938 \\ \hline 185 \end{array}$$

so that the remainder is one hundred and eighty-five (185).

The reason for the Rule will appear from the following considerations.

We cannot take 8 units from 3 units; we therefore add 10 units to the 3 units, which are thus increased to 13 units; and taking 8 units from 13 units we have 5 units left; we therefore place 5 under the column of units; but having added 1 ten units to the upper number, we must add the same number of units (1 ten units) to the lower number, so that the difference between the two numbers may not be altered; and adding 1 ten units to the 3 ten units in the lower number, we obtain 4 tens or forty instead of 3 tens or 30.

Again, we cannot take 4 tens from 2 tens; we therefore add 10 tens or 1 hundred to the 2 tens, which thus becomes 12 tens or 120; and then taking 4 tens or 40 from 12 tens or 120, we have 8 tens or 80 remaining; we therefore place 8 under the columns of tens; but having added 1 hundred to the upper number, we must add 1 hundred to the lower number for the reason given above; and adding 1 hundred to the 9 hundreds in the lower number, we obtain 10 hundreds or 1000 instead of 900.

Again, we cannot take 10 hundreds from 1 hundred, and we therefore add 10 hundreds or 1 thousand to the 1 hundred, which thus becomes 11 hundreds or 1100; and taking 10 hundreds or 1000 from

SUBTRACTION.

11 hundreds or 1100, we have 1 hundred or 100 left; we therefore place 1 under the column of hundreds; but having added 10 hundreds or 1 thousand to the upper number, we must add 1 thousand to the lower number for the reason given above; and adding 1 thousand to the 4 thousands in the lower number, we obtain 5 thousands or 5000;

5000 taken from 5000 leaves 0;

therefore the whole difference or remainder is 185.

The above Example might have been worked thus, putting down at full length the local values of the figures:

$$5123 = 5000 + 100 + 20 + 3$$
$$= 4000 + 1000 + 100 + 20 + 3$$
$$= 4000 + 1000 + 100 + 10 + 10 + 3$$
$$= 4000 + 1000 + 110 + 13$$

(collecting the first 10 with the 100, and the second 10 with the 3),

$$4938 = 4000 + 900 + 30 + 8.$$

Therefore subtracting the columns, thousands from thousands, &c., we get the remainder or difference

$$= 100 + 80 + 5$$
$$= 185$$

NOTE. The truth of all results in Subtraction may be proved by adding the less number to the difference or remainder; if this sum equals the larger number, the result obtained by subtraction may be presumed to be correct.

Ex. III.

Examples in Simple Subtraction.

(1) 1000000
    100101

(2) 400357261
     99988877

(3) 89487182
   15790293

(4) Find the difference between 6543756 and 412848; 7863927 and 826957; 303233334 and 192001222.

ARITHMETIC.

(5) How much greater is 164326289 than 48476798?

.................................10000001000 than 7077070077?

.................................7559030640021 than 6990040005679?

(6) Take two thousand and nine, from ten thousand and ninety-six; three thousand and eight, from seven thousand, nine hundred and forty-four.

(7) Required the difference between four and four millions; also between one hundred millions and three hundred thousand.

(8) Subtract five hundred and eighty-four thousand and seventy-six, from fifteen millions, one hundred thousand and three.

(9) The revenue of Newfoundland for the year 1866 was $716287.97; the Expenditure, $662763.15. How much did the Revenue exceed the Expenditure?

(10) For the year 1866 the Imports into New Brunswick were $10000794; the Exports, $8186185. How much more was imported than exported.

(11) The imports into Nova Scotia for the years 1865, 1866 were respectively; $14381662, $14381095. How much less was imported during the latter than the former year.

---

19. The following method of expressing numbers was used by the Romans, and it is still in occasional, though not in common use among ourselves. They represented the number one by the character I; five by V; ten by X; fifty by L; one hundred by C; five hundred by D or I⊃; one thousand by M or CI⊃.

All other numbers were formed by a combination of the above characters, subject to the following Rules:

1st. When a character was *followed* by one of *equal or less* value, the whole expression denoted the *sum* of the values of the single characters; for instance, II stood for 2; III for 3; VI for 6; VIII for eight; LV for 55; LXXVII for 77; CCXI for 211.

2d. When a character was *preceded* by one of *less* value, the whole expression denoted the *difference* of the values of the single characters; for instance IV stood for 5—1, or 4; IX for 10—1, or 9; XIX for 10+10—1, or 19; XL for 50—10, or 40; XC for 100—10, or 90.

3d. Every ⊃ annexed to I⊃ increased the value of the latter tenfold; for instance, I⊃⊃ stood for 5000; I⊃⊃⊃ for 50000; and so forth. And every C prefixed and ⊃ annexed to CI⊃ increased the value of the latter tenfold; for instance, CCI⊃⊃ stood for 10000; CCCI⊃⊃⊃ for 100000; and so forth.

4th. A line drawn over a character or characters increased the value of the latter a thousandfold; for instance, $\overline{\text{V}}$ stood for 5000; $\overline{\text{C}}$ for 100000; $\overline{\text{IX}}$ for 9000; and so forth.

It follows then that either XXXXVI or XLVI will represent 46; and that either M.DCCC.LIV, or CI⊃.I⊃CCCLIV, or $\overline{\text{I}}$.DCCCLIIII will represent 1854.

Ex. IV.

(1) Express in Roman characters, thirty, forty-eight, fifty-nine; 222; 600; 1843.

(2) Express in words, and also in Arabic figures, the values of XXIII; LXIX; CCXVIII; $\overline{\text{VI}}$; CLDCIII; $\overline{\text{MM}}$ C.

MULTIPLICATION.

20. MULTIPLICATION is a short method of finding the sum of any given number repeated as often as there are units in another given number; thus, when 3 is multiplied by 4, the number produced by the multiplication is the sum of 3 repeated four times, which sum is equal to $3+3+3+3$ or 12.

The number to be repeated or added to itself, is called the MULTIPLICAND.

The number which shows how often the multiplicand is to be repeated, is called the MULTIPLIER.

The number found by multiplication is called the PRODUCT.

The multiplicand and multiplier are sometimes called "FACTORS," because they are factors or makers of the product.

21. Multiplication is of two kinds, SIMPLE and COMPOUND. It is termed Simple Multiplication, when the multiplicand is either an abstract number, or a concrete number of one denomination.

It is termed Compound Multiplication, when the multiplicand contains numbers of more than one denomination, but all of the same kind

22. The sign ×, placed between two numbers, signifies that the numbers are to be multiplied together.

23. The following table ought to be learned correctly:

1	2	3	4	5	6	7	8	9	10	11	12
2	4	6	8	10	12	14	16	18	20	22	24
3	6	9	12	15	18	21	24	27	30	33	36
4	8	12	16	20	24	28	32	36	40	44	48
5	10	15	20	25	30	35	40	45	50	55	60
6	12	18	24	30	36	42	48	54	60	66	72
7	14	21	28	35	42	49	56	63	70	77	84
8	16	24	32	40	48	56	64	72	80	88	96
9	18	27	36	45	54	63	72	81	90	99	108
10	20	30	40	50	60	70	80	90	100	110	120
11	22	33	44	55	66	77	88	99	110	121	132
12	24	36	48	60	72	84	96	108	120	132	144

In the above Table, the second line from the top shows the product of each of the numbers, 1, 2, 3, 4, &c., 11, 12, in the first line, when multiplied by 2; the several products being placed under the respective numbers of the line above, from the multiplication of which they arise: the third line shows the several products, when the figures in the first line are respectively multiplied by 3; and so on.

NOTE.—One of the factors, namely, the multiplier, must necessarily be an "abstract number"; since it would be absurd to speak of 6 shillings multiplied by 4 shillings. We can multiply 6 shillings by 4, *i. e.*, we can find how many shillings there are in four times six shillings; but there is no meaning in six shillings multiplied by 4 shillings.

SIMPLE MULTIPLICATION.

24. Rule. Place the multiplier under the multiplicand, units under units, tens under tens, and so on. Multiply each figure of the multiplicand, beginning with the units, by the figure in the units' place of the multiplier (by means of the table given for Multiplication); set down and carry as in Addition. Then multiply each figure of the multiplicand, beginning with the units, by the figure in the tens' place of the multiplier, placing the first figure so obtained under the tens of the line above, the next figure under the hundreds, and so on. Proceed in the same way with each succeeding figure of the multiplier. Then add up all the results thus obtained, and the sum will be the required product.

25. A number which cannot be separated into factors, which are respectively greater than unity, is called a PRIME number. Thus 3, 5, 7, 11, 13 are prime numbers.

26. A number which can be separated into factors respectively greater than unity, or which, in other words, is produced by multiplying together two or more numbers respectively greater than unity, is called a COMPOSITE number. Thus 4 which $=2\times 2$, 6 which $=2\times 3$, 8 which $=2\times 2\times 2$, are composite numbers; because they are composed or consist of the product of two or more numbers, each of which is greater than unity.

27. If more than two factors have to be multiplied together, as $2\times 4\times 9$, it is termed CONTINUED MULTIPLICATION, and since $2\times 4=8$, and $8\times 9=72$, and $\therefore 2\times 4\times 9=72$, we shall of course obtain the same result, whether we multiply any number by 72, or by its factors 2, 4, and 9, by continued multiplication; and so of any other number.

28. Numbers which have no common factor greater than unity, are said to be PRIME to one another. Thus the numbers 3, 5, 8, 11, are prime to each other.

NOTE. If the multiplier does not exceed 12, the multiplication can be effected easily in one line, by means of the Table given above.

Ex. Multiply 7654 by 397.

Proceeding by the Rule given above, we obtain.

$$
\begin{array}{r}
7654 \\
397 \\
\hline
53578 \\
68886 \\
22962 \\
\hline
3038638
\end{array}
$$

The reason for the Rule will appear from the following considerations.

When 7654 is to be multiplied by 7, we first take 4 seven times, which by the Table gives 28, *i. e.*, 8 units and 2 tens; we therefore place down 8 in the units' place and carry on the 2 tens; again, 5 tens taken 7 times gives 35 tens, to which add 2 tens, and we obtain 37 tens, or 7 tens and 3 hundreds; we put down 7 in the tens' place, and carry on 3 hundreds: again, 6 hundreds taken 7 times give 42 hundreds, to which add 3 hundreds, and we obtain 45 hundreds, or 4 thousands and 5 hundreds; we put down 5 in the hundreds' place, and carry on the 4 thousands: again, 7 thousands taken 7 times give 49 thousands, to which we add the 4 thousands, thus obtaining 53 thousands, which we write down.

Next, when we multiply 7654 by the 9, we in fact multiply it by 90; and 4 units taken 90 times give 360 units, or 3 hundreds, 6 tens, and 0 units; therefore, omitting the cypher, we place the 6 under the tens' place, and carry on the 3 to the next figure, and proceed with the operation as in the line above.

When we multiply 7654 by the 3, we in fact multiply it by 300; and 4 multiplied by 300 gives 1200, or 1 thousand, 2 hundreds, 0 tens, and 0 units; therefore, omitting the cyphers, we place the first figure 2 under the hundreds' place, and proceed as before. Then adding up the three lines of figures which we have just obtained, we obtain the product of 7654 by 397.

SIMPLE MULTIPLICATION. 25

The above Example might have been worked thus, putting down t full length the local values of the figures:

$$7654 = 7\times1000 + 6\times100 + 5\times10 + 4$$
$$397 = 3\times100 + 9\times10 + 7$$

$$ 49\times1000 + 42\times100 + 35\times10 + 28$$
$$63\times10000 + 54\times1000 + 45\times100 + 36\times10$$
$$21\times100000 + 18\times10000 + 15\times1000 + 12\times100$$

$$21\times100000 + 81\times10000 + 118\times1000 + 99\times100 + 71\times10 + 28$$

which =

$$20\times10000 + 1\times100000$$
$$+ 8\times100000 + 1\times10000$$
$$+ 1\times100000 + 1\times10000 + 8\times1000$$
$$+ 9\times1000 + 9\times100$$
$$+ 7\times100 + 1\times10$$
$$+ 2\times10 + 8$$

$$ 2000000 + 10\times100000 + 2\times10000 + 17\times1000 + 16\times100 + 3\times10 + 8$$
$$= 2000000 + 1000000 + 2\times10000 + \overline{10\times1000 + 7\times1000} + 10\times100 + 6\times100 + 3\times10 + 8$$
$$= 3000000 + 2\times10000 + 1\times10000 + 7\times1000 + 1\times1000 + 6\times100 + 3\times10 + 8$$
$$= 3000000 + 3\times10000 + 8\times1000 + 600 \times 30 + 8$$
$$= 3000000 + 30000 + 8000 + 600 + 30 + 8$$
$$= 3038638$$

29. If the multiplier or multiplicand, or both, end with cyphers, we may omit them in the working; taking care to affix to the product as many cyphers as we have omitted from the end of the multiplier or multiplicand, or both. Thus, if 263 be multiplied by 6200, and 570 be multiplied by 3200, we have

```
    263              570
   6200             3200
   ----             ----
    526              114
   1578              171
   ----             ----
 1630600          1824000
```

The reason is clear; for in the first case, when we multiply by
2

the 2, in fact we multiply by 200; and 3 multiplied by 200 gives 600; in the second case, the 7 multiplied by the 2 is the same as 70 multiplied by 200; and 70 multiplied by 200 gives 14000.

30. If the MULTIPLIER contain any cypher in any other place, then, in multiplying by the different figures of the multiplier we may pass over the cypher; taking care, however, when we multiply by the next figure, to place the first figure arising from that multiplication under the third figure of the line above instead of the second figure. The reason of this is clear: for, if we were multiplying by 206, when we multiply by the 6 we take the multiplicand 6 times, when we multiply by the 2, we really take the multiplicand, not 20 times, but 200 times.

31. When two numbers are to be multiplied together, it is a matter of indifference, so far as the product is concerned, which of them be taken as the multiplicand or multiplier; in other words, the product of the first multiplied by the second, will be the same as the product of the second multiplied by the first.

Thus, $2 \times 4 = 2+2+2+2 = 8$,
$4 \times 2 = 4+4 = 8$;

therefore the results are the same, that is, $2 \times 4 = 4 \times 2$.

That the product of one number multiplied by another, will be equal to the product of the latter multiplied by the former, may perhaps appear more clearly from the following mode of showing this equality in the case of the numbers 3 and 5.

$$3 = 1+1+1;$$

$$\therefore 3 \times 5 = (1+1+1)+(1+1+1)+(1+1+1)+(1+1+1)+(1+1+1)$$

$$\left.\begin{array}{l} =1+1+1 \\ +1+1+1 \\ +1+1+1 \\ +1+1+1 \\ +1+1+1 \end{array}\right\} = 15.$$

Now, if we regard the *ones* from left to right, there are 3 *ones* taken 5 times; if we regard them taken from top to bottom, we have 5 *ones* repeated 3 times: and the number of ones in each case is the

SIMPLE MULTIPLICATION.

same, *i. e.*, $3 \times 5 = 5 \times 3$; and so in the case of any two other numbers multiplied together.

32. The accuracy of results in Multiplication is often tested by the following method, which is termed "CASTING OUT THE NINES": add together all the figures in the multiplicand, divide their sum by 9, and set down the remainder, then divide the sum of the figures in the multiplier by 9, and set down the remainder; multiply these remainders together, and divide their product by 9, and set down the remainder: if this remainder be the same as the remainder which results after dividing the product, or the sum of the digits in the product, of the multiplicand and multiplier by 9, the operation is very probably right; but if different, it is sure to be wrong.

This test depends upon the fact that "if any number and the sum of its digits be each divided by 9, the remainders will be the same"; the proof of which may be shown thus:

$$100 = 99 + 1,$$

where the remainder must be one, whether 100, or the sum of the digits in 100, viz., 1, be divided by 9, since 99 is divisible by 9 without a remainder.

Similarly,
$$200 = 2 \times 99 + 2,$$
$$300 = 3 \times 99 + 3,$$
$$400 = 4 \times 99 + 4,$$
$$500 = 5 \times 99 + 5,$$
&c., &c.

Hence it appears that if 100, 200, 300, 400, 500, &c., be each divided by 9, and the sum of the digits making up the respective numbers be also divided by 9, the two remainders in each case will be the same.

Also the number
$$532 = 500 + 30 + 2$$
$$= 5 \times 100 + 3 \times 10 + 2$$
$$= \overline{5 \times 99 + 5} + \overline{3 \times 9 + 3} + 2;$$

whence it appears that if the parts 5×100, 3×10, and 2, which make up the entire number, be each divided by 9, the remainders will be 5, 3, 2 respectively; and therefore the remainder, when 532 is divided by 9, will clearly be the same, as when $5 + 3 + 2$ is divided by 9.

To explain why the test holds, let us take as an example 533 multiplied by 57.

$$\begin{array}{r} 533 \\ 57 \\ \hline 3731 \\ 2665 \\ \hline 30381 \end{array}$$

Now $\qquad 533 = 9 \times 59 + 2 = 531 + 2$
$\qquad\qquad 57 = 9 \times 6 + 3 = 54 + 3.$

It is clear, since 531 contains 9 without a remainder, that 531×57 contains 9 without a remainder; therefore the remainder which is left after dividing the product of 533 and 57 by 9, must be the same as the remainder which is left after dividing the product of 2 and 57 by 9.

Again, since the product of 57 and $2 = (54+3) \times 2$, and the product of 54 and 2 when divided by 9 leaves no remainder, therefore the remainder which is left after dividing the product of 533 and 57 by 9, must be the same as the remainder left after dividing the product of 3 and 2 by 9, *i. e.*, after dividing the product of the remainders which are left after the division of the multiplicand and multiplier respectively by 9.

Now on dividing either 30381, or the sum of its digits, which is 15, by 9, the remainder left is 6, and 3×2 divided by nine also leaves 6 as remainder. Therefore we conclude that 30381 is the correct product of 533 and 57.

NOTE. If an error of 9, or any of its multiples, be committed, or if cyphers be introduced or omitted, the results will nevertheless agree, and so the error in these cases remains undetected.

Ex. V.

(1) 87298
  46

(2) 16097
  59

(3) 296897
  83

(4) 69284
  90

(5) 840607
  80

(6) 175
  189

(7) 6298
  769

(8) 5423
  603

SIMPLE MULTIPLICATION.   29

(9) 25607   (10) 78847   (11 672934684
        5004.   8803   765
     ———   ———   ———

(12) Find the product of 234578 by 18, by 29, and also by 53; of 924846 by 67, by 95, and also by 430; 2846067 by 206, by 1008, and also by 907; 8409631 by 21711, by 7009, by 8435, and also by 7980.

(13) Find the product of 1754 and 9306; of 47506 and 4500; of 149570 and 15790; of 554768 and 39314; of 815085 and 20048; of 123456789 and 987654321: and of 57298492692 and 700809050321.

(14) Multiply 9487352 by 4731246; 4342760 by 599999; 17376872 by 7399078; 38015732 by 400700065; 574585614865 by 2837154309.

(15) Multiply six hundred and fifty thousand and ninety, by three thousand and eight; also seventy-six millions, eight thousand, seven hundred and sixty-five, by nine millions, nine thousand and nine.

(16) Find the continued product of 12, 17, and 19; of 3781, 3782, and 3783; and of 6565, 6786, and 9898.

(17) Multiply 20470 by 1030, and 2958 by 476, explaining the reason of each step in the process.

The following *abbreviations* in Multiplication may be noticed.

83. *To multiply a number by* 5

RULE. Multiply the number by 10, and divide by 2.

Ex. Multiply 8763 by 5.

$$5 = \frac{10}{2}; \therefore 8763 \times 5 = 8763 \times \frac{10}{2} = \frac{87630}{2} = 43815$$

74. *To multiply a number* (1) *by* 25; (2) *by* 125.

RULE. Multiply the number in case (1) by 100, and divide by 4; in case (2), by 1000, and divide by 8.

Ex. Multiply (1) 839 by 25, (2) 7563 by 125.

(1) $25 = \frac{100}{4}; \therefore 839 \text{ by } 25 = 839 \times \frac{100}{4} = \frac{83900}{4} = 20975$

(2) $125 = \frac{1000}{8}; \therefore 7563 \text{ by } 125 = 7563 \times \frac{1000}{8} = \frac{7563000}{8} = 945375$

85. *To multiply a number* (1) *by* 15; (2) *by* 35; (3) *by* 45; (4) *by* 55.

30 ARITHMETIC.

Rule. Multiply the number in case (1) by 30; in case (2) by 70 in case (3) by 90: in case (4) by 110, and divide the product in each case by 2.

Ex. Multiply (1) 728 by 15; (2) 837 by 35; (3) 678 by 45.

(1) $15 = \dfrac{30}{2}$; $\therefore 728 \times 15 = 728 \times \dfrac{30}{2} = \dfrac{21840}{2} = 10920$

(2) $35 = \dfrac{70}{2}$; $\therefore 837 \times 35 = 837 \times \dfrac{70}{2} = \dfrac{58590}{2} = 29295$

(3) $45 = \dfrac{90}{2}$; $\therefore 678 \times 45 = 678 \times \dfrac{90}{2} = \dfrac{61020}{2} = 30510$

36. *To multiply a number* (1) *by* 75; (2) *by* 175; (3) *by* 225; (4) *by* 275.

Rule. Multiply the number in case (1) by 300; in case (2) by 700; in case (3) by 900; in case (4) by 1100, and divide in each case by 4.

Ex. Multiply (1) 973 by 75; (2) 687 by 175; (3) 978 by 225; (4) 1314 by 275.

(1) $75 = \dfrac{300}{4}$; $\therefore 973 \times 75 = 973 \times \dfrac{300}{4} = \dfrac{291900}{4} = 72975$

(2) $175 = \dfrac{700}{4}$; $\therefore 687 \times 175 = 687 \times \dfrac{700}{4} = \dfrac{480900}{4} = 120225$

(3) $225 = \dfrac{900}{4}$; $\therefore 978 \times 225 = 978 \times \dfrac{900}{4} = \dfrac{880200}{4} = 220050$

(4) $275 = \dfrac{1100}{4}$; $\therefore 1314 \times 275 = 1314 \times \dfrac{1100}{4} = \dfrac{1445400}{4} = 361350$

37. *To multiply a number by any number of nines.*

Rule. Multiply the number by the same power of 10, as is indicated by the number of nines; subtract the multiplicand from the product, and the remainder is the required result.

Ex. 1. Multiply 789786 by 999.

$999 = 10^3 - 1$; $\therefore 789786 \times 999 = 789786 \, (10^3 - 1)$
$= 789786000 - 789786 = 788996214$

SIMPLE DIVISION.   31

Ex. 2. Multiply 2686734 by 99999.

$99999 = 10^5 - 1$; $\therefore 2686734 \times 99999 = 2686734 \, (10^5 - 1)$
$= 268673400000 - 2686734 = 268670713266$

Similarly in any other case.

DIVISION.

38. DIVISION is the method of finding how often one number, called the DIVISOR, is contained in another number, called the DIVIDEND. The result is called the QUOTIENT.

39. Division is of two kinds, Simple and Compound. It is called Simple Division, when the dividend and divisor are, both of them, either abstract numbers, or concrete numbers of one and the same denomination.

It is called Compound Division, when the dividend, or when both divisor and dividend contain numbers of different denominations, but of one and the same kind.

40. The sign $\div$, placed between two numbers, signifies that the first is to be divided by the second.

41. In Division, if the dividend be a concrete number, the divisor may be either a concrete number or an abstract number, and the quotient will be an abstract number or a concrete number, according as the divisor is concrete or abstract. For instance, 5 shillings taken 6 times give 30 shillings, therefore 30 shillings divided by 5 shillings give the abstract number 6 as quotient; and 30 shillings divided by 6 give the concrete number 5 shillings as quotient.

SIMPLE DIVISION.

42. RULE. Place the divisor and dividend thus:

divisor) dividend (quotient.

Take off from the left-hand of the dividend the least number of figures which make a number not less than the divisor; then find by the Multiplication Table, how often the first figure on the left-hand side of the divisor is contained in the first figure, or the first two figures, on the left-hand side of the dividend, and place the figure which denotes this number of times in the quotient; multiply the divisor by this figure, and bring down the product, and subtract it from the number which was taken off at the left of the dividend; then bring down the next figure of the dividend, and place it to the right of the remainder, and proceed as before; if the divisor be greater than any of these remainders, affix a cypher to the quotient, and bring down the next figure from the dividend to the right of the remainder, and proceed as before. Carry on this operation till all the figures of the dividend have been thus brought down, and the quotient, if there be no remainder, will be thus determined, or if there be a remainder, the quotient and the remainder will be thus determined.

NOTE 1. If any product be greater than the number which stands above it, the last figure in the quotient must be changed for one of smaller value: but if any remainder be greater than the divisor, or equal to it, the last figure of the quotient must be changed for a greater.

NOTE 2. If the divisor does not exceed 12, the division can easily be effected by means of the Multiplication Table.

Ex. Divide 2338268 by 6758.

Proceeding by the Rule given above, we obtain

$$6758) 2338268 (346$$
$$20274$$
$$\overline{}$$
$$31086$$
$$27032$$
$$\overline{}$$
$$40548$$
$$40548$$

Therefore the quotient is 346.

The reason for the Rule will appear from the following considerations.

The divisor represents six thousand, seven hundred and fifty-eight; the first five figures on the left-hand side of the dividend represent two millions, three hundred and thirty-eight thousand, and two hundred.

Now the divisor is contained in this 300 times; and $6758 \times 300 = 2027400$, or omitting the two cyphers at the end for convenience in

SIMPLE DIVISION.

working, we properly place the 4 under the 2 in the line above; we subtract the product thus found, and we obtain a remainder of 3108, which represents three hundred and ten thousand, and eight hundred. Bring down the 6 by the Rule; this 6 denotes 6 tens or 60, but the cypher is omitted for the reason above stated; the number now represents three hundred and ten thousand, eight hundred and sixty; 6758 is contained 40 times in this, and 6758 × 40 = 270320; we omit the cypher at the end as before, and subtract the 27032 from the 31086; and after subtraction the remainder is 4054, which represents forty thousand five hundred and forty. Bring down the 8 by the Rule, and the number now represents forty thousand, five hundred and forty-eight; 6758 is contained 6 times exactly in this number.

Therefore 346 is the quotient of 2338268 by 6758.

The above example worked without omitting the cyphers would have stood thus:

$$6758) 2338268 (300 + 40 + 6$$
$$2027400$$
$$\overline{310868}$$
$$270320$$
$$\overline{40548}$$
$$40548$$

hence it appears that the divisor is subtracted from the dividend 300 times, and then 40 times from what remains, and then 6 times from what then remains, and there being now no remainder, 6758 is contained exactly 346 times in 2338268.

The truth of the above method might have been shown as follows:

$$2338268 = 2027400 + 270320 + 40548$$
$$6758) 2027400 + 270320 + 40548 (300 + 40 + 6$$
$$2027400$$
$$\overline{ + 270320}$$
$$+ 270320$$
$$\overline{ + 40548}$$
$$+ 40548$$

Ex. 2. Divide 56438971 by 4064.

$$
\begin{array}{r}
4064\,)\,56438971\,(\,13887 \\
4064 \\
\hline
15798 \\
12192 \\
\hline
36069 \\
32512 \\
\hline
35577 \\
32512 \\
\hline
30651 \\
28448 \\
\hline
2203
\end{array}
$$

therefore 4064 is contained in 56438971, 13887 times, with the remainder 2203.

43. *If the divisor terminate with cyphers, the process can be abridged by the following Rule.*

RULE. Cut off the cyphers from the divisor, and as many figures from the right-hand of the dividend, as there are cyphers so cut off at the right-hand end of the divisor; then proceed with the remaining figures according to the Rule, Art. (42); and to the last remainder annex the figures cut off from the dividend for the total remainder.

Ex. Divide 537523 by 34

Proceeding by the Rule,

$$
\begin{array}{r}
34,00\,)\,5375,23\,(\,158 \\
34 \\
\hline
197 \\
170 \\
\hline
275 \\
272 \\
\hline
3
\end{array}
$$

therefore 3400 is contained in 537523, 158 times with remainder 323.

SIMPLE DIVISION.   35

The Reason for the Rule will appear from the following considerations.

537523 is 5375 hundreds and 23, of which 537500 contains 3400, 158 times with a remainder 300 over; and as 23 does not contain 3400 at all, the quotient will evidently be 158, with remainder 300 + 23, or 323.

NOTE. The same rule applies when the divisor and dividend both terminate with cyphers.

44. When the divisor is a composite number, and made up of two factors, neither of which exceeds 12, the dividend may be divided by one of the factors in the way of Short Division, and then the result by the other factor; if there be a remainder after each of these divisions, the true remainder will be found by multiplying the second remainder by the first divisor, and adding to the product the first remainder.

Ex. Divide 56732 by 45.

$$45 \begin{cases} 9 \\ 5 \end{cases} \begin{array}{|l} 56732 \\ \hline 6303 - 5 \\ \hline 1260 - 3 \end{array}$$

the total remainder is 9 × 3 + 5, or 27 + 5 = 32.

Therefore the quotient arising from the division of 56732 by 45 is 1260, with a remainder 32.

The reason for the above Rule is manifest from the following considerations.

> 6303 is 5 times 1260 together with 3,
> and 56732 is 9 times 6303 together with 5,
>   or is 9 times (5 times 1260 + 3) together with 5,
>   or is 45 times 1260 + 28 + 5,
>   or is 45 times 1260 + 32.

45. The truth of all results in Division may be proved by multiplying the DIVISOR and QUOTIENT together, adding to the product the remainder, if there be any; the result (if the work is correct) will be the DIVIDEND.

ARITHMETIC.

46. Since the product of the divisor and quotient equals the DIVIDEND less the remainder; therefore, the accuracy of questions in division may be tested by the process of casting out the nines, pointed out in Art. (32).

Ex. VI.

Examples in Simple Division.

(1) 14683059 ÷ 27.
(2) 817286228 ÷ 44.
(3) 54906734 ÷ 59.
(4) 6848734752 ÷ 96.
(5) 70865432 ÷ 87.
(6) 649305745 ÷ 55.
(7) 28894545 ÷ 123.
(8) 433418175 ÷ 615.
(9) 1674918 ÷ 189.
(10) 31884740 ÷ 779.
(11) 536819741 : 907.
(12) 111111111111 ÷ 50160.
(13) 8235460800 ÷ 1440.
(14) 57380625 ÷ 7575.
(15) 353008972662 ÷ 5406.
(16) 599961567212 ÷ 2468.
(17) 26799534687 ÷ 7890000.
(18) 57111104051 ÷ 3851.
(19) 10000000000000000 ÷ 1111, and also by 11111.
(20) 634394567 ÷ 164600.
(21) 67157148372 ÷ 90009.
(22) 1220225292 ÷ 200563.
(23) 7428927415293 ÷ 8496427.
(24) 60435674536815 ÷ 79094451.
(25) 65358547823 ÷ 5578.
(26) 3968901531620 ÷ 687637943.

(27) Divide 152181255 by 3854, and explain the process.

(28) Divide 143255 by 4093. Explain the operation, and show that it is correct.

(29) Divide 203534191 by 72.

(30) The remainder is 613, quotient 78936, divisor 873. Find the dividend.

(31) The dividend is 855856651, the quotient 86783, the remainder 2705. Find the divisor.

(32) The distance between Liverpool and Quebec is 3060 miles; the usual length of a voyage by a Montreal Ocean steamship is 11 days. Find the number of miles which the vessel goes per hour.

(33) The length of the Rideau Canal, is 126 miles; cost of building, $4,380,000; Length of Welland Canal, 51 miles; cost $7,000,000. Find, 1st, cost of each per mile; 2d, difference of cost per mile.

(34) The number of miles open for traffic on the Grand Trunk Railway is 1377; the cost for building and equipping the road, $94,405,914; number of miles open on the Great Western is 852; cost

SIMPLE DIVISION. 37

for building and equipping, $24,777,430. Find, 1st, cost of each per mile; 2d difference of cost per mile.

NOTE. In the above exercise, whenever the Divisor is a composite number, divide, 1st, by Long Division and then by its factors, and show that the results in both cases coincide.

The following *abbreviations* in Division may be noticed.

To Divide a number by 5.

RULE. Multiply the number by 2, and divide the product by 10.

Ex. Divide 637 by 5.

$$5 = \frac{10}{2} \; ; \; \therefore \; \frac{637}{5} = \frac{637}{\frac{10}{2}} = \frac{637 \times 2}{10} = \frac{1274}{10} = 127\tfrac{4}{10}$$

48. **To divide a number** (1) *by* 25; *by* 125.

RULE. Multiply the number in case (1) by 4, and divide the product by 100; in (2) by 8, and divide the product by 1000.

Ex. Divide (1) 541 by 25, and (2) 5600741 by 125.

(1) $25 = \frac{100}{4} \; ; \; \therefore \; \frac{541}{25} = \frac{541}{\frac{100}{4}} = \frac{541 \times 4}{100} = \frac{2164}{100} = 21\tfrac{64}{100}$

(2) $125 = \frac{1000}{8} \; ; \; \therefore \; \frac{5600741}{125} = \frac{5600741}{\frac{1000}{8}} = \frac{5600741 \times 8}{1000}$

$$= \frac{44805928}{1000} = 44805\tfrac{928}{1000}$$

49. **To divide a number** (1) *by* 15; (2) *by* 35; (3) *by* 45; (4) *by* 55.

RULE. Multiply the number in each case by 2, and divide the product in case (1) by 30, in (2) by 70, in (3) by 90, in (4) by 110.

Ex. Divide (1) 683 by 45; (2) 5603 by 35.

(1) $45 = \frac{90}{2} \; ; \; \therefore \; \frac{683}{45} = \frac{683}{\frac{90}{2}} = \frac{683 \times 2}{90} = \frac{1366}{90} = 15\tfrac{3}{15}$

(2) $35 = \frac{70}{2} \; ; \; \therefore \; \frac{5603}{35} = \frac{5603}{\frac{70}{2}} = \frac{5603 \times 2}{70} = \frac{11206}{70} = 160\tfrac{3}{35}$

50. *To divide a number* (1) *by* **75**; (2) *by* 175; (3) *by* 225; (4) *by* 275.

Rule. Multiply the number in each case by 4, and divide the product in (1) by 300, in (2) by 700, in (3) by 900, in (4) by 1100.

Ex. Divide (1) 2697 by 75; (2) 23647 by 275.

(1) $75 = \dfrac{300}{4}$; $\therefore \dfrac{2697}{75} = \dfrac{2697}{\frac{300}{4}} = \dfrac{2697 \times 4}{300} = \dfrac{10788}{300} = 35\tfrac{21}{25}$

(2) $275 = \dfrac{1100}{4}$; $\therefore \dfrac{23647}{275} = \dfrac{23647}{\frac{1100}{4}} = \dfrac{23647 \times 4}{1100} = \dfrac{94588}{1100} = 85\tfrac{273}{275}$

To divide a number by any number of nines.

Rule. Divide the given number by the same power of 10 as is indicated by the number of nines; repeat the same operation as often as necessary with each successive quotient obtained; add all these quotients together; their sum is the quotient required.

Ex. Divide 2897637 by 9999.

$$\begin{array}{r} 289\cdot7637 \\ \cdot02897637 \\ \cdot000002897637 \\ \hline 289\cdot792679267637 \end{array}$$

Note 1. If the sum of the partial remainders should be the same as the divisor in any example (*i. e.* a number of nines), it is plain that there is no remainder, but that one should be added to the integral part.

Note 2. By carrying on the operation, as in the given example, the digits which recur very soon appear; for instance, as *in the example*, 9267, so that the answer above might be written 289·7̇9267̇.

Ex. VII.

Miscellaneous Questions and Examples on the foregoing Articles.

I.

(1) Explain the principle of the common system of numerical notation. Multiply 603 by 48, and give the reasons for the several steps.

(2) **Write** at length the meaning of 9090909, **and of 90909.** Find their **sum** and difference, and explain fully **the** processes employed.

(3) A person, whose age is 73, was 37 years old **at the birth of his** eldest son ; what is the son's age ?

(4) Explain the meaning of the **terms** " vinculum ", " bracket "; and of the signs $+, -, =, \therefore, \times$.

Find the value of the following expression :

$$15 \times 37153 - 73474 - 67152 \div 4 + 40734 \times 2.$$

(5) By the **census of 1861, the population of Ontario was** found to be **1396091** ; of Quebec, **1111566** ; of New Brunswick, 252017, of Nova Scotia, 330857 ; of Prince Edward Island, 80857 ; of Newfoundland **(1857), 124288** ; British Columbia and **Vancouver's** Island, 34816 ; **Rupert's Land, 101000. Find the whole population of the** above named **provinces.**

II.

(1) Define "**a Unit**", " Number ", "**Arithmetic** ". **What is the** difference between Abstract **and** Concrete numbers ?

(2) **The annual** deaths in **a town being 1 in** 45, and in **the country 1** in 50, in how many **years will the number of** deaths **out of 18675 persons** living **in the town, and 79250 persons living in the country amount together to 10000 ?**

(3) **Define "** Notation " " **Numeration** "; express **in numbers seven** hundred quadrillions four hundred **and nine** trillions.

(4) Find **the value of**

$494871 - 94853 + (45079 - 3177) - (54312 - 3987) - (1763 + 231) \div 379 \times 379.$

(5) What **number** divided **by 528 will give 36 for the quotient,** and leave 44 **as a remainder.**

III.

(1) Define Multiplication and **Division. Shew that the product** of two **numbers** is the same in whatever order the operation is performed.

(2) The **Iliad contains** 15683 lines, and the Æneid contains 9892 **lines** ; how many days **will it take a boy to read through both of them,** at the rate of **eighty-five lines a day ?**

(3) The dividend is 813215640, the quotient 62513, the remainder 46536; what is the divisor?

(4) Explain the meaning of the sign ÷, and find the value of
(7854−4913)×3 − (20374−12530)÷53 − 6+(395456−2364)÷556.

(5) At a game of cricket A, B, and C together score 108 runs, B and C together score 90 runs, and A and C together score 51 runs; find the number of runs scored by each of them.

IV.

(1) Define Addition and Subtraction. What is meant by a prime number? When are numbers said to be prime to each other? Give examples.

Explain the rule of *carrying* in the addition of numbers; exemplify it in the addition of 3864, 4768, and 15938.

(2) A father was 21 years old when his eldest son was born; how old will his son be when he is 50 years old, and what will be the father's age when the son is 50 years old.

(3) Write in figures one hundred millions, one hundred thousand, one hundred and one; and in words 1010101010. Express in figures M.DCCC.XL.

(4) Explain the short method of multiplying and dividing a given number by any number of nines; exemplify by the number 8795678 being separately multiplied and divided by 9999.

(5) The estimated population of the British American Provinces for the year 1870, is as follows: Ontario, 2047334; Quebec, 1387884; New Brunswick, 319398; Nova Scotia, 389343; Prince Edward Island, 97246; Newfoundland, 133000; British Columbia, 60000; Rupert's Land, 115000. Find the total estimated population of the above provinces for the year 1870.

V.

(1) Multiply 478 by 146, and test the result by casting out the nines. In what cases does this method of proof fail? Divide 4843 by 99, and prove the correctness of the operation by any test you please.

(2) What number multiplied by 86 will give the same product as 163 by 430?

(3) In the city of Montreal, for every two persons who speak English only, three speak French only, and seven both English and French; and the whole population is 120000. How many speak English only, French only, and both English and French?

(4) A gentleman dies, and leaves his property thus: 10000 dollars to his widow; 15000 dollars to his eldest son, on the condition of his giving to a school-library 350 dollars; 5500 dollars to each of his four younger sons; 3750 dollars to each of his three daughters; 4563 dollars to different societies; and 599 dollars in legacies to his servants. What amount of property did he die possessed of?

(5) The quotient arising from the division of 9281 by a certain number is 17, and the remainder is 373. Find the divisor.

VI.

(1) Explain briefly the Roman method of Notation. Express 1563 and 9000 in Roman characters.

(2) Explain the terms "factor", "product"; "quotient"; show by an example how the process of Division can be abridged, if the divisor terminate with cyphers.

(3) The remainder of a division is 97, the quotient 665, and the divisor 91 more than the sum of both. What is the dividend?

(4) Express in words the numbers 270130 and 26784; also write down in figures the number ten thousand two hundred and thirty-four; and find the least number which added to the last number will make it divisible by 8.

(5) A gentleman, whose age is 60, has two sons and a daughter; his age equals the sum of the ages of his children; two years since his age was double that of his eldest son; the sum of the ages of the father and the eldest son is seven times as great as that of the youngest son; find the ages of the children.

SECTION II.

CONCRETE NUMBERS.

TABLES.

52. Our operations hitherto have been carried on with regard only to abstract numbers, or concrete numbers of one denomination. It is evident that if concrete numbers were all of one denomination; if, for instance, shillings were the only units of money, yards of length, years of time, and so on, such numbers would be subject to the common rules for abstract numbers. Again, if the concrete numbers were of different denominations, and those denominations differed from each other by 10 or multiples of 10, then all operations with such concrete numbers could be carried on by the rules which have been given for whole numbers. But generally with concrete numbers such a relation does not hold between the different denominations, and therefore it is necessary to commit to memory tables, which connect the different units of money together, the different units of length together, the different units of time together, and so on.

We shall now put down some of the most useful of these tables, with a few brief remarks on each.

MONEY TABLES.

CANADIAN CURRENCY.

53. The Silver Coins are: a 5 cent piece.
          a 10 " "
          a 20 " "
          a 25 " "
          a 50 " "

100 cents makes one dollar, or $1.

NOTE. The cent (ct.), which is made of bronze, is one inch in diameter, and 100 cents weigh one pound avoirdupois. The Canadian silver coinage is of the same degree of fineness as that of Great Britain.

The copper coinage is not, according to the present law, a legal tender for more than 20 cents; nor is the silver coinage for more than $10; the gold coinage of Great Britain being the standard of this country.

CONCRETE NUMBERS.

HALIFAX OR OLD CANADIAN CURRENCY.

64.
- 2 Farthings make 1 Half-penny......½d.
- 2 Half-pence ...... 1 Penny ....1d.
- 12 Pence .... 1 Shilling............1s.
- 5 Shillings........... 1 Dollar..............$1.
- 4 Dollars.............. 1 Pound..............£1.

NOTE 1. The farthing is written thus, ¼d; and three farthings thus, ¾d.

ENGLISH OR STERLING CURRENCY.

- 2 Farthings make 1 Half-penny, or ½d.
- 2 Half-pence .... 1 Penny............1d.
- 12 Pence .... 1 Shilling............1s.
- 20 Shillings .... 1 Pound ....£1.

NOTE 2. The sterling pound or sovereign = $4.86⅔ Canadian Currency.

Pounds, shillings, pence, and farthings were formerly denoted by £, s, d, q respectively, these letters being the first letters of the Latin words, *libra*, *solidus*, *denarius*, and *quadrans*, the Latin names of certain Roman coins or sums of money. £. s. d are still the abbreviated forms for pounds, shillings, and pence respectively.

The following coins are in common use in England:

COPPER COINS.

A Farthing the coin of least value.
A Half-penny = 2 Farthings.
A Penny = 4 Farthings.

SILVER COINS.

Three penny-piece = 3 Pence.
Four penny-piece = 4 Pence.
A Six pence = 6 Pence.
A Shilling = 12 Pence.
A Florin = 2 Shillings.
A Half-Crown = { 2 shillings and 6 Pence
A Crown = 5 Shillings.

GOLD COINS.

A Half sovereign = 10 Shillings.
A Sovereign = 20 Shillings.

The following coins have been in use at various periods in England, but with the exception of the first two, which are used under different names, they are now obsolete.

SILVER COINS.

A Groat = 4 Pence.
A Tester = 6 Pence.

GOLD COINS.

	£.	s.	d.
A Noble	= 0	6	8
An Angle	= 0	10	0
A Half-guinea	= 0	10	6
A Mark or Merk	= 0	13	4
A Guinea	= 1	1	0
A Carolus	= 1	3	0
A Jacobus	= 1	5	0
A Moidore	= 1	7	0

ARITHMETIC.

Note 1. The office at which coin is made and stamped, so as to pass or become current for legal money, is called *the Mint*.

The *standard* of gold coin in Great Britain and Ireland is 22 parts of *pure gold* and 2 parts of *copper*, melted together. From a pound Troy of standard gold there are coined at the Mint $46\frac{2 8}{4 0}$ sovereigns or £46. 14s. 6d.; therefore the Mint price of gold is $\frac{1}{12}$ of £46. 14s. 6d. or £3. 17s. $10\frac{1}{2}d.$ per ounce standard (12 ounces Troy=1 pound Troy).

The *standard* of silver coin is 37 parts of *pure silver* and 3 parts of *copper*. From a pound Troy of standard silver are coined 66 shillings. Therefore the Mint price of silver is 5s. 6d. per ounce standard.

In the copper coinage, 24 pence are coined from 1 pound Avoirdupois of copper. Therefore 1 penny should weigh $\frac{1}{24}$th of a pound Avoirdupois.

UNITED STATES CURRENCY.

55.
10 Mills (m)............make 1 Cent......ct.
10 Cents........................ 1 Dime......D.
10 Dimes .... 1 Dollar......$.
10 Dollars .... 1 Eagle......E.

Note 2. The dollar, it seems, was originally a German coin, said to be derived from *Dale*, the name of the town where it was first coined.

Cent, most likely from the Celtic *Cant*, meaning a hundred.

MEASURES OF WEIGHT.

TABLE OF TROY WEIGHT.

56. This table derives its name probably from *Troyes* in France, the first city in Europe where it was adopted. It seems to have been brought thither from Egypt. It has also been derived from *Troynovant*, the monkish name for London. It is used in weighing gold, silver, diamonds, and other articles of a costly nature; also in determining specific gravities; and generally in philosophical investigations.

TABLES—WEIGHT.

The different units are grains (written grs.), pennyweights (dwts.) ounces (oz.), and pounds (lbs. or ℔.) and they are connected thus:

```
24 Grains ............ make 1 Pennyweight ... 1 dwt.
20 Pennyweights ............ 1 Ounce ............ 1 oz.
12 Ounces .................. 1 Pound ............ 1 lb. or ℔.
```

NOTE 1. As the origin of weights, a grain of wheat was taken from the middle of the ear, and being well dried, was used as a weight, and called '*a grain*.'

NOTE 2. Diamonds and other precious stones are weighed by '*Carats*, each carat weighing about $3\frac{1}{5}$ grains. The term 'carat' applied to gold has a relative meaning only; any quantity of pure gold, or of gold alloyed with some other metal, being supposed to be divided into 24 equal parts (carats); if the gold be pure, it is said to be 24 carats fine; if 22 parts be pure gold and 2 parts alloy, it is said to be 22 carats fine.

Standard gold is 22 carats fine; jewellers' gold is 18 carats fine.

TABLE OF APOTHECARIES' WEIGHT.

57. Apothecaries' weight only differs from Troy weight in the subdivisions of the pound, which is the same in both. This table is used in mixing medicines. The different units are grains (grs.), scruples (℈), drams (℥), ounces (℥), pounds (lbs. or ℔.), and they are connected thus:

```
20 Grains......... make 1 Scruple ...... 1 sc. or 1 ℈.
 3 Scruples ............. 1 Dram ........ 1 dr. or 1 ℥.
 8 Drams ............... 1 Ounce ........ 1 oz. or 1 ℥.
12 Ounces ............... 1 Pound ........ 1 lb. or ℔.
```

TABLE OF AVOIRDUPOIS WEIGHT.

58. Avoirdupois weight derives its name from *Avoirs* (goods or chattels) and *Poids* (weight). It is used in weighing all heavy articles, which are coarse and drossy, or subject to waste, as butter, meat, and the like, and all objects of commerce, with the exception of medicines,

gold, silver, and some precious stones. The different units are drams (drs.) ounces (oz.), pounds (lbs.), quarters (qrs.), hundredweights (cwts.), tons (tons), and they are connected thus:

16 Drams..........make	1 Ounce..............	1 oz.
16 Ounces..................	1 Pound..............	1 lb.
25 Pounds..................	1 Quarter............	1 qr.
4 Quarters	1 Hundredweight.	1 cwt.
20 Hundredweights.....	1 Ton	1 Ton.

In general, 1 Stone (1 st.) = 14 lbs. Avoirdupois, but for butchers' meat or fish, 1 Stone = 8 lbs.; 1 Firkin of Butter = 56 lbs.; 1 Fodder of Lead = 19½ cwt.; 1 Great Pound of Silk = 24 ounces; 1 Pack of Wool = 240 pounds.

    1 lb. Avoirdupois weighs 7000 grains Troy;
    1 lb. Troy   weighs 5760 grains Troy.

MEASURES OF LENGTH.

TABLE OF LINEAL MEASURE.

59. In this measure, which is used to measure distances, lengths, breadths, heights, depths, and the like, of places or things:

8 Barley-corns (in length) make	1 Inch, which is written	1 in.
12 Inches	1 Foot,.......................	1 ft.
3 Feet	1 Yard,.......................	1 yd.
6 Feet	1 Fathom,..................	1 fth.
5½ Yards	1 Rod, Pole, or Perch, ...	1 po.
40 Poles (220 yds.)	1 Furlong, ...	1 fur.
8 Furlongs	1 Mile, ...	1 m.
3 Miles.................................	1 League, ...	1 lea.
69½ Miles	1 Degree,....................	1 deg. or 1°

NOTE. A grain of Barley, or a Barley-corn, is supposed to have been the original element of Lineal Measure.

The following measurements may be added, as useful in certain cases:

4 inches make 1 Hand (used in measuring horses),
22 Yards make 1 Chain
100 Links make 1 Chain $\Big\}$ used in measuring land,

a Palm = 3 inches, a Span = 9 inches, a Cubit = 18 inches,
a Pace = 5 feet, 1 Geographical mile = $\frac{1}{60}$th of a degree,
a Line = $\frac{1}{12}$th of an inch.

TABLE OF CLOTH MEASURE.

60. In this measure, which is used by linen and woollen drapers:

$2\frac{1}{4}$ inches make 1 Nail.
4 Nails.......... 1 Quarter.... 1 qr.
4 Quarters.... 1 Yard....... 1 yd.
5 Quarters..... 1 English Ell.
6 Quarters..... 1 French Ell.
3 Quarters..... 1 Flemish Ell.

MEASURES OF SURFACE

TABLE OF SQUARE MEASURE.

61. This measure is used to measure all kinds of surface or superficies, such as land, paving, flooring, in fact everything in which length and breadth are to be taken into account.

A SQUARE is a four-sided figure, whose sides are equal, each side being perpendicular to the adjacent sides. See figure below.

A square inch is a square, each of whose sides is an inch in length a square yard is a square, each of whose sides is a yard in length.

144 Square Inches make 1 Square Foot... 1 sq. ft. or 1 ft.
9 Square Feet .... 1 Square Yard... 1 sq. yd. or 1 yd.
$30\frac{1}{4}$ Square Yards .... 1 Square Pole... 1 sq. po. or 1 po.
40 Square Poles .... 1 Square Rood.. 1 ro.
4 Roods .... 1 Acre .... 1 ac.

48   ARITHMETIC.

$$\begin{aligned}
25000 \text{ Square Links} &= 1 \text{ Rood.} \\
100000 \ \ldots\ldots\ldots\ldots\ldots &= 1 \text{ Acre.} \\
10 \ \ldots\ldots \text{Chains} &= 1 \text{ Acre.} \\
4840 \ \ldots\ldots \text{ Yards} &= 1 \text{ Acre.} \\
640 \ \ldots\ldots \text{ Acres} &= 1 \text{ Square Mile.}
\end{aligned}$$

Note. This table is formed from the table for lineal measure, by multiplying each lineal dimension by itself.

The truth of the above table will appear from the following considerations.

Suppose *AB* and *AC* to be lineal yards placed perpendicular to each other.

Then by definition *ABCD* is a square yard. If *AE*, *EF*, *FB*, *AG*, *GH*, *HC* = 1 lineal foot each, it appears from the figure that there are 9 squares in the square yard, and that each square is 1 square foot.

The same explanation holds good of the other dimensions.

The following measurement may be added:

$$\text{A Rod of Brickwork} \ldots\ldots = 272\tfrac{1}{4} \text{ Square Feet.}$$

(*The work is supposed to be 14 in., or rather more than a brick-and-a-half thick.*)

TABLE OF SOLID OR CUBIC MEASURE.

62. This measure is used to measure all kinds of solids, or figures which consist of three dimensions, length, breadth, and depth or thickness.

A CUBE is a solid figure contained by six equal squares; for instance, a die is a cube. A cubic inch is a cube whose side is a square inch. A cubic yard is a cube whose side is a square yard.

$$\begin{aligned}
1728 \text{ Cubic Inches}\ldots\ldots\ldots\ldots\ldots \text{make } &1 \text{ Cubic Foot, or 1 c. ft.} \\
27 \text{ Cubic Feet} \ldots\ldots\ldots\ldots\ldots\ldots &1 \text{ Cubic Yard. or 1 c. yd.} \\
40 \text{ Cubic Feet of Rough Timber or} & \\
50 \text{ Cubic Feet of Hewn Timber}\ldots\ldots &1 \text{ Load.} \\
42 \text{ Cubic Feet} \ldots\ldots\ldots\ldots\ldots\ldots &1 \text{ Ton of Shipping.} \\
128 \text{ Cubic Feet of Fire-wood}\ldots\ldots &1 \text{ Cord.} \\
16 \text{ Cubic Feet of Fire-wood} \ldots\ldots &1 \text{ Cord-foot.}
\end{aligned}$$

The truth of the first part of the above table will appear from the following considerations.

If AB, AC, and AD be perpendicular to each other, and each of them a lineal yard in length, then the figure DE is a cubic yard.

Suppose DH a lineal foot, and $HKLM$ a plane drawn parallel to side DC.

By the table Art. 61, there are 9 square feet in side DC. There will therefore be 9 cubic feet in the solid figure DL.

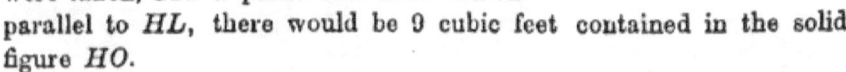

Similarly if another lineal foot HN were taken, and a plane NO were drawn parallel to HL, there would be 9 cubic feet contained in the solid figure HO.

Similarly, there would be 9 cubic feet in the solid figure NE.

Therefore, there are 27 cubic feet in the solid figure DE, or 1 cubic yard.

Note. A pile of wood 4 feet high, 4 feet wide, and 8 feet long, makes a cord.

MEASURES OF CAPACITY.

TABLE OF WINE MEASURE.

63. In this measure, by which wines and all liquids, with the exception of malt liquors and water, are measured,

    4 Gills................make 1 Pint......... 1 pt.
    2 Pints.... .... 1 Quart ...... 1 qt.
    4 Quarts .... 1 Gallon...... 1 gal.
    63 Gallons.................... 1 Hogshead.. 1 hhd.
    2 Hogsheads .... 1 Pipe......... 1 pipe.
    2 Pipes .... 1 Tun......... 1 tun.

TABLE OF ALE AND BEER MEASURE.

64. In this measure, by which all malt liquors and water are measured:

    2 Pints........................ 1 Quart ...... 1 qt.
    4 Quarts .... 1 Gallon...... 1 gal.

9 Gallons.................. 1 Firkin........... 1 fir.
18 Gallons.................. 1 Kilderkin...... 1 kil.
36 Gallons.................. 1 Barrel........... 1 bar.
1½ Barrels, or 54 Gallons..... 1 Hogshead...... 1 hhd.
2 Hogsheads .... 1 Butt............. 1 butt.
2 Butts .... 1 Tun .... 1 tun.

TABLE OF DRY MEASURE.

65.   2 Pints ....make 1 Quart........... 1 qt.
      4 Quarts .... 1 Gallon.......... 1 gal.
      2 Gallons..................... 1 Peck .... 1 pk.
      4 Pecks .... 1 Bushel .... 1 bu.
     36 Bushels .... 1 Chaldron ...... 1 ch.

66.   34 Pounds make 1 Bushel of Oats.
     48 Pounds......... 1 Bushel of B'kwheat, Barley or Timothy.
     50 Pounds......... 1 Bushel of Flax Seed.
     56 Pounds......... 1 Bushel of Rye or Indian Corn.
     60 Pounds......... 1 Bushel of Wheat, Potatoes, Peas, Beans,
                           Onions, or Red Clover Seed.

Note 1. Grains are sold by the cental (100 lbs.), or by parts thereof.

MISCELLANEOUS TABLE.

67.   12 Units....................make 1 Dozen.
      12 Dozen....................... 1 Gross.
      12 Gross .... 1 Great Gross.
      20 Units .... 1 Score.
      24 Sheets of Paper .... 1 Quire.
      20 Quires....................... 1 Ream.
     100 Pounds .... 1 Quintal.
     196 Pounds .... 1 Barrel of Flour.
     200 Pounds .... 1 Barrel of Pork or Beef.

Note 2. A sheet folded into two leaves is called a folio, into 4 leaves a quarto, into 8 leaves an octavo, into 16 leaves a 16mo, into 18 leaves an 18mo, &c.

MEASURES OF TIME.

TABLE OF TIME.

68.   1 Second is written thus 1″.

    60 Seconds .... make 1 Minute .... 1′.
    60 Minutes .... 1 Hour .... 1 hr.
    24 Hours .... 1 Day .... 1 day.
    7 Days .... 1 Week .... 1 wk.
    4 Weeks, or 28 days ...... 1 Lunar Month .... 1 mo.
    365 Days .... 1 Civil or common year 1 yr.

A year is divided into 12 months, called Calendar Months, the number of days in each of which may be easily remembered by means of the following lines :

> Thirty days hath September,
> April, June, and November:
> February hath twenty-eight alone,
> And all the rest have thirty-one :
> But leap-year coming once in four,
> February then has one day more.

NOTE 3.—A civil or common year = 52 wks., 1 day.
        A leap-year   = 366 days.

A day, or rather a *mean solar day*, which is divided in 24 equal portions, called *mean solar hours*, is the standard unit for the measurement of time, and it is the mean or average time which elapses between two successive transits of the Sun across the meridian of any place.

The time between the Sun's leaving a certain point in the *Ecliptic* and its return to that point consists of 365·242218 mean *solar days*, or 365 days, 5 hours, 48 minutes, 47½ seconds, very nearly, and is called a *solar year*. Therefore the *civil* or *common* year, which contains 365 days, is about ¼th of a day less than the *solar* year ; and this error

52   ARITHMETIC.

would, of course, in time be very considerable, and cause great confusion.

Julius Cæsar, in order to correct this error, enacted that every 4th year should consist of 366 days; this was called *Leap* or *Bissextile year*. In that year February had 29 days, the extra day being called "the *Intercalary*" day.

But the solar year contains 365·242218 days, and the Julian year contains 365·25 or $365\frac{1}{4}$ days.

Now $365·25 - 365·242218 = ·007782.$

Therefore in one year, taken according to the Julian calculation, the Sun would have returned to the same place in the Ecliptic ·007782 of a day before the end of the Julian year.

Therefore in 400 years the Sun would have come to the same place in the Ecliptic ·007782 × 400 or 3·1128 days before the end of the Julian year; and in 1257 years would have come to the same place, ·007782 × 1257 or 9·7819, or about ten days before the end of the Julian year. Accordingly, the vernal equinox which, in the year 325 at the council of Nice, fell on the 21st of March, in the year 1582 (that is, 1257 years later), happened on the 11th of March; therefore Pope Gregory caused 10 days to be omitted in that year, making the 15th of October immediately succeed the 4th, so that in the next year the vernal equinox again fell on the 21st of March; and to prevent the recurrence of the error, ordered that, for the future, in every 400 years 3 of the leap years should be omitted, viz. those which complete a century, the numbers expressing which century, are *not* divisable by 4; thus 1600 and 2000 are leap years, because 16 and 20 are exactly divisable by 4; but 1700, 1860, and 1900 are not leap years, because 17, 18, and 19 are not exactly divisible by 4.

This Gregorian style, which is called the *new style*, was adopted in England on the 2nd of September 1752, when the error amounted to 11 days.

The Julian calculation is called the *old style*; thus old Michaelmas and old Christmas take place 12 days after New Michaelmas and New Christmas.

In Russia, they still calculate according to the *old style*, but in the other countries of Europe the new style is used. Sir Harris Nicolas in his Chronology gives the dates at which the new style was adopted in different countries. Of course, it was almost immediately adopted by most of the Roman Catholic courts of Europe.

TABLE OF ANGULAR MEASURE.

69.   1 Second is written 1 sec. or 1".
 60 Seconds make 1 Minute.......... 1 min. or 1'.
 60 Minutes .... 1 Degree .... 1 deg. or 1°.
 90 Degrees .... 1 Right Angle.... 1 rt. ang. or 90°.

The circumference of every circle is considered to be divided into 360 equal parts, each of which is often called a degree, as it subtends an *angle* of 1° at the centre of the circle.

70. An Act of Parliament "FOR ASCERTAINING AND ESTABLISHING UNIFORMITY OF WEIGHTS AND MEASURES," in Great Britain and Ireland, came into operation on the first of January, 1826.

It is thereby enacted,

First. That the *brass Standard Yard of* 1760, then in custody of the Clerk of the House of Commons, shall be the *Imperial Standard Yard,* (the brass being at the temperature of 62° by Fahrenheit's thermometer); and that this Imperial Standard Yard, shall be the unit or only standard measure of extension, wherefrom or whereby all other measures of extension whatsoever, whether the same be lineal, superficial or solid, shall be derived, computed, and ascertained; and that the *thirty-sixth* part of this yard shall be *an inch.*

Now the length of a *Pendulum* vibrating *seconds* in the latitude of London, in a vacuum, and at the level of the sea, is found to be 39·1393 such inches, *i. e.* 39 such inches and 1393 ten thousandths of another such inch.

This affords the means of recovering the Imperial Standard Yard should it be lost. In fact, the brass Standard Yard of 1760 was destroyed or rendered useless by the fire at the House of Commons in 1834.

Secondly. That the *brass weight of one pound Troy of the year* 1758, then in the custody of the same officer, shall continue the unit or

Standard Measure of Weight, from which all other weights shall be derived, computed and ascertained; that 5760 grains shall be contained in the Imperial Standard Troy Pound, and 7000 such grains in the Avoirdupois Pound.

Now the weight of a *cubic inch* of distilled water is 252·458 grains Troy, the barometer being at 30 inches and the thermometer at 62°. This affords the means of recovering the Imperial Standard Pound should it be lost. In fact, the brass weight of 1758 was destroyed or lost at the above-mentioned fire.

3d. That the *Standard Measure of capacity* for Liquids and Dry Goods shall be "the *Imperial Standard Gallon*," containing 10 Pounds Avoirdupois weight of distilled water, weighed in air at a temperature of 62° Fahrenheit's thermometer, and the barometer being at 30 inches.

Now this weight fills 277·274 cubic inches, therefore the Imperial Standard Gallon contains 277·274 cubic inches.

The *Imperial Bushel*, consisting of *eight* gallons, will consequently be 2218·192 cubic inches.

REDUCTION.

71. When a number is expressed in one or more denominations, the method of finding its value in one or more other denominations is REDUCTION. Thus, £1 is of the same value of 240d., and 7s. 1½d. is of the same value as 342 farthings, and conversely: the method or process by which we find this to be so, is REDUCTION.

72. First. *To express a number of a higher denomination or of higher denominations in units of a lower denomination.*

RULE. Multiply the number of the highest denomination in the proposed quantity by the number of units of the next lower denomination contained in one unit of the highest, and to the product add the number of that lower denomination, if there be any in the proposed quantity.

Repeat this process for each succeeding denomination, till the required one is arrived at.

Ex. 1. How many cents in $127.15?

REDUCTION. 55

Proceeding by the Rule given above,

$127.15
100
───────
12700 + 15 = 12715 cents.
or $127.15 = 12715 cents.

Reasons for the process.
Since 100 cents make one dollar;
∴ $127 = (127 × 100) cts. = 12700 cts.
∴ $127.15 = 12700 cts. + 15 cts.
 = 12715 cents.
or $127.15 = 12715 cents.

Ex. 2. Reduce 27 acres, 1 rood, 32 poles, to poles.

```
acres.  rood.  poles.
 27  .   1  .   32
  4   (add the 1 rood)
 ────
 109  ro.
  40  (add the 32 poles)
 ────
 4392  poles.
```

Ex. 3. How many inches in 106 miles, 6 fur., 25 poles, 2½ yards?

```
miles.   fur.   pol.   yds.
 106  .   6  .  25  .   2½
   8   (add the 6 fur.)
 ─────
  854   fur.
   40   (add the 25 po.)
 ─────
 34185   poles
    5½   (add the 2½ yards.)
 ───────
 170927½
 170922½  (product of the ½)
 ───────
 188020   yards
      3
 ───────
 564060   feet
     12
 ───────
 6768720  inches
```

73. Second. *To express a number of lower denomination or denominations in units of a higher denomination.*

Rule. Divide the given number by the number of units which connect that denomination with the next higher, and the remainder, if any, will be the number of surplus units of the lower denomination.

Carry on this process, till you arrive at the denomination required.

Ex. 1. In 17392 cents, how many dollars and cents?

By the rule,

$$100 \begin{cases} 10 \\ 10 \end{cases} \begin{array}{l} 17392 \\ \overline{1739 \text{--} 2} \\ \overline{\$173 \text{--} 92 \text{ cts.}} \end{array}$$

Reason for the Rule.

100 cents = $1.

∴ 17392 cents ÷ 100 = $173 + 92 cents.

∴ 17392 cents = $173.92 cents.

NOTE.—From the above example, we see that by cutting off the last 2 figures on the right of any number of cents, gives the dollars, and the figures so cut off will be the cents.

Ex. 2. Reduce 49 acres, 28 poles, 10 yards, 8 feet, 112 inches, to inches. Prove the result.

```
    ac.    po.   yds.  ft.   in.
    49  .  28  .  10  .  8  . 112
     4
    ───
    196    ro.
     40    (add the 28 po.)
    ───
    7868   poles
      30¼  (add the 10 yds.)
    ──────
    236050
      1967 (product of ¼)
    ──────
    238017 yards
         9 (add the 8 ft.)
    ──────
    2142161 feet
        144 (add the 112 in.)
    ──────
    8568756
    8568644
    2142161
    ──────
    308471296 inches
```

REDUCTION

Proof
$$144 \begin{cases} 12 \\ 12 \end{cases} \begin{array}{|l} \text{sq. in.} \\ 308471296 - 4 \\ \hline 25705941 - 9 \end{array} \Big\} 112 \text{ sq. in.}$$

$$\begin{array}{r|l} 9 & 2142161 - 8 \text{ sq. ft.} \\ \hline & 238017 \end{array}$$

Now, since $30\frac{1}{4}$ or $\frac{121}{4}$ sq. yds = 1 sq. po., we multiply by 4, which reduces the sq. yds. into quarters of sq. yds., and then divide that result by 121, or 11 × 11, which brings it into sq. poles.

$$\begin{array}{r} 238017 \\ 4 \end{array}$$

$$121 \begin{cases} 11 \\ 11 \end{cases} \begin{array}{|l} 952068 - 7 \\ \hline 86551 - 3 \end{array} \Big\} \begin{array}{l} 40 \text{ quarters of sq. yds.} \\ \text{or } 10 \text{ sq. yds} \end{array}$$

$$\begin{array}{r|l} 4,0 & 786,8 - 28 \text{ sq. po.} \\ \hline 4 & 196 \\ \hline & 49 \end{array}$$

Therefore in 308471296 sq. in., there are 49 ac., 28 sq. po., 10 sq. yds., 8 sq. ft., 112 sq. in.

Ex. VIII.

Reduce and show by opposite process that your result is correct:

(1) $878.28 to cents; and $1027.87 to cents.

(2) £57 to pence; and £15.12s. to pence.

(3) 8s. $4\frac{1}{2}d.$ to half pence; and £1. 0s. $3\frac{3}{4}d.$ to farthings.

(4) £83. 15s. $6\frac{1}{4}d.$ to farthings; and £393. 0s. $11\frac{1}{2}d.$ to half-pence.

(5) 738 half-crowns to farthings; and 570 crowns to fourpenny pieces.

(6) Find the number of pounds in 5673542 farthings, and prove the truth of the result.

(7) How many half-crowns, how many sixpences, and how many fourpences, are there in 25 pounds?

Reduce, verifying the result in each case, the following:

ARITHMETIC.

(8) 59 lbs., 7 oz., 14 dwts., 19 grs., to grains; and 37400157 grs. to lbs.

(9) 56332005 scrs. to lbs. Troy: and 536 lbs. to drams and scruples.

(10) 7 tons, 15 cwt., 2 qrs., 16 lbs. to ounces; and 7563241 drs. to tons.

(11) 5838297 oz. to tons; and 33 tons, 17 cwt., 3 qrs., 27 lbs., 15 drs. to drams (cwt.=112 lbs.).

(12) 17 lbs., 2 ℥, 2 ℈ to grains; and 34678 grs. Apoth. to oz. Troy.

(13) 3 m., 7 fur., 8 po. to yards; and 573 miles to inches.

(14) 1364428 in. to leagues; and 74 m., 3 fur., 4 yds to inches.

(15) 4 lea., 2 m., 2 in. to barleycorns; and 50 m., 3 po. to yards.

(16) 7 fur., 200 yds. to chains; and 6 cubits, 1 span to feet.

(17) 84 yds., 1 qr to nails; and 56 Eng. ells, 1 qr. to nails.

(18) 83 Fr. ells, 3 qrs. to nails; and 73 Fl. ells, 1 qr. to nails.

(19) 35 ac., 2 ro. to poles; and 56 ac., 2 ro. to yards.

(20) 3 ro., 37 po., 26 yds. to inches; and 3 ac., 30 po. to feet.

(21) 15 ac., 3 ro. to links; and 50000 po. to acres.

(22) 29 cub. yds. to feet; and 158279 cub. in. to yds.

(23) 17 cub. yds., 1001 cub. in. to inches; and 26 cub. yds., 19 cub. ft. to inches.

(24) 563 gals. to pints; and 365843 gills to gallons.

(25) 760 bus., 3 pks. to quarts; and 2875646 quarts to bus.

(26) 250 chaldrons to bushels; and 186043 pks. to chaldrons.

(27) 56 reams, 19 quires to sheets; and 52073 sheets of paper to reams.

(28) 36 wks., 5 d., 17 hrs., to seconds; and 1 mo. of 30 days, 23 hrs. 59 sec. to seconds.

(29) How many barrels, gallons, quarts, and pints are there in 1336381 half-pints?

(30) One year being equivalent to 365 days, 6 hours, find how many seconds there are in 27 years, 245 days.

(31) From 9 o'clock P. M., Aug. 5, 1852, to 6 o'clock A. M., March 3, 1853, how many hours are there, and how many seconds?

(32) In Great Britain and Ireland there are 121838 square miles; in British North America, 3389345 square miles; in the United States of America, 3306000 square miles; how many acres in each of those counties?

COMPOUND ADDITION.

74. Compound Addition is the method of collecting several numbers of the same kind, but containing different denominations of that kind, into one sum.

Rule. Arrange the numbers, so that those of the same denomination may be under each other in the same column, and draw a line below them.

Add the numbers of the lowest denomination together, and find by Reduction how many units of the next higher denomination are contained in this sum.

Write the remainder, if any, under the column just added, and carry the quotient to the next column.

Proceed thus with all the columns.

Ex. 1. Add together $37.95, $36.87, $97.48.

By the Rule,

$37.95
36.87
97.48
―――
$172.30

The sum of the right-hand column is 20; write 0 under that column, and carry 2 to the next: the sum of the next column together with the 2 carried is 23; write 3 under that column and carry 2 to the next, and so on; the same way as was done in the Simple Rules, and for the same reason.

Ex. 2. Add together £2. 4s. 7½d., £3. 5s. 10¼d., £15. 15s., and £33. 12s. 11½d.

Proceeding by the Rule given above,

```
   £  .  s.  .  d.
   2  .  4  .  7½
   3  .  5  . 10¼
  15  . 15  .  0
  33  . 12  . 11½
  ―――――――――――――
 £54  . 18  .  5¼
```

Reason for the above process.

The sum of 2 farthings, 1 farthing and 2 farthings, = 5 farthings,

= 1 penny, and 1 farthing; we therefore put down $\frac{1}{4}$, that is, one farthing, and carry 1 penny to the column of pence. Then

$$(1+11+10+7)d. = 29d. = (12\times 2+5)d.$$

or 2 shillings, and 5 pence; we therefore put down $5d.$, and carry on the 2 to the column of shillings.

Then $(2+12+15+5+4)s. = 38s. = (20\times 1+18)s.$ £1., and 18*s.*; we therefore put down 18*s.*, and carry on the 1 pound to the column of pounds. Then (1+33+15+3+2) pounds = £54.

Therefore the result is £54. 18*s.* $5\frac{1}{4}d.$

NOTE. The method of proof is the same as that in Simple Addition.

Ex. 3. Add together 34 tons, 15 cwt., 1 qr., 14 lbs.; 42 tons, 3 cwt. 18 lbs.; 18 tons, 19 cwt., 3 qrs.; 7 cwt., 6 lbs.; 2 qrs., 19 lbs.; and 3 tons, 7 lbs.

	tons.	cwt.	qrs.	lbs.
	34	15	1	14
	42	3	0	18
	18	19	3	0
	0	7	0	6
	0	0	2	19
	3	0	0	7
Ans.	99	6	0	14

Ex. IX.

(1)	£.	s.	d.
	5	17	$10\frac{1}{2}$
	36	0	11
	7	3	$4\frac{1}{4}$
	73	19	$8\frac{3}{4}$
	30	14	$5\frac{1}{4}$

(2)	£.	s.	d.
	63	15	$2\frac{1}{4}$
	83	8	$9\frac{3}{4}$
	41	0	$11\frac{3}{4}$
	6	7	$10\frac{1}{2}$
	76	17	$1\frac{3}{4}$

(3)	£.	s.	d.
	528	14	$11\frac{3}{4}$
	854	19	4
	578	18	$9\frac{1}{2}$
	507	0	$0\frac{3}{4}$
	859	14	$11\frac{1}{2}$

(4)	tons.	cwt.	qrs.	lbs.
	16	17	2	24
	13	10	0	20
	17	15	2	19
	84	0	3	22
	11	11	1	11

(5)	oz.	drs.	sc.	grs
	22	3	2	19
	56	0	1	10
	3	2	2	11
	15	6	1	9
	79	4	1	10

(6)	ac.	ro.	po.
	82	2	24
	18	3	14
	20	1	27
	56	0	0
	45	3	30

COMPOUND ADDITION. 61

(7) Find the sum of £3966. 16s. 9¼d., £2. 11s. 7¼d., £3795. 0s. 2¼d. £37. 17s. 0¾d., £48. 0s. 0¾d., and £59000. 14s. 6¼d.; also of £6491, £3651. 10s. 3¼d., £8000. 0s. 11¾d., £5510. 19s. 10½d., £50430. 12s. 1¼d., £316. 14s. 5¾d., and £4850. 18s. 4d.; also of £306217. 13s. 9¾d., £55. 0s. 9d., £450812. 15s. 2¼d., £9837. 1s. 5½d., and £2939. 3s. 11¾d.; and prove the result in each case.

(8) Add together 2 lbs., 9 oz., 1 dwt., 23 grs.; 8 lbs., 6 oz., 4 dwt., 20 grs.; 1 lb., 10 oz., 5 dwt., 12 grs.; 14 lbs., 11 oz., 14 dwts., 19 grs.; and 21 lbs., 8 oz., 13 dwt., 11 grs.: also 22 lbs., 7 dwt., 15 grs.; 15 lbs., 11 oz., 18 grs.; 34 lbs., 9 oz., 12 dwt.; 74 lbs., 1 oz., 1 dwt., 20 grs.; and 46 lbs., 11 oz,, 16 dwt., 19 grs.: also 1740 oz., 9 dwt., 19 grs.; 4179 oz., 11 dwt., 14 grs.; 8497 oz., 12 dwt., 22 grs.; 5629 oz., 19 dwt., 17 grs.; and 1038 oz., 4 dwt., 14 grs.: verify each result.

(9) Add together 10 lbs., 8 oz., 4 drs., 1 scr.; 66 lbs., 10 oz., 2 drs., 19 lbs., 9 oz., 3 drs., 2 scr.; 55 lbs., 6 drs.; and 79 lbs., 11 oz., 4 drs., 1 scr.: also 13 lbs., 6 oz., 7 drs., 2 scr., 17 grs.; 19 lbs., 11 oz,, 1 scr., 18 grs.; 36 lbs., 3 oz., 2 scr., 19 grs.; 6 oz., 7 drs., 7 grs.; and 176 lbs., 96 grs.: explain the process in each case.

(10) Find the aggregate of 1 cwt., 2 qrs., 26 lbs., 10 oz.; 11 cwt., 18 lbs., 9 oz.; 13 cwt., 3 qrs., 17 lbs., 14 oz.; 7 cwt., 1 qr., 25 lbs., 9 oz.; and 19 cwt., 2 qrs., 19 lbs., 14 oz.: also of 306 tons, 15 cwt., 2 qrs., 15 lbs.; 731 tons, 6 cwt., 3 qrs., 24 lbs; 279 tons, 7 cwt., 10 lbs.; 896 tons, 9 cwt., 1 qr., 17 lbs.; and 10 cwt., 2 qrs., 16 lbs.: also of 23 tons, 12 cwt., 15 lbs., 12 oz.; 58 tons, 17 cwt., 1 qr., 10 oz.: 67 tons, 3 qrs., 15 oz.; 19 cwt., 27 lbs.; and 3 tons, 13 lbs., 13 oz.: prove the results (cwt. = 112 lbs.).

(11) Find the sum of 26 m., 7 fur., 23 po., 3 yds.; 22 m., 5 fur., 27 po., 5 yds.; 37 m., 4 fur., 3 yds.; 86 m., 6 fur., 38 po., 3 yds.; and 25 m., 1 fur., 29 po., 2½ yds.; also of 14 m., 7 fur., 23 po., 2½ yds., 2 ft., 11 in.; 12 m., 5 fur., 1 yd., 2 ft., 3 in.; 27 m., 2 fur., 13 po., 3½ yds., 1 ft., 10 in.; 36 m., 6 fur., 33 po., 4½ yds., 2 ft., 6 in.; and 75 m., 1 fur., 21 po., 3 yds., 1 ft., 7 in.: also of 2 lea., 1 m., 3 fur., 103 yds., 67 lea., 3 fur., 157 yds.; 11 lea., 1 m., 98 yds.; 9 lea., 2 m., 5 fur., 87 yds.; and 34 lea., 2 m., 7 fur., 198 yds.

(12) Find the sum of 43 yds., 2 qrs., 3 na.; 37 yds., 2 qrs., 1 na.; 23 yds., 3 qrs., 2 na.; 41 yds., 2 qrs., 2 na.; and 38 yds., 2 qrs., 3 na.: and of 11 Eng. ells, 2 qrs., 3 na.; 13 Eng. ells, 2 qrs., 1 na.; 39 Eng. ells,

4 qrs., 2 na.; 37 Eng. ells, 4 qrs., 3 na.; and 79 Eng. ells, 3 na.: and prove each result.

(13) Find the sum of 25 ac., 2 ro., 16 po.; 30 ac., 2 ro., 25 po; 26 ac., 2 ro., 35 po.; 63 ac., 1 ro., 31 po.; and 34 ac., 2 ro., 29 po.: also of 5 ac., 2 ro., 15 po., 25¼ sq. yds., 101 sq. in.; 9 ac., 1 ro., 35 po., 12½ sq. yds., 87 sq. in.; 42 ac., 3 ro., 24 po., 23¾ sq. yds., 57 sq. in.; 12 ac., 2 ro., 5 pa., 13¾ sq. yds., 23 sq. in.; and 17 ac., 24 po., 30 sq. yds., 113 sq. in.: explain each process.

(14) Find the sum of 3 c. yds., 23 c. ft., 171 c. in.; 17 c. yds., 17 c. ft., 31 c. in.; 28 c. yds., 26 c. ft., 1000 c. in.; and 34 c. yds., 23 c. ft., 1101 c. in.

(15) Add together 39 gals., 3 qts., 1 pt.; 48 gal., 2 qts., 1 pt.; 56 gals., 1 pt.; 74 gals., 3 qts.; and 84 gals., 3 qts., 1 pt.: also 2 pipes, 42 gals., 3 qts.; 36 gals., 1 qt.; 5 pipes, 48 gals.; 12 pipes, 53 gals., 3 qts.; and 27 pipes, 2 qts., of wine: also 19 hhds., 10 gals., 3 pts.; 29 hhds., 50 gals., 7 pts.; 116 hhds., 46 gals., 5 pts.; 2 hhds., 2 pts.; and 235 hhds., 1 bar., 3 qts., of beer.

(16) Add together $19.28, $27.35, $37.39, $216.16, $152.93, $225.17, and $23.19; also $2795.28, $3878.15, $737.35, $6797.27, $9689.21, $5293.78, $69256.36, $52678.38, $27812.15.

(17) Add together 4 mo., 3 w., 5 d., 23 h., 46 m.; 5 mo., 1 d., 17 h., 57 m.; 6 mo., 2 w., 1 h.; 1 w., 6 d., 23 h., 59 m.; and 11 mo., 1 w., 58 m.: also 7 yrs., 28 w., 3 s.; 26 yrs., 5 w., 5 d.; 58 yrs., 6 d., 23 h., 59 s.; 43 w., 23 h., 50 m., 12 s.; and 124 yrs., 14 w., 19 h., 37 s.

(18) When B was born, A's age was 2 yrs., 9 m., 3 w., 4 d.; when C was born, B's age was 13 yrs., and 3 d.; when D was born, C's age was 9 mo., 2 w., 3 d., 23 h.; when E was born, D's age was 6 yrs., 11 mo., 23 h.; when F was born, E's age was 7 yrs., 3 w., 5 d., 15 h. What was A's age on F's 5th birth-day?

COMPOUND SUBTRACTION.

75. **Compound Subtraction** is the method of finding the difference between two numbers of the same kind, but containing different denominations of that kind.

COMPOUND SUBTRACTION.

RULE. **Place the less number below the greater,** so that the numbers of the same denomination may be under each other in the same column, and draw a line below them.

Begin at the right hand, and subtract if possible each number of the lower line from that which stands above it, and set the remainder underneath.

But when any number in the lower line is greater than the number above it, add to the upper one as many units of the same denomination as make one unit of the next higher denomination; subtract as before, and carry one to the number of the next higher denomination in the lower line.

Proceed thus throughout the columns.

Ex. 1. From $2782.25 take $1783.29.

$2782.25
$1783.29
―――――
$998.96

This example is worked in the same way as Simple Subtraction.

Ex. 2. Subtract £88. 18s. 8½d, from £146. 19s. 6¼d.

Proceeding by the Rule given above.

£.	s.	d.
146 .	19 .	6¼
88 .	18 .	8½
£58 .	0 .	9¾

Reason for the above process.

Since ½d. is greater than ¼d. we add to ¼d. 4 farthings or 1 penny, thus raising it to 5 farthings; and when 2 farthings are subtracted from 5 farthings, we have 3 farthings left; we therefore write down ¾d.; and in order to increase the lower number equally with the upper number, we add 1 penny to the 8 pence.

Now 9 pence cannot be taken from 6 pence; we therefore add 12 pence or 1s. to 6 pence, thus raising the latter to 18d.: we take the 9d. from 18d., and put down the remainder 9d.; then adding 1s. to 18s., the latter becomes 19s.: 19s. taken from 19s. leave no remainder: we then subtract £88. from £146., as though they were abstract numbers.

It is manifest that in this process, whenever we add to the upper line we also add a number of the same value to the lower line, so that the final difference is not altered.

Ex. 3. Subtract 106 lbs., 11 oz., 16 dwt., from 144 lbs., 8 oz. 14 dwts.

lb.		oz.		dwt.
144	.	8	.	14
106	.	11	.	16
37	.	8	.	18

Ex. X.

	£.		s.		d.			£.		s.		d.
(1)	343	.	18	.	5¼		(2)	663	.	5	.	11½
	11	.	18	.	5¾			349	.	19	.	9¾

	cwt.		qr.		lbs.		oz.			fur.		po.		yds.
(3)	63	.	0	.	18	.	1	(4)		14	.	34	.	5
	58	.	1	.	12	.	10			1	.	38	.	4

	ac.		ro.		po.			qrs.		bus.		pk.		gal.
(5)	63	.	1	.	29		(6)	64	.	3	.	1	.	0
	57	.	2	.	38			8	.	5	.	3	.	1

(7) Subtract £456. 15s. 11¾d. from £534. 13s. 10½d.; and prove the result.

Find the difference between the following numbers, and verify the results:

(8) 5836 lbs., and 4976 lbs., 7 oz., 13 dwt., 19 grs.

(9) 26 tons, 2 qrs., 23 lbs., and 19 tons, 3 cwt., 3 qrs., 18 lbs.

(10) 144 lbs., 9 oz., 4 drs., 1 sc., and 129 lbs., 7 drs., 3 sc.

(11) 418 yds., 1 qr., 1 na., and 387 yds., 3 qrs., 3 na.

(12) 15 yds., 1 ft., 5 in., and 13 yds., 2 ft. 7 in.

(13) 13 m., 6 fur., 35 po., 3½ yds., and 12 m., 38 po., 4 yds.

(14) 3 ro., 28 po., 27 sq. yds., 7 sq. ft., and 1 ro., 39 po., 28¼ sq. yds., 8 sq. ft.

COMPOUND MULTIPLICATION. 65

(15) 37 cub. yds., 18 cub. ft., 857 cub. in., and 35 cub. yds., 24 cub. ft. 1280 cub. in.

(16) 203 tuns, 19 gals., 3 qts., 1 pt., of wine, and 187 tuns, 1 hhd., 29 gals., 2 qts.

(17) 83 bar., 2 fir., 7 gals., of beer, and 77 bar., 2 fir., 8 gals., 29 qts.

(18) 216 yrs., 9 mo., 2 w., 4 d., and 217 yrs.

(19) The latitude of the Provincial "University of Toronto" at Toronto is 43°, 39′, 24″, north, that of St. Paul's at London (England) is 51°, 30′, 49″ north. Find the difference of their latitude.

(20) What sum added to £947. 19s., 7¾d., will make £1000?

(21) A furnished house is worth $17935.50; unfurnished, it is worth $5978.50. By how much does the value of the furniture exceed the value of the house?

COMPOUND MULTIPLICATION.

76. COMPOUND MULTIPLICATION is the method of finding the amount of any proposed compound number, that is, of any number composed of different denominations, but all of the same kind, when it is repeated a given number of times.

RULE. Place the multiplier under the **lowest denomination of the** multiplicand.

Multiply the **number of** the lowest denomination **by the** multiplier, **and find the number of units** of the next denomination contained in this first product; if there be a remainder, write it down; for the second product, **multiply the number of the next denomination in the** multiplicand by **the multiplier, and** after adding to it the above-mentioned **number of units, proceed with the result as with the first** product.

Carry this operation through with all the different denominations of the multiplicand.

Ex. Multiply $212.13 by 12.

$212.13
    12
―――
$2545.56

In this example we do the same as in simple Multiplication, observing to place the point separating the dollars and cents in its proper place.

Ex. 2. Multiply 56. 4s. 6½d. by 5.
Proceeding by the Rule given above,

£.	s.	d.
56 .	4 .	6½
		5
£281 .	2 .	8½

Reason for the above process.

½d. multiplied by 5 is the same as $(\frac{1}{2}+\frac{1}{2}+\frac{1}{2}+\frac{1}{2}+\frac{1}{2})d.=5$ half-pence $=2\frac{1}{2}d.$; we therefore write down ½d., and carry 2d. to the denomination of pence:

6d. multiplied by $5=30d.$; therefore $(2+6\times5)d.=32d.=(2\times12+8)d.=2s.+8d.$; we therefore write down 8d., and carry 2s. to the denomination of shillings:

4s. multiplied by $5=20s.$; therefore $(2+4\times5)s.=22s.=(20+2)s.==£1+2s.$; we therefore write down 2s., and carry £1 to the denomination of pounds:

Now by Simple Multiplication $£56\times5=£280$; therefore $£(1+56\times5)=£(1+280)=£281.$

Therefore the total amount is £281. 2s. 8½d.

77. When the multiplier exceeds 12, a convenient method is to split the multiplier into factors, or into factors and parts: thus $15=3\times5$; $17=3\times5+2$; $23=4\times5+3$; $240=4\times6\times10$: and so on.

Ex. Multiply £55. 12s. 9¼d. by 23.

£.	s.	d.	
55 .	12 .	9¼	
		4	
222 .	11 .	1	=value of £55. 12s. 9¼d. multiplied by 4.
		5	
1112 .	15 .	5	=value of £222. 11s. 1d. multiplied by 5, or of £55. 12s. 9¼d. multiplied by (4×5), or 20.
166 .	18 .	3¾	=value of £55. 12s. 9¼d. multiplied by 3.
£1279 .	13 .	8¾	=value of £55. 12s. 9¼d multiplied by (20+3), or 23.

NOTE. For an example, when the multiplier is a large number, see the Elementary Arithmetic, p. 60.

COMPOUND DIVISION. 67

NOTE. **When** the multiplicand contains farthings, if one of the factors of the multiplier be even, it will often be advantageous **to use** it first, as the farthings **may disappear.**

Ex. XI.

Multiply
(1) $217.35 separately by 8 and 14.
(2) £7. 19s. 7½d. separately by 10 and 12.
(3) £721. 0s. 5¼d. separately by 81 and 96.
(4) £2579. 0s. 0¾d. separately by 147, 155, 474, and 2331.
(5) 86 lbs., 7 oz., 16 dwt., 11 grs. separately by 8 and 36.
(6) 3 tons, 24 lbs., 13 oz. separately by 11 and 76.
(7) 45 lbs., 7 oz., 3 drs., 2 sc. separately by 12 and 68.
(8) 67 yds., 1 qr., 2 na. separately by 9 and 53.
(9) 70 yds., 2 ft., 10 in. separately by 7 and 29.
(10) 16 ac., 3 ro., 38 po., 27 yds., 2 ft. by 11.
(11) 380 ac., 3 ro., 32 po. separately by 12 and **106.**
(12) 57 gals., 3 qts., separately by 10 and 257.
(13) 76 qrs., 5 bus., 2 pks. separately by 13 and 240.
(14) 5 wks., 6 d., 18 h., 14 m. separately by 11 and 339.
(15) 84 hhds., 43 gals., 1 pt. of wine separately by 27 and 364.
(16) 43 bar., 13 gals., 1 qt., 1 pt. of beer separately by 39 and 764.
(17) A person buys 67 lambs at £1. 0s. 9½d. each; 73 sheep at £2. 2s. 11¼d. each; 12 cows at the average of £37. 0s. 2¾d. for every 3 of them; and 17 horses at 37 guineas each; the expenses of getting them all home amount to 17½ guineas. What money must he draw from his bankers to pay for the whole outlay?
(18) There are 7 chests of drawers: in each chest there are 18 drawers; and in each drawer 8 divisions; and in each division there is placed $25.25. How much money is deposited in the chests?

COMPOUND DIVISION.

78. COMPOUND DIVISION is the method of dividing a compound number, that is, a number composed of several denominations, but all of the same kind, into as many equal parts as the divisor contains units; and also of finding how often one compound number is contained in another of the same kind.

ARITHMETIC.

When the Divisor is an abstract number, either larger or not larger than 12.

RULE. Place the numbers as in Simple Division; then find how often the divisor is contained in the highest denomination of the dividend; put this number down in the quotient; multiply as in Simple Division and subtract.

If there be a remainder, reduce the remainder to the next inferior denomination, adding to it the number of that denomination in the dividend, and repeat the division.

Carry on this process through the whole dividend.

Ex. 1. Divide £199. 6s. 8d. by 130.

Proceeding by the Rule given above,

```
       £.    s.   d.                 The work is usually written thus:
130) 199 .  6  . 8 (1£.                 £    s.   d.
     130                         130) 199 .  6 . 8 (1£. 10s. 8d.
     ───                              130
      69                              ───
      20 (add the 6s.)                 69
     ───                               20
130) 1386 (10s.                       ────
     1300                             1386
     ────                              1300
       86                              ────
       12 (add the 8d.)                 86
     ────                               12
130) 1040 (8d.                        ────
     1040                             1040
```

Therefore the answer is £1. 10s. 8d.

Reason for the above process.

We first subtract £1 taken 130 times, from £199. 6s. 8d., and there remains £69. 6s. 8d.

Now, £69: 6s. 8d. = 1386s. 8d.; from this amount we subtract 10s. taken 130 times, and there remains 86s. 8d.

Again, 86s. 8d. = 1040d.; from this amount we subtract 8d. taken 130 times, and nothing remains.

Therefore £1. 10s. 8d. is contained 130 times in £199. 6s. 8d.

COMPOUND DIVISION. 69

Ex. 2. Divide £1076. 4s. 3¼d. by 527.

```
              £.    s.   d.
      527)  1076 .  4 .  3¼   (2£. 0s. 10d, 247/527.
            1054
            ────
              22
              20 (add the 4s.)
             ────
             444  (0s.
              12 (add the 3d.)
             ────
      527)  5331  (10d.
             527
             ───
              61
               4 (add the 3q.)
             ────
      527)   247  (0q.
```

Therefore the result is £2 0s. 10d., and there remain 247 farthings to be divided by 527, which division will clearly not give so much as one farthing.

Therefore the quotient is £2. 0s. 10d. 0 247/527 q.

Ex. XII.

(1) £386. 16s. 5¼d. ÷ 11. (2) £473. 14s. 6d. ÷ 12.
(3) 459 lbs., 4 oz., 5 dwt., 22 grs. ÷ 29.
(4) 15511 lbs., 3 oz., 6 drs., 2 sc. ÷ 68. (5) £1288. 1s. 8d. ÷ 754.
(6) 2 fur., 10 po., 1 yd., 1 ft., 10 in. ÷ 35. (7) £165. 15s. 8¼d. ÷ 139.
(8) £2728 ÷ 744. (9) 1738 c. yds., 1236 c. in. ÷ 798.
(10) £37. 3s. 1d. ÷ 74. (11) 266 tuns, 33 gals. ÷ 102.
(12) £492710. 1s. 8d. ÷ 6352. (13) $61411 ÷ 217.
(14) £1746 ÷ 2737. (15) £130264. 9s. 6d. ÷ 9416.
(16) 1288 cwt., 4 lbs. ÷ 75. (17) 178 cwt., 3 qrs., 14 lbs. ÷ 53.
(18) 206 mo. of 28 days, 4 d. ÷ 26. (19) 684 d., 8 h., 9 m, ÷ 47.
(20) 15 cwt., 27 lb., 11 oz. ÷ 456. (21) 76 cwt, ÷ 963.
(22) 75 ac., 3 ro., 39 po. ÷ 26. (23) 13 ac., 1 ro. ÷ 147.
(24) 91 yds., 2 qrs., 1 na. ÷ 903. (25) 97 qrs., 3 bus., 3 pks. ÷ 107
(26) $455455 ÷ 637.

79. It may sometimes be found convenient to break up the divisor into factors: thus,

ARITHMETIC.

Ex. 3. Divide £131. 2s. 8½d. by 48, and also by its factors 6 and 8, and show that the results coincide.

$$
\begin{array}{r}
\text{£.} \quad s. \quad d. \\
48)\ 131\ \ 2\ .\ 8\tfrac{1}{2}\ \ (2\text{£.}\ 14s.\ 7\tfrac{1}{2}d.\ \tfrac{34}{48}.\\
96 \\ \hline
35 \\
20 \\ \hline
48)\ 702\ \ (14s. \\
48 \\ \hline
222 \\
192 \\ \hline
30 \\
12 \\ \hline
48)\ 368\ \ (7d. \\
336 \\ \hline
32 \\
4 \\ \hline
48)\ 130\ \ (2q. \\
96 \\ \hline
34
\end{array}
$$

Now, dividing by the factors 6 and 8 we get

$$
48\begin{cases} 6 \\ 8 \end{cases}
\begin{array}{|ccc}
\text{£.} & s. & d. \\
131 & 2 & 8\tfrac{1}{2} \\ \hline
21 & 17 & 1\tfrac{1}{4}\ \text{rem. } 4 \\ \hline
2 & 14 & 7\tfrac{1}{2}\ \tfrac{34}{48}\ \text{rem. } 5
\end{array}
$$

Therefore the true remainder
$$= (5 \times 5 + 4)q. = 34q.,$$

and since our divisor is 48, we write the remainder; thus $\tfrac{34}{48}$.

Therefore the quotient is £2. 14s. 7½d. $\tfrac{34}{48}q.$

80. *When the divisor and dividend are* both compound numbers of the same kind.

RULE. Reduce both numbers to the same denomination: divide as in Simple Division, and the result will be the answer required.

Ex. How often is 5s. 3¾d. contained in £15. 8s. 9d.?

Proceeding by the above Rule,

$$
\begin{array}{r}
s. \quad d. \\
5\ .\ 3\tfrac{3}{4} \\
12 \\ \hline
63 \\
4 \\ \hline
255
\end{array}
\qquad
\begin{array}{r}
\text{£.} \quad s. \quad d. \\
15\ .\ 18\ .\ 9. \\
20 \\ \hline
318 \\
12 \\ \hline
3825 \\
4 \\ \hline
15300
\end{array}
$$

$$
255)\ 15300\ (60 \\
\ 1530
$$

Therefore 60 is the answer.

MISCELLANEOUS EXAMPLES WORKED OUT. 71

Reason for the above process.
$$5s.\ 3\tfrac{3}{4}d. = 255 \text{ farthings.}$$
$$£15.\ 18s.\ 9d. = 15300 \text{ farthings.}$$
and 255 farthings subtracted 60 times from 15300 farthings leaves no remainder.

Ex. XIII.

(1) £2. 12s. 3d. ÷ 1s. 4½d.
(2) £55. 18s. 10¼d. ÷ £2. 8s. 7¾d.
(3) £160. 4s. 8¼d. ÷ £1. 10s. 6¼d.
(4) £401. 4s. 3d. ÷ £2. 11s. 5¼d.
(5) 44 cwt., 2 qrs., 11 lbs. ÷ 1 cwt., 2 qrs., 17 lbs.
(6) 272 yds., 1 qr. ÷ 7 yds., 2 qrs., 1 na.
(7) 9487 bus., 2 pks. ÷ 143 bus., 3 pks.
(8) 1416 ac., 2 ro., 16 po., ÷ 4 ac., 3 ro., 27 po.
(9) 57 lea., 1 m., 956 yds. ÷ 7 fur., 87 yds., 1 ft., 5 in.
(10) $63729 ÷ $873.

Miscellaneous Examples, depending on Arts. (71—80) *worked out.*

Ex. 1. A person bought 500 yards of cloth at $3.78 a yard, and retailed it at $3.90 a yard : what was his profit?

His profit on 1 yard = 12 cents; for $3.90 − $3.78 = 12 cents. therefore his whole profit = (12 × 500) cts.
$$= \$60$$

Ex. 2. A spring of water, which yields 75 gallons an hour, supplies 600 families: how much water may each family use daily?

The daily supply of water = (75 × 24) gallons;

therefore each family may use daily $\dfrac{75 \times 24}{600}$ gals., or 3 gals.

Ex. 3. How many revolutions will a wheel, which is 4 yards in circumference, make in 3 miles?
$$3 \text{ miles} = (3 \times 1760) \text{ yards} = 5280 \text{ yards,}$$
and since the wheel passes over 4 yards in one revolution;

$\dfrac{5280}{4}$ or 1320 = number of revolutions required.

Ex. 4. How many guineas, sovereigns, half-crowns, and shillings and of each an equal number, are there in £1246?

Now, 1 guinea + 1 sovereign + 1 half-crown + 1 shilling

$$= (42+40+5+2) \text{ sixpences}$$
$$= 89 \text{ sixpences};$$

and £1246 = (1246 × 20 × 2) sixpences = 49840 sixpences;

the question therefore is reduced to this: How often are 89 sixpences contained in 49840 sixpences?

$$\text{Number required} = \frac{49840}{89} = 560$$

Ex. 5. How much water must be added to a cask containing 60 gallons of spirit at $3 a gallon, to reduce the price to $1.92 a gallon?

Cost of cask = ($3) × 60 = $180 = 18000 cents.

$1.92 = 192 cents

therefore $\frac{18000}{192}$, or $93\frac{3}{4}$ = the number of gallons which the cask must contain, in order that its contents may be sold at $1.92 a gallon.

Therefore $(93\frac{3}{4} - 60)$, or $33\frac{3}{4}$ = the number of gallons of water which have to be added.

Ex. 6. A traveller walks 22 miles a day, and after he has gone 84 miles, another follows him at the rate of 34 miles a day; in what time will the second traveller overtake the first?

The second traveller has to walk over 84 miles more than the first before he can overtake him.

Each day he walks (34 − 22) or 12 miles more than the first;

therefore $\frac{84}{12}$ or 7 is the number of days required.

Ex. 7. A mixture is made of 8 gallons of spirits at 12s. 10d. a gallon, 7 gallons at 10s. 6d. a gallon, and 10 gallons at 9s. 1d. a gallon; at what price per gallon must the mixture be sold, 1st, that the seller may neither gain nor lose by his bargain; 2nd, that he may gain £1. 13s. by it.

MISCELLANEOUS EXAMPLES WORKED OUT. 73

1st. If he is neither to gain nor lose, he must sell 1 gallon for $\frac{£13.\ 7s.}{25}$; which, worked out, gives 10s. $8\frac{4}{25}d.$ as the price required.

2nd. If he is to gain £1. 13s.

25 gallons must be sold for £13. 7s. + £1. 13s., or £15; therefore, 1 gallon must be sold for $\frac{£15}{25}$; which, worked out, gives 12s. as the price required.

Ex. 8. Divide $49 among 5 men, 6 women, and 7 boys; giving each woman twice as much as each boy, and each man thrice as much as each woman.

Since each woman's share = twice each boy's share,
therefore 6 women's shares = 12 boys' shares.

Again, since each man's share = thrice each woman's share,
therefore, 5 men's shares = 15 women's shares,
= 30 boys' shares,

but **5 men's shares + 6 women's shares + 7 boys' shares = $49**,
or **30 boys' shares + 12 boys' shares + 7 boys' shares = $49**,
or 49 boys' shares = $49

Therefore, each **boy's share** = $$\frac{49}{49}$ = 1.

Therefore, each **woman's** share = $2.
.... each **man's** share = $6.

NEW DECIMAL COINAGE.

81. The House of Commons of the Dominion of Canada **intends to** adopt a coinage based on the decimal system, (as the present is), but of which the unit shall be the sovereign; thus having the same coinage as is proposed to be used in England. The coins proposed for such a system **are pounds,** florins, cents, **mils.** The table would stand thus:

    10 mils (m.) make 1 cent, 1 c.
    10 cents .... 1 florin, 1 fl.
    10 florins .... 1 pound, £1.

82. In such a system, much of the labour of reducing superior to inferior denominations, and the converse, is done away with; **for we**

4

could at once say, £24. 3 fl. 7 c. 2 m.= 24372 m.   Since by performing the operation of reducing at length, we obtain

$$
\begin{array}{cccc}
£. & \text{fl.} & \text{c.} & \text{m.} \\
24 & .\ 3 & .\ 7 & .\ 2
\end{array}
$$

        10

240+3, or 243 fl.
              10

        2430+7, or 2427 c.
                    10

            24370+2, or 24372 m.

or we might say £24. 3 fl. 7 c. 2 m.= £24·372.
    Similarly, £24. 3 fl. 7 c. 2 m.= 243·72 fl., or = 2437·2 c.
        Conversely 24372 mils = £24. 3 fl. 7 c. 2 m.
for, **proceeding by Rule** (Art. 73), we get

$$
\begin{array}{r|l}
10 & 24372 \\ \hline
10 & 2437\text{--}2\ \text{m.} \\
10 & 243\text{--}7\ \text{c.} \\ \hline
& 24\text{--}3\ \text{fl.}
\end{array}
$$

      hence 24372 m.= £24. 3 fl. 7 c. 2 m.

Again, £254. 5¼ c.= £254. 5·5 c.
             100

      25400 c.+5·5 c.
       = 25405·5 c.
       = 254055 m.

Also, £254. 5¼ fl.= £254. 5·25 fl.
            10

     2540 fl.+5·25. fl.
    = 2545·25 fl.
    = 25452·5 c.
    = 254525 m.

Ex. XIV.

Reduce, expressing in each successive inferior denomination and verifying each result:

(1) £15. 6 fl. to mils, and 6 fl. 3 c. 2 m. to mils.

(2) £30. 9½ fl. to mils, and £96. 1 fl. 2 c. 9 m. to mils.
(3) £18. 6½ c. to mils, and 9¼ fl. to mils.
(4) £10. 1 m. to mils, and £46. 2½ c. to mils.

83. The addition, subtraction, multiplication, and division of money would be easily performed, as will be evident from the following examples.

Ex. 1 Find the sum of £18. 6 fl. 3 c. 5 m.; 9 fl. 9 m.; £24. 1 m.; 3 c. 2 m.; 5¼ fl.

	m.		£.
£18. 6 fl. 3 c. 5 m.=	18635,	or =	18·635,
9 fl. 9 m.=	909,	or =	·909,
£24. 1 m.=	24001,	or =	24·001,
3 c. 2 m.=	32,	or =	·032,
5¼ fl.=	525,	or =	·525,
	44102 m.,	or =	£44·102,

each of which results = £44. 1 fl. 2 m.

Ex. 2. From £16. 3 c. 2 m., subtract £14. 4 fl. 9 m.

	m.		£.
£16. 3 c. 2 m.=	16032.	or =	16·032,
£14. 4 fl. 9 m.=	14409,	or =	14·409,
	1623 m,	or =	£1·623,

each of which results = £1. 6 fl. 2 c. 3 m

Ex. 3. Multiply £16. 3 c. 2 m. by 23.

£16. 3 c. 2 m.= 16032 m., or = £16·032.

m.	£
16032	16·032
23	23
48096	48096
32064	32064
368736 m.	£368·736

each of the above results = £368. 7 fl. 3 c. 6 m.

Ex. 4. Divide £368. 7 fl. 3 c. 6 m. by 23.

76   ARITHMETIC.

In other words, divide 368736 m. by 23, or £368·736 by 23.

```
         m.      m³                    £.      £.¹
     23) 368736 (16032             23) 368·736 (16·032
         23                            23
         ---                           ---
         138                           138
         138                           138
         ---                           ---
           73                            73
           69                            69
           ---                           ---
           46                            46
           46                            46
```

each of the above results = £16. 0 fl. 3c. 2 m.

Note. Similar advantages would result from the use of a decimal system in weights and measures.

Ex. XV.

Add together

(1) £76. 8 fl. 5 c. 3 m.; £27. 9 fl. 9 m.; £84. 1 c.; £56. 8 fl. 6 c. 2 m.; £19. 1 m.

(2) £252. 2½ fl.; £300. 2½ c.; 4½ fl.; 5½ c.

Find the difference between

(3) £19. 5 fl., and £16. 3 fl. 9 c.

(4) £20, and £19. 9 fl. 9 c. 9 m.

(5) £5. 5½ fl., and £4. 4½ c.

Multiply

(6) £76. 8 fl. 3 m. separately by 5 and 63.

(7) 9 fl. 2½ c. separately by 18 and 1008.

(8) £150. 5 m. separately by 2005 and 18576.

Divide

(9) £194. 5 fl. 7 c. 5 m. by 5.

(10) £10764. 2 fl. 4 m. by 11.

(11) £342136. 8 fl. by 7380.

Note. For the method of reducing the old *Canadian Currency* into the *Decimal Currency* and the contrary, see the Elementary Arithmetic, p. 65.

Ex. XVI.

Miscellaneous Questions and Examples on Arts. (71–83).

NOTE. Where the contrary is not expressed, a year is supposed to consist of 365 days, and a cwt. of 100 lbs.

I.

(1) Explain the meaning of the term 'Reduction.' Reduce 27 cwt. 2 qrs. 2 lbs. into parcels, each weighing 8 lbs.

(2) What is the standard of the gold, and silver, and copper coinage in this country? According to the present law in Canada, for what sums respectively are copper and silver legal tenders?

(3) What is meant by 'Compound Multiplication'? Can concrete numbers of the same or different kinds be multiplied together? Give the reason. What is the cost of school accommodation for 13750 children at $9.12. each?

(4) A person bought 1763 yards of cloth at $1.27 per yard, and retailed it at $1.66 per yard: what was his profit?

(5) A person's weekly income is £14, and his quarterly expenditure is £128. 10s.; how much will he have saved at the end of 8 years? (supposing a year to consist of 52 weeks).

(6) What quantity of water must I add to a pipe of wine, which cost $450, to reduce its price to $2.50 a gallon?

II.

(1) Explain the meaning of 'Compound Division': what different cases are there of it? If £1844. 2s. 8¼d. be divided equally among 49 persons, how much will each receive?

(2) A house and its furniture are worth $32324.58; but the house is worth 8 times as much as the furniture; what is the house worth?

(3) Define 'a square', 'a cube'; show clearly by a figure how many cubic feet there are in a cubic yard. Reduce 4203239040 cub. in. to cub. yds.; and find how many grains of wheat there are in a load, if a pint contains 7000 grains.

(4) Divide £3. 13s. 9d. between two persons, so that one shall receive half as much again as the other.

(5) If I bottle off two-thirds of 2 pipes of wine into quarts and the rest into pints, how many dozens of each shall I have?

(6) A servant's wages are £10. 8s. a year: how much ought he to receive for 7 weeks (supposing a year to consist of 52 weeks)?

III.

(1) What are the different uses to which **Troy** weight and Avoirdupois weight are respectively applied? Express 56 lbs. Avoirdupois in lbs. &c. Troy.

(2) A factor bought 56 pieces of stuff for $7535.36 at $1.16 a yard: how many yards were there in each piece?

(3) Goods are bought at $6\frac{1}{2}d.$ per lb., and the cost of carriage is $1\frac{1}{4}d.$ per lb.; they are sold at £4. 10s. per cwt.: what is the gain or loss per cwt.?

(4) What is meant by a 'mean solar day'? How does the 'solar' year differ from the 'civil' year? State clearly the methods which have been adopted to correct the error arising therefrom.

(5) A gentleman laid up in the year 1851 $1411.56, having spent daily $7.80: what was his income in that year?

(6) Divide 198 dollars among 4 persons, so that the second may have twice as much as the first, the third 3 times as much as the second, and the fourth 4 times as much as the third.

IV.

(1) What are the standards of weight and capacity in Canada, and how are they fixed.

(2) Two persons buy postage-stamps at 12 a shilling; one retails them at 11 for a shilling, and the other at 13d. for a dozen; compare the gains on selling the same number of stamps.

(3) A hundred sovereigns all equally light, are worth ninety-five pounds; what is the value of each in shillings?

(4) Find
1. The sum of £27. 3 c. 9 m.; £560. $2\frac{1}{2}$ fl.; £30. 3 c. 7 m.
2. The quotient of £405. 5 fl. 3 c. 6 m. by 16.

(5) A person lays out $208.64 in spirits at $1.28 a gallon; some of which leaked out in the carriage; however, he sold the remainder for $259.20 at the rate of $1.80 a gallon: how many gallons leaked out?

(6) If a piece of ground contain 24 acres, and an inclosure of 17 acres, 3 roods be taken out of it, how many perches are there in the remainder?

(7) How many hours have elapsed since the birth of Christ to the year 1852, supposing each year to consist of 365 days, 6 hours?

MISCELLANEOUS QUESTIONS AND EXAMPLES.

V.

(1) Explain how the statute defines 'a yard', with reference to a natural standard of length. Find the corresponding linear unit, when an acre is one hundred thousand square units.

(2) How many barley-corns will reach round the earth, supposing the circumference of it to be 25000 miles?

(3) If a single article cost 86 cents, how many dozens can be bought for $415.20.

(4) How many times will a pendulum vibrate in 24 hours, which vibrates 5 times in 2 seconds?

(5) If the sum paid for 247 gallons of spirit, amount, together with the duty, to $859.56; and the duty on each gallon be $\frac{1}{5}$ part of its original cost; what is the duty per gallon.

(6) 12 persons on a journey each spend £23. 4 c. 6 m. in board and lodging; 6 of them agree to pay the travelling expenses, the share of each amounting to £18. 1 m. Find the amount of expenditure during the journey.

VI.

(1) What is the meaning of the word 'carat' as applied to gold, and as applied to diamonds? How many 'carats' fine is standard gold? If from 2793461 lbs Troy of gold there be coined £130524465. 4s. 6d., find the value of each lb.

(2) A wheel makes 514 revolutions in passing over 1 mile, 467 yards, 1 foot: what is its circumference?

(3) A grocer buys a hogshead of sugar, containing half a ton, for $90, and retails it at 11 cents per lb.; how much money does he make?

(4) A merchant buys 10 gallons of spirit at 12s. a gallon; 15 gallons at 14s. 6d. a gallon; and 18 gallons at 15s. 9d. a gallon: what will be the price of a gallon of the mixture, so that he may gain £2. 5s. 6d. on his outlay?

(5) A gentleman distributed $198 among 12 men, 16 women, and 30 children; to every man he gave twice as much as to a woman, and to every woman three times as much as to a child: what did each receive?

(6) A merchant expends £1636, 5s. on equal quantities of wheat at £2. 2s. a quarter, barley at £1. 1s. a quarter, and oats at 14s. a quarter: what quantity of each will he have?

VII.

(1) How many minutes are there in the 10 years, of which the first is 1852?

(2) Divide 425 tons, 15 cwt., 2 qrs., 12 lbs., by 27: and 1361 m., 4 fur., 28 po., by 28.

(3) Two boys run a race of 1 mile, one of them gains 5 feet in every 110 yards; how far will the other be left behind at the end of the race?

(4) Light travels at the rate of 192000 miles a second: how many days will it be in coming to us from the star α Centauri, supposed to be 20 trillions of miles distant?

(5) Divide \$480.60 equally among 45 people; supposing 20 of them to have received their portions, and 10 of the remaining 25 to have given up their portions to the other 15, how much would each of the 15 receive?

(6) A father left his oldest son \$24000 more than he left his second son, and the second son \$7560 more than the third; to the third he left \$60480: what was the oldest son's portion; and what sum did the father leave to his 3 sons.

VIII.

(1) Can you attach any meaning (1) to the multiplication of 6s. 8d. by £1. 2s 3d., (2) to the division of 1 yard, 2 feet, 3 inches, by 6 feet, 8 inches? State reasons for your answer.

(2) A carriage-load is found to weigh 1 ton, 3 cwt., 1 qr., and it consists of 315 equal packages, what is the weight of each?

(3) A person counts on the average 7000 shillings in an hour: what sum will he count in 67 days, if he work 9 hours a day?

(4) A gentleman's average daily expenditure for the year 1852 is \$9.63; and this allows him to lay by £50 at the end of the year: what is his income?

(5) Show how to perform the following operations: (1) the addition of £896. 5 fl. 4 c. 7 m.; £391. 5 fl. 3 c. 8 m.: £23. 9c. 6m.: (2) the subtraction of the second sum from the first; and (3) the multiplication of the third by 248; reading off each result.

(6) I hire a house at £90 a year; which is assessed in the rate-book at $\frac{4}{5}$ths of its rent; I agree to pay the rates upon it, viz., 3 poor rates of 9d., 10d., and 1s. 2d. respectively in the £, a church-rate of 8d. in the £, and a paving-rate of 1s 7d; in the £: what is the whole annual cost of the house?

MISCELLANEOUS QUESTIONS AND EXAMPLES

IX.

(1) Explain the calendar as now in use. On June 21 of 1851, the Duke of Wellington had lived 30,000 days. Find the day and year of his birth.

(2) The fore-wheel of a carriage is 10 feet in circumference, and the hind-wheel is 16 feet: how many revolutions will one make more than the other in 100 miles?

(3) Sound travels at the rate of 1142 feet a second: if a gun be discharged at the distance of $4\frac{1}{2}$ miles, how long will it be, after seeing the flash, before I hear the report.

(4) How many times will a clock, which chimes the quarters, strike and chime in 1854?

(5) How long will a person be in walking from Toronto to Brampton, a distance of 16 miles, when he takes 110 steps of $2\frac{1}{4}$ feet every minute?

(6) A manufacturer employs 60 men and 45 boys, who respectively work 10 and 14 hours per day during 5 days of the week, and half the time on the remaining day; each man receives 6d per hour, and each boy, 2d. per hour: what is the amount of wages paid in the year? (a year = 52 weeks).

X.

(1) A gentleman sent a tankard to his silversmith, which weighed 100 oz., 16 dwts., and ordered him to make it into spoons, each weighing 2 oz., 16 dwts.: how many spoons did he receive?

(2) A gentleman's estate, for the 5 years ending with 1849, yielded £1227. 15s.; how much could he spend one day with another, so as to lay by 135 guineas?

(3) The length of a year being $365\frac{1}{4}$ days, and that of a lunar month being $29\frac{1}{2}$ days, how many lunar months are there in 19 years?

(4) What is the value of a talent of silver, if silver be worth $1.20 per oz., and a talent consists of 1000 shekels, each weighing 219 grains?

(5) A merchant bought 7 pieces of cloth, each 27 yards, for $266.88; and sold 56 yards at $1.27 per yard; at what must he sell the remainder per yard in order to gain $17.04 on the whole?

(6) A certain number of men, twice as many women, and three times as many boys earned in 5 days £7. 15s.; each man earned 1s. 6d., each woman 10d., and each boy 8d. a day. How many were there of each?

SECTION III.

GREATEST COMMON MEASURE,

84. A *MEASURE* of any given number is a number which will divide the number exactly, *i. e.*, without a remainder. Thus, 2 is a measure of 6, because 2 is contained 3 times exactly in 6.

When one number is a measure of another, the former is said to measure the latter.

85. A MULTIPLE of any given number is a number which contains it an exact number of times. Thus, 6 is a multiple of 2.

86. A COMMON MEASURE of two or more given numbers is a number which will divide each of the given numbers exactly: thus, 3 is a common measure of 18, 27, and 36.

The GREATEST COMMON MEASURE, (G. C. M.), of two or more given numbers, is the greatest numbers which will divide each of the given numbers exactly: thus, 9 is the greatest common measure of 18, 27, and 36.

87. *If a number measure each of two others, it will also measure their sum* or *difference; and also any multiple of either of them.*

Thus, 3 being a common measure of 9 and 15, will measure their sum, their difference, and also any multiple of either 9 or 15.

The sum of 9 and $15 = 9 + 15 = 24 = 3 \times 8$;

therefore 3 measures their sum 24.

The difference of 15 and $9 = 15 - 9 = 6 = 2 \times 3$;

therefore 3 measures their difference 6.

Again 36 is a multiple of 9, and $36 = 3 \times 12$; therefore 3 measures this multiple of 9; and similarly any other multiple of 9.

Again, 75 is a multiple of 15: and $75 = 3 \times 25$; therefore 3 measures this multiple of 15; and similarly any other multiple of 15.

88. *To find the greatest common measure of two numbers.*

RULE. Divide the greater number by the less; if there be a remainder, divide the first divisor by it; if there be still a remainder,

GREATEST COMMON MEASURE. 83

divide the second divisor **by this** remainder, and so on; always dividing the last preceding divisor **by the last** remainder, till nothing remains. The last divisor will **be the greatest** common measure required.

Ex. Required the greatest common measure of 475 and 589.

Proceeding by the Rule given above,

```
475) 589 (1
     475
     ───
     114) 475 (
          456
          ───
          19) 114 (6
              114
              ───
                0
```

therefore 19 is the greatest common measure of 475 and 589.

Reason for the above process.

Any number which **measures 589 and 475,** also measures their difference, or $589 - 475$ or 114, Art. (87); also measures any multiple of 114, and therefore 4×114, or 456, Art. (87); and any number which measures 456 and 475, also measures their difference, or $475 - 456$, or 19; and no number greater than 19 can measure the original numbers 589 and 475; for it has just been shown that any number which measures them must also measure 19.

Again, 19 itself will measure 589 and 475.
For 19 measures 114 (since $114 = 6 \times 19$);
therefore 19 measures 4×114, or 456, Art. (87);
therefore 19 measures $456 + 19$, or 475, Art. (87);
therefore 19 measures $475 + 114$, or 589;
therefore since **19** measures them both, and no number greater than 19 can measure them both,

19 is their greatest common measure.

89. *To find the greatest common measure of three or more numbers.*

RULE. Find the greatest common measure of the first two numbers; then the greatest common measure of the common measure so found and the third number; **then** that of the common measure last found and the fourth number, and so on. The last common measure so found will be the greatest common measure required.

Ex. Find the greatest common measure of 16, 24, and 18.
Proceeding by the Rule given above,

$$\begin{array}{r}16)\,24\,(1\\16\\\hline 8)\,16\,(2\\16\\\hline\end{array}$$

therefore 8 is the greatest common measure of 16 and 24.

Now to find the greatest common measure of 8 and 18,

$$\begin{array}{r}8)\,18\,(2\\16\\\hline 2)\,8\,(4\\8\\\hline 0\end{array}$$

therefore 2 is the greatest common measure required.

Reason for the above process.

It appears from Art. (87) that every number, which measures 16 and 24, measures 8 also;

therefore every number, which measures 16, 24, and 18, measures 8 and 18.

therefore the greatest common measure of 16, 24, and 18, is the greatest common measure of 8 and 18.

But 2 is the greatest common measure of 8 and 18;

therefore 2 is the greatest common measure of 16, 24, and 18.

Ex. XVII.

Find the greatest common measure of

(1) 16 and 72.　　(2) 30 and 75.　　(3) 63 and 99.
(4) 55 and 121.　　(5) 128 and 324.　　(6) 120 and 320.
(7) 272 and 425.　　(8) 394 and 672.　　(9) 720 and 860.
(10) 825 and 960.　　(11) 775 and 1800.　　(12) 856 and 936.
(13) 176 and 1000.　　(14) 1236 and 1632.　　(15) 6409 and 7395.
(16) 689 and 1573.　　(17) 1729 and 5850.　　(18) 5210 and 5718.
(19) 2023 and 7581.　　(20) 468 and 1266.　　(21) 2484 and 2628.
(22) 3444 and 2268.　　(23) 5544 and 6652.　　(24) 4067 and 2573.

LEAST COMMON MULTIPLE. 85

(25) 10395 and 16819.
(26) 80934 and 110331.
(27) 1242 and 2323.
(28) 13536 and 23148.
(29) 42237 and 75582.
(30) 285714 and 999999.
(31) 10353 and 14877.
(32) 271469 and 30599.
(33) 14, 18, and 24.
(34) 16, 24, 48, and 74.
(35) 13, 52, 416, and 78.
(36) 837, 1134, and 1347.
(37) 805, 1311, and 1978.
(38) 28, 84, 154, and 343.
(31) 504, 5292 and 1520.
(40) 396, 5184, and 6914.

LEAST COMMON MULTIPLE.

90. A **Common Multiple** of two or more given numbers is a number which will contain each of the given numbers an exact number of times without a remainder. Thus, 144 is a common multiple of 3, 9, 18, and 24.

The **Least Common Multiple** (L. C. M.), of two or more given numbers is the least number which will contain each of the given numbers an exact number of times without a remainder. Thus, 72 is the least common multiple of 3, 9, 18, and 24.

91 *To find the least common multiple of two numbers.*

RULE. Divide their product by their greatest common measure: the quotient will be the least common multiple of the numbers.

Ex. Find the least common multiple of 18 and 30.

Proceeding by the Rule given above,

$$
\begin{array}{r}
18)\ 30\ (1 \\
18 \\ \hline
12)\ 18\ (1 \\
12 \\ \hline
6)\ 12\ (2 \\
12 \\ \hline
0
\end{array}
$$

therefore 6 is the greatest common measure **of 18 and 30.**

$$18 \times 30 \div 6 = 90$$

therefore 90 is the least common multiple of 18 and 30.

Reason for the above process.

$$18 = 3 \times 6, \text{ and } 30 = 5 \times 6.$$

Since **3 and 5** are prime factors, it is clear that 6 is the greatest common measure of 18 and 30; therefore their least common multiple must contain 3, 6, and 5, as factors.

Now every multiple of 18 must contain 3 and 6 as factors; and every multiple of 30 must contain 5 and 6 as factors; therefore every number, which is a multiple of 18 and 30, must contain 3, 5 and 6 as factors; and the least number which so contains them is $3 \times 5 \times 6$, or 90.

Now, $90 = (3 \times 6) \times (5 \times 6)$, divided by 6,
$= 18 \times 30$, divided by 6,
$= 18 \times 30$, divided by the greatest common measure of 18 and 30.

92. Hence it appears that the least common multiple of two numbers, which are prime to each other, or have no common measure but unity is their product.

93. *To find the least common multiple of three or more numbers.*

RULE. Find the least common multiple of the first two numbers; then the least common multiple of that multiple and the third number, and so on. The last common multiple so found will be the least common multiple required.

Ex. Find the least common multiple of 9, 18, and 24.

Proceeding by the Rule given above,

Since 9 is the greatest common measure of 18 and 9, their least common multiple is clearly 18.

Now, to find the least common multiple of 18 and 24.

$$\begin{array}{r} 18)\ 24\ (1 \\ 18 \\ \hline 6)\ 18\ (3 \\ 18 \\ \hline 0 \end{array}$$

therefore 6 is the greatest common measure of 18 and 24;
therefore the least common multiple of 18 and 24 is equal to (18×24) divided by 6,

$$24 \times 18 \div 6 = 72$$

therefore 72 is the least common multiple required.

Reason for the above process.

Every multiple of 9 and 18 is a multiple of their least common multiple 18; therefore every multiple of 9, 18, and 24 is a multiple of

18 and 24; and thererefore the least common multiple of 9, 18, and 24 is the least common multiple of 18 and 24: but 72 is the least common multiple of 18 and 24; therefore 72 is the least common multiple of 9, 18, and 24.

94. *When the least common multiple of several numbers is required, the most convenient practical method is that given by the following Rule.*

RULE. Arrange the numbers in a line from left to right, with a comma placed between every two. Divide those numbers which have a common measure by that common measure and place the quotients so obtained and the undivided numbers in a line beneath, separated as before. Proceed in the same way with the second line, and so on with those which follow, until a row of numbers is obtained in which there are no two numbers which have any common measure greater than unity. Then the continued product of all the divisors and the numbers in the last line will be the least common multiple required.

NOTE. It will in general be found advantageous to begin with the lowest prime number 2 as a divisor, and to repeat this as often as can be done; and then to proceed with the prime numbers, 3, 5, &c., in the same way.

Ex. Find the least common multiple of 18, 28, 30, and 42.

Proceeding by the Rule given above,

$$\begin{array}{c|cccc} 2 & 18, & 28, & 30, & 42 \\ 2 & 9, & 14, & 15, & 21 \\ 3 & 9, & 7, & 15, & 21 \\ 7 & 3, & 7, & 5, & 7 \\ \hline & 3, & 1, & 5, & 1 \end{array}$$

therefore the least common multiple required
$$= 2 \times 2 \times 3 \times 7 \times 3 \times 5 = 1260.$$

Reason for the above process.

Since $18 = 2 \times 3 \times 3$; $28 = 2 \times 2 \times 7$: $30 = 2 \times 3 \times 5$; $42 = 2 \times 3 \times 7$; it is clear that the least common multiple of 18 and 28 must contain as a factor $2 \times 2 \times 3 \times 3 \times 7$; and this factor itself is evidently a common multiple of $2 \times 3 \times 3$, or 18, and of $2 \times 2 \times 7$, or 28; now the least number which contains $2 \times 2 \times 3 \times 3 \times 7$ as a factor, is the product of these numbers; therefore $2 \times 2 \times 3 \times 3 \times 7$ is the least common multiple of 18 and 28; also it is clear that the least common multiple of 18, 28 and 30, or of $2\times2\times3\times3\times7$ and 30, or of $2\times2\times3\times3\times7$ and $2\times3\times5$ must contain as a factor $2\times2\times3\times3\times7\times5$, and this factor itself is evidently a common multiple of $2\times3\times3$ or 18, $2\times2\times7$ or 28, and $2\times3\times5$ or 30; hence it follows as before that $2\times2\times3\times3\times7\times5$ is the least common multiple of 18, 28, and 30; again the least common multiple of $2\times2\times3\times3\times7\times5$ and 42, or of $2\times2\times3\times3\times7\times5$ and $2\times3\times7$ must contain $2\times2\times3\times3\times7\times5$ as a factor, and this factor, as before, is evidently itself a common multiple of 18, 28, 30, and 42; now the least number which contains $2\times2\times3\times3\times7\times5$ as a factor, is the product of these numbers.

Therefore this product, or 1260, is the least common multiple required.

NOTE. The above method is sometimes shortened by rejecting in any line, any number, which is exactly contained in any other number in the same line; for instance, if it be required to find the least common multiple of 2, 4, 8, 16, 10, and 48; the numbers 2, 4, 8, 16, since each of them is exactly contained in 48, may be left out of consideration, and 240, the least common multiple of 10 and 48, will evidently be the least common multiple required.

Ex. XVIII.

Find the least common multiple of
(1) 16 and 24.   (2) 36 and 75.   (3) 7 and 15.
(4) 28 and 35.   (5) 319 and 407.   (6) 333 and 504.
(7) 2961 and 799.   (8) 7568 and 9504.   (9) 4662 and 5476.
(10) 6327 and 23997.   (11) 5415 and 30105.
(12) 15863 and 21489.
(13) 12, 8, and 9.   (14) 8, 12, and 16.
(15) 6, 10, and 15.   (16) 8, 12, and 20.
(17) 27, 24, and 15.   (18) 12, 51, and 68.
(19) 19, 29, and 38.   (20) 24, 48, 64, and 192.
(21) 63, 12, 84, and 14.   (22) 5, 7, 9, 11, and 15.
(23) 6, 15, 24, and 25.   (24) 12, 18, 30, 48, and 60.
(25) 15, 35, 63, and 72.   (26) 9, 12, 14, and 210.
(27) 54, 81, 63, and 14.   (28) 24, 10, 32, 45, and 25.
(29) 1, 2, 3, 4, 5, 6, 7, 8, and 9.
(30) 7, 8, 9, 18, 24, 72, and 144.
(31) 12, 20, 24, 54, 81, 63, and 14.
(32) 225, 255, 289, 1023, and 4095.

SECTION IV

FRACTIONS.

95. Let unity be represented by the line AB, which we will consider to be 1 yard in length.

Suppose AB to be divided into 3 equal parts AD, DE, EB; then one of such parts AD is a foot or one-third part of the yard, and it is denoted thus $\frac{1}{3}$ (read *one-third*); two of them AE, or two feet, thus $\frac{2}{3}$ (read *two-thirds*); three of them AB, or three feet, or the whole yard thus, $\frac{3}{3}$ or 1.

If another equal portion BF of a second yard BC, divided in the same manner as the first, be added, then AF, or four feet, is denoted thus $\frac{4}{3}$; and so on.

Such expressions, representing any number of the equal parts of a unit, *i. e.* of a quantity which is denoted by 1, are called BROKEN NUMBERS or FRACTIONS.

96. A FRACTION denotes one or more of the equal parts of a unit; it is expressed by two numbers placed one above the other with a line between them; the lower number is called the DENOMINATOR (Den^r.), and shews into how many equal parts the unit is divided; the upper is called the NUMERATOR (Num^r.), and shews how many of such parts are taken to form the fraction.

97. A Fraction also represents the quotient of the numerator by the denominator.

Thus $\frac{2}{3} = 2 \div 3$; for we obtain the same result, whether we divide one unit, AB or 1 yard, into 3 equal parts AD, DE, EB, each = 1 ft. or 12 in., and take 2 of such parts AE (represented by $\frac{2}{3}$), = 12 in. × 2 = 24 in., or divide 2 units, AC or 2 yards, into 3 equal parts, AE, EF, FC, each = 2 ft. or 24 in., and take 1 of such parts AE; which is equal to $\frac{1}{3}$ part of AC or 2 units, or = 2 ÷ 3. Hence $\frac{2}{3}$ and $2 \div 3$ have the same meaning.

98. When fractions are denoted in the manner above explained, they are called VULGAR FRACTIONS.

Fractions, whose denominators are composed of 10, or 10 multiplied by itself, any number of times, are often denoted in a different manner and when so denoted, they are called DECIMAL FRACTIONS.

VULGAR FRACTIONS.

99. In treating of the subject of Vulgar Fractions, it is usual to make the following distinctions:

(1) A PROPER FRACTION is one whose numerator is less than the denominator; thus, $\frac{3}{4}, \frac{4}{5}, \frac{6}{7}$, are proper fractions.

(2) An IMPROPER FRACTION is one whose numerator is equal to or greater than the denominator; thus, $\frac{5}{5}, \frac{6}{5}, \frac{7}{5}$, are improper fractions.

(3) A SIMPLE FRACTION is one whose numerator and denominator are simple integer numbers; thus, $\frac{1}{3}, \frac{4}{5}$, are simple fractions.

(4) A MIXED NUMBER is composed of a whole number and a fraction; thus, $5\frac{1}{6}, 7\frac{3}{4}$ are mixed numbers, representing respectively 5 units, together with $\frac{1}{6}$th of a unit; and 7 units, together with $\frac{3}{4}$ths of a unit.

(5) A COMPOUND FRACTION is a fraction of a fraction; thus, $\frac{1}{2}$ of $\frac{3}{4}$, $\frac{5}{6}$ of $\frac{7}{8}$ of $\frac{9}{10}$ are compound fractions.

(6) A COMPLEX FRACTION is one which has either a fraction or a mixed number in one or both terms of the fraction; thus, $\frac{\frac{3}{5}}{\frac{6}{7}}, \frac{2\frac{1}{2}}{3}, \frac{3}{4\frac{1}{2}}, \frac{2\frac{1}{2}}{5\frac{1}{6}}$, $\frac{\frac{3}{4} \text{ of } \frac{1}{2}}{2\frac{1}{4}}$ are complex fractions.

100. It is clear from what has been said, that every integer may be considered as a fraction whose denominator is 1; thus, $5 = \frac{5}{1}$, for the unit is divided into 1 part comprising the whole unit, and 5 of such parts, that is 5 units, are taken.

101. *To multiply a fraction by a whole number.*

RULE. Multiply the numerator of the fraction by the whole number. Thus, $\frac{2}{7} \times 3 = \frac{6}{7}$.

Reason for the above process.

In $\frac{2}{7}$ the unit is divided in 7 equal parts, and 2 of those parts are taken: whereas in $\frac{6}{7}$ the unit is divided into 7 equal parts, and 6 of those parts are taken; *i. e.* 3 times as many parts are taken in $\frac{6}{7}$ as are taken in $\frac{2}{7}$, the value of each part being the same in each case.

VULGAR FRACTIONS.   91

Ex. XIX.

(1) Multiply $\frac{5}{13}$ separately by 3, 9, 12, 36.
(2) Multiply $\frac{11}{64}$ separately by 7, 15, 21, 45.

102. *To divide a fraction by a whole number.*

RULE. Multiply the denominator of the fraction by the whole number.

Thus, $\dfrac{2}{7} \div 3 = \dfrac{2}{7 \times 3} = \dfrac{2}{21}$

Reason for the above process.

In the fraction $\frac{2}{7}$, the unit is divided into 7 equal parts, and 2 of those parts are taken; in the fraction $\frac{2}{21}$, the unit is divided into 21 equal parts, and of 2 such parts are taken; but since each part in the latter case is equal to one-third of each part in the former case, and the same number of parts are taken in each case, it is clear that $\frac{2}{21}$ represents one-third part of $\frac{2}{7}$, or $\frac{2}{7} \div 3$.

Ex. XX.

(1) Divide $\frac{5}{8}$ separately by 2, 3, 4, 5, 10.
(2) Divide $\frac{15}{287}$ separately by 11, 20, 25, 45.

103. *If the numerator and denominator of a fraction be both multiplied or both divided by the same number, the value of the fraction will not be altered.*

Thus, if the numerator and denominator of the fraction $\frac{2}{7}$ be multiplied by 3, the fraction resulting will be $\frac{6}{21}$, which is of the same value as $\frac{2}{7}$.

Reason for the above process.

In the fraction $\frac{2}{7}$ the unit is divided into 7 equal parts, and 2 of those parts are taken; in the fraction $\frac{6}{21}$ the unit is divided in 21 equal parts, and 6 of such parts are taken. Now there are 3 times as many parts taken in the second fraction as there are in the first fraction; but 3 parts in the second fraction are only equal to 1 part in the first fraction; therefore the 6 parts taken in the second fraction equal the 2 parts taken in the first fraction; therefore $\frac{2}{7} = \frac{6}{21}$.

104. Hence it follows that a whole number may be converted into a vulgar fraction with any denominator, by multiplying the number

by the required denominator for the numerator of the fraction, and placing the required denominator underneath;

for $6 = \frac{6}{1}$.

and to convert it into a fraction with a denominator 5 or 14, we have

$$6 = \frac{6}{1} = \frac{6 \times 5}{1 \times 5} = \frac{30}{5},$$

$$6 = \frac{6}{1} = \frac{6 \times 14}{1 \times 14} = \frac{84}{14}.$$

Ex. XXI.]

Reduce (1) 7, 9, 11, to fractions with denominators 3, 7, and 22 respectively; and (2) 26, 109, 117, and 125, to fractions with denominators 2, 5, 13, 23, and 35 respectively.

105 *Multiplying the numerator of a fraction by any number, is the same in effect as dividing the denominator by it, and conversely.*

For if the numerator of the fraction $\frac{6}{8}$ be multiplied by 4, the resulting fraction is $\frac{24}{8}$; and if the denominator be divided by 4, the resulting fraction is $\frac{6}{2}$.

Now the fraction $\frac{24}{8}$ signifies that unity is divided into 8 equal parts, and that 24 such parts are taken; these are equivalent to 3 units: also $\frac{6}{2}$ signifies that unity is divided into 2 equal parts, and that 6 such parts are taken; these are equivalent to 3 units: hence $\frac{24}{8}$ and $\frac{6}{2}$ are equal. The proof of their equality may also be put in this form: that since the unit, in the case of the second fraction, is only divided into 2 equal parts, each part in that case is 4 times as great as each part in the case of the first fraction, where the unit is divided into 8 equal parts; and therefore 4 parts in the case of the first fraction are equal to 1 part in the case of the second; or the 24 parts denoted by the first are equal to the 6 denoted by the second; or, in other words, the fractions $\frac{24}{8}$ and $\frac{6}{2}$ are equal.

Again, if we divide the numerator of the fraction $\frac{6}{8}$ by 2, the resulting fraction is $\frac{3}{8}$; and if we multiply the denominator by 2, the resulting fraction is $\frac{6}{16}$.

Now, $\frac{3}{8}$ signifies that the unit is divided into 8 equal parts, and that 3 of such parts are taken; and $\frac{6}{16}$ signifies that the unit is divided into 16 equal parts, and that 6 of such parts are taken; but each part in $\frac{3}{8}$ is equal to 2 parts in $\frac{6}{16}$; and therefore $\frac{3}{8}$ is of the same value as $\frac{2 \times 3}{16}$, or $\frac{6}{16}$.

VULGAR FRACTIONS.

106. *To represent an improper fraction as a whole or mixed number.*

RULE. Divide the numerator by the denominator: if there be no remainder, the quotient will be a whole number; if there be a remainder put down the quotient as the integral part, and the remainder is the numerator of the fractional part, and the given denominator as the denominator of the fractional part.

Ex. Reduce $\frac{25}{5}$ and $\frac{35}{6}$ to whole or mixed numbers.

By the Rule given above,

$$\frac{25}{5} = 5,\text{ a whole number}$$
$$\frac{35}{6} = 5\tfrac{5}{6}.$$

Reason for the above process.

Since $\dfrac{25}{5} = \dfrac{5 \times 5}{5} = \dfrac{5}{5} \times 5$, (Art. 101)

and since $\frac{5}{5}$ signifies that the unit is divided into 5 equal parts, and that 5 of those parts are taken, which 5 parts are equal to the whole unit or **1**; there $\frac{25}{5} = \frac{5}{5} \times 5 = 1 \times 5$, or 5.

Again, $\dfrac{35}{6} = \dfrac{30+5}{6} = \dfrac{6 \times 5 + 5}{6}$

which equals $\dfrac{6 \times 5}{6}$ together with $\frac{5}{6}$, that is, $=5$ together with $\frac{5}{6}$, by what has been said above; or, as it is written, $5\tfrac{5}{6}$.

Ex. XXII.

Express the following improper fractions as **mixed or whole numbers**:

(1) $\frac{15}{4}$.   (2) $\frac{77}{7}$.   (3) $\frac{49}{9}$.   (4) $\frac{113}{9}$.
(5) $\frac{63}{7}$.   (6) $\frac{143}{12}$.   (7) $\frac{587}{13}$.   (8) $\frac{183}{55}$.
(9) $\frac{751}{24}$.   (10) $\frac{5801}{63}$.   (11) $\frac{848}{108}$.   (12) $\frac{5876}{157}$.
(13) $\frac{10000}{111}$.   (14) $\frac{231750}{153}$.   (15) $\frac{14264}{236}$.   (16) $\frac{95950}{999}$.
(17) $\frac{25713}{1168}$.   (18) $\frac{633623}{1741}$.   (19) $\frac{878911}{36125}$.   (20) $\frac{46325}{3724}$.

107. *To reduce a mixed number to an improper fraction.*

RULE. Multiply the integer by the denomenator of the fraction, and to the product add the numerator of the fractional part; the result will be the required numerator, and the denominator of the fractional part the required denominator.

ARITHMETIC.

Ex. Convert $2\frac{4}{7}$ into an improper fraction.

Proceeding by the Rule given above,

$$2\frac{4}{7} = \frac{2 \times 7 + 4}{7} = \frac{18}{7}.$$

Reason for the above process.

$2\frac{4}{7}$ is meant to represent the integer 2 with the fraction $\frac{4}{7}$ added to it.

But 2 is the same as $\frac{2 \times 7}{7}$ or $\frac{14}{7}$; and therefore $2\frac{4}{7}$ must be the same as $\frac{14}{7}$ increased by $\frac{4}{7}$, or as $\frac{18}{7}$; for $\frac{18}{7}$ denotes that unity is divided into 7 equal parts, and represents 14 such parts together with 4 such parts.

Ex. XXIII.

Reduce the following mixed numbers to improper fractions:

(1) $2\frac{1}{3}$.   (2) $5\frac{3}{7}$.   (3) $4\frac{5}{9}$.   (4) $7\frac{2}{3}$.
(5) $25\frac{11}{12}$.   (6) $43\frac{7}{15}$.   (7) $25\frac{1}{13}$.   (8) $14\frac{13}{15}$.
(9) $2003\frac{1}{4}$.   (10) $857\frac{11}{15}$.   (11) $57\frac{31}{43}$.   (12) $13\frac{8}{21}$.
(13) $3\frac{25}{227}$.   (14) $26\frac{201}{232}$.   (15) $164\frac{113}{141}$.   (16) $106\frac{119}{837}$.
(17) $157\frac{42}{137}$.   (18) $172\frac{605}{850}$.   (19) $427\frac{5}{107}$.   (20) $100\frac{113}{726}$.

108. *To reduce a compound fraction to its equivalent simple fraction.*

RULE. Multiply the several numerators together for the numerator of the simple fraction, and the several denominators together for its denominator.

Ex. Convert $\frac{3}{5}$ of $\frac{7}{8}$ into simple fraction.

Proceeding by the Rule given above,

$$\frac{3}{5} \text{ of } \frac{7}{8} = \frac{3 \times 7}{5 \times 8} = \frac{21}{40}.$$

Reason for the above process.

By $\frac{3}{5}$ of $\frac{7}{8}$, we mean $\frac{3}{5}$ths of that part of unity which is denoted by $\frac{7}{8}$: thus if unity be divided in 8 equal parts, and 7 of these taken, and if each of these be again divided into 5 equal parts, and 3 of each set of parts be taken, then each of the parts will be one-fortieth part of

VULGAR FRACTIONS. 95

the original unit, and the number of parts **taken will be** 3×7, or 21·
the result therefore is $\dfrac{21}{40}$ or $\dfrac{3 \times 7}{5 \times 8}$: that is,

$$\frac{3}{5} \text{ of } \frac{7}{8} = \frac{3 \times 7}{5 \times 8}.$$

NOTE. In reducing compound fractions to simple ones, we may strike out factors **common to one of the** numerators and one of the denominators: for this **is in fact** simply dividing the numerator and denominator of the fraction **by the same number.** Art. (103).

Thus $\frac{3}{5}$ of $2\frac{1}{12}$ of $1\frac{1}{15} = \frac{3}{5}$ of $\frac{25}{12}$ of $\frac{16}{15}$.

$$= \frac{3 \times 25 \times 16}{5 \times 12 \times 15} = \frac{3 \times 5 \times 5 \times 4 \times 4}{5 \times 3 \times 4 \times 3 \times 5} = \frac{4}{3}$$

(striking out the factors 3, 5, 5, 4 from the numerator and denominator).

Ex. XXIV.

Reduce the following **compound fractions to simple** ones:
(1) $\frac{2}{3}$ of $\frac{4}{5}$. (2) $\frac{6}{7}$ of $\frac{9}{10}$. (3) $\frac{2}{5}$ of $\frac{3}{4}$.
(4) $\frac{5}{6}$ of $1\frac{1}{12}$. (5) $\frac{7}{8}$ of $\frac{5}{6}$ of 7. (6) $\frac{5}{6}$ of $\frac{3}{7}$ of $\frac{2}{3}$ of $\frac{1}{7}$ of 28.
(7) $\frac{5}{11}$ of $2\frac{1}{2}$ of $\frac{5}{7}$ of $10\frac{1}{2}$. (8) $\frac{7}{9}$ of $12\frac{1}{2}$ of $\frac{4}{5}$ of $\frac{5}{9}$ of $\frac{3}{8}$ of 9.
(9) $\frac{5}{18}$ of $\frac{7}{3}$ of $\frac{9}{10}$ of $\frac{3}{4}$ of $\frac{3}{10}$ of 2 of $\frac{8}{27}$.
(10) $\frac{5}{7}$ of $\frac{3}{8}$ of $\frac{9}{7}$ of $70\frac{2}{5}$ of $\frac{9}{40}$ of $1\frac{7}{11}$ of 147.

109. A FRACTION is in its LOWEST TERMS, when its numerator and denominator are PRIME to each other.

NOTE. **When the numerator and** denominator of a fraction are not prime to each other, they have (Art. 25) a common factor greater than unity. If we divide each of **them by this, there** results a fraction *equal* to the former, but of which the **terms, that** is, the numerator and denominator are less, **or** *lower* than those of the original fraction: and it may be considered to be the same fraction in *lower* terms. When the numerator **and** denominator **of** a fraction are PRIME to each other, that is, have **no common factor greater** than unity, it is clear that its terms cannot **be** made lower by division of this kind, and on this account the fraction **is** said to be in its LOWEST TERMS.

110 *To reduce a fraction to its lowest terms.*

RULE. Divide the numerator and denominator by their greatest common measure.

ARITHMETIC.

Ex. 1. Reduce $\frac{6465}{7335}$ to its lowest terms.

The G. C. M. of 6465 and 7335 is 15 : 431, 489 are the quotients of the numerator and denominator, respectively divided by the G.C.M. 15; therefore the fraction in its lowest terms $= \frac{431}{489}$.

Reason for the above process.

If the numerator and denominator of a fraction be divided by the same number, the value of the fraction is not altered (Art. 99.); and the greatest number which will divide the numerator and denominator is their greatest common measure.

Note. Sometimes it is unnecessary to find the greatest common measure, as it is easier to bring the fraction to its lowest terms by successive divisions of the numerator and denominator by common factors, which are easily determined by inspection.

Ex. 2. Reduce $\frac{540}{750}$ to its lowest terms,

$\frac{540}{750} = \frac{54}{75}$, dividing numerator and denominator by 10.

$= \frac{18}{25}$, dividing numerator and denominator by 3.

Ex. XXV.

Reduce each of the following fractions to its lowest terms:

(1) $\frac{4}{8}$. (2) $\frac{10}{15}$. (3) $\frac{14}{21}$. (4) $\frac{18}{45}$.
(5) $\frac{56}{63}$. (6) $\frac{54}{81}$. (7) $\frac{63}{135}$. (8) $\frac{66}{121}$.
(9) $\frac{216}{312}$. (10) $\frac{247}{323}$. (11) $\frac{272}{425}$. (12) $\frac{330}{726}$.
(13) $\frac{825}{2700}$. (14) $\frac{630}{936}$. (15) $\frac{324}{612}$. (16) $\frac{936}{2368}$.
(17) $\frac{5184}{6912}$. (18) $\frac{3444}{3556}$. (19) $\frac{7845}{96780}$. (20) $\frac{2472}{3264}$.
(21) $\frac{625}{6000}$. (22) $\frac{81}{1672}$. (23) $\frac{1653}{2973}$. (24) $\frac{10265}{14371}$.
(25) $\frac{4301}{05897}$. (26) $\frac{55247}{74841}$. (27) $\frac{6093}{9171}$. (28) $\frac{10813}{22860}$.
(29) $\frac{25194}{58176}$. (30) $\frac{374192}{576080}$. (31) $\frac{114135}{220661}$. (32) $\frac{128352}{233365}$.

111. *To reduce fractions to equivalent ones with a common denominator.*

Rule. Find the least common multiple of the denominators: this will be the common denominator. Then divide the common multiple so found by the denominator of each fraction, and multiply each quotient so found into the numerator of the fraction which belongs to it for the new numerator of that fraction.

VULGAR FRACTIONS. 97

NOTE. 1. If the given fractions be in their *lowest* terms, the above rule will reduce them to others having the *least* common denominator; if the *least* common denominator be required, the given fractions should be reduced to their lowest terms before the rule be applied.

Ex. Reduce $\frac{5}{12}, \frac{9}{16}, \frac{11}{24}, \frac{17}{33}$, into equivalent fractions with a common denominator.

The least common multiple of the denominators $= 528$; therefore the fractions become respectively,

$$\frac{5 \times 44}{12 \times 44}, \frac{9 \times 33}{16 \times 33}, \frac{11 \times 22}{24 \times 22}, \frac{17 \times 16}{33 \times 16}, \text{ or}$$

$$\frac{220}{528}, \frac{297}{528}, \frac{242}{528}, \frac{272}{528}.$$

Reason for the above process.

The least common multiple of the denominators of the given fractions will evidently contain the denominator of any one of the fractions an exact number of times. If both the numerator and denominator of that fraction be multiplied by that number, the value of the fractions will not be altered (Art. 103); and the denominator will then be equal to the least common multiple of all the denominators. If this be done with all the fractions, they will evidently be, in like manner, reduced to others of the same value, and having the least common multiple of all the denominators for the denominator of each fraction.

NOTE 2. If the denominators have no common measure, we must then multiply each numerator into all the denominators, except its own, for a new numerator for each fraction, and all the denominators together for the common denominator.

Ex. Reduce $\frac{1}{5}, \frac{2}{7}, \frac{1}{9}$, to equivalent fractions with a common denominator.

The least common multiple of the denominators

$$= 5 \times 7 \times 9;$$

therefore the fractions become

$$\frac{1 \times 7 \times 9}{5 \times 7 \times 9}, \frac{2 \times 5 \times 9}{7 \times 5 \times 9}, \frac{1 \times 5 \times 7}{9 \times 5 \times 7}, \text{ or}$$

$$\frac{63}{315}, \frac{90}{315}, \frac{35}{315}.$$

Ex. XXVI.

Reduce the fractions in each of the following sets to equivalent fractions, having the least common denominator:

(1) $\frac{1}{2}, \frac{2}{3},$ and $\frac{4}{5}$.
(2) $\frac{2}{3},$ and $\frac{7}{8}$.
(3) $\frac{4}{5}, \frac{3}{4},$ and $\frac{5}{8}$.
(4) $\frac{2}{3},$ and $\frac{5}{27}$.
(5) $\frac{3}{7}, \frac{5}{14},$ and $\frac{11}{28}$.
(6) $\frac{1}{2}, \frac{3}{4},$ and $\frac{5}{8}$.
(7) $\frac{7}{8}, \frac{11}{12},$ and $\frac{17}{18}$.
(8) $\frac{5}{12}, \frac{7}{16},$ and $\frac{13}{24}$.
(9) $\frac{5}{6}, \frac{9}{10},$ and $\frac{11}{14}$.
(10) $\frac{2}{3}, \frac{3}{5}, \frac{5}{6},$ and $\frac{7}{10}$.
(11) $\frac{2}{3}, \frac{3}{4}, \frac{5}{8},$ and $\frac{7}{8}$.
(12) $\frac{4}{7}, \frac{4}{11}, \frac{4}{13},$ and $\frac{2}{3}$.
(13) $\frac{3}{5}, \frac{35}{80},$ and $\frac{14}{200}$.
(14) $\frac{7}{12}, \frac{6}{7}, \frac{22}{63},$ and $\frac{13}{84}$.
(15) $\frac{7}{9}, \frac{5}{14}, \frac{13}{18}, \frac{3}{22},$ and $\frac{7}{36}$.
(16) $\frac{1}{3}, \frac{7}{5}, \frac{5}{6}, \frac{13}{14}, \frac{2}{8},$ and $\frac{17}{32}$.
(17) $\frac{3}{5}, \frac{4}{6}, \frac{7}{27}, \frac{8}{81}, \frac{16}{243},$ and $\frac{31}{729}$.
(18) $\frac{n}{10}, \frac{n}{100}, \frac{n}{1000},$ and $\frac{n}{10000}$.
(19) $\frac{31}{60}, \frac{17}{90}, \frac{13}{25}, \frac{1}{100},$ and $\frac{5}{6}$.
(20) $\frac{31}{84}, \frac{11}{28}, \frac{63}{63},$ and $\frac{1}{12}$.

NOTE. Whenever a comparison has to be made between fractions, in respect of their magnitudes, they must be reduced to equivalent ones with a common denominator; because then we shall have the unit divided, in the case of each fraction so obtained, into the same number of equal parts; and the respective numerators will show us how many of such parts are taken in each case; or which is the greatest fraction, which the next, and so on.

Ex. Compare the values of $\frac{5}{27}, \frac{11}{24}, \frac{5}{6}, \frac{4}{15},$ and $\frac{3}{5}$.

The least common multiple of the denominators $= 1080$; therefore the fractions become

$$\frac{5 \times 40}{27 \times 40}, \frac{11 \times 45}{24 \times 45}, \frac{5 \times 180}{6 \times 180}, \frac{4 \times 72}{15 \times 72}, \frac{3 \times 216}{5 \times 216}, \text{ or}$$

$$\frac{200}{1080}, \frac{495}{1080}, \frac{900}{1080}, \frac{288}{1080}, \frac{648}{1080};$$

therefore $\frac{5}{6}$ is the greatest, $\frac{3}{5}$ the next, $\frac{11}{24}$ the next, $\frac{4}{15}$ the next, and $\frac{5}{27}$ the least.

Ex. XXVII.

Compare the values of

(1) $\frac{3}{5}, \frac{6}{8},$ and $\frac{7}{10}$.
(2) $\frac{1}{2}, \frac{3}{4}, \frac{5}{6},$ and $\frac{7}{8}$.
(3) $\frac{1}{2}$ of $\frac{3}{4}, \frac{7}{12},$ and $\frac{4}{5}$ of $\frac{6}{7}$.
(4) $\frac{5}{12}, \frac{7}{16}, \frac{19}{24},$ and $\frac{21}{25}$.
(5) $\frac{3}{7}, \frac{5}{14}, \frac{9}{22}, \frac{1}{15},$ and $\frac{21}{25}$.
(6) $\frac{3}{4}, \frac{25}{32}, \frac{7}{16}, \frac{7}{10},$ and $\frac{27}{40}$.
(7) $\frac{2}{7}$ of $\frac{5}{8}$ of $4, \frac{2}{11}$ of $\frac{3}{8}$ of $5, \frac{1}{5}$ of $\frac{1}{2}$ of $4\frac{1}{4},$ and $\frac{14}{25}$.

ADDITION OF VULGAR FRACTIONS. 99

(8) $\frac{15}{4}$, $3\frac{1}{3}$, and $\frac{2}{7}$ of $9\frac{2}{5}$. (9) $\frac{6}{7}$, $\frac{13}{28}$, $\frac{14}{9}$, $\frac{3}{8}$, and $\frac{33}{56}$.
(10) $\frac{8}{9}$, $\frac{3}{11}$, $\frac{7}{18}$, $\frac{9}{22}$, and $\frac{5}{36}$. (11) $5\frac{1}{76}$, $1\frac{13}{152}$, $1\frac{1}{38}$, $\frac{101}{418}$, and $\frac{709}{418}$.
(12) $\frac{15}{4}$, $3\frac{1}{3}$, $\frac{2}{7}$ of $9\frac{2}{5}$, and $\frac{2}{7}$ of $\frac{3}{3}$ of $\frac{4}{5}$.

Find the greatest and least of the fractions
(13) $\frac{3}{4}$, $\frac{7}{12}$, $\frac{2}{3}$, $\frac{1}{3}$, and $\frac{1}{2}$. (14) $1\frac{1}{12}$, $\frac{29}{30}$, $1\frac{7}{18}$, $1\frac{7}{16}$, and $1\frac{7}{18}$.

ADDITION OF VULGAR FRACTIONS.

112. Rule. Reduce the fractions to equivalent ones with the least common denominator; add all the new numerators together, and under their sum write the common denominator.

Ex. Find the sum of $\frac{7}{15}$, $\frac{10}{21}$, and $\frac{16}{35}$.

Proceeding by the Rule given above,

The least common multiple of the denominators = 105; therefore the sum of the fractions is

$$\frac{49}{105} + \frac{50}{105} + \frac{48}{105} = \frac{49+50+48}{105} = \frac{147}{105} = 1\frac{2}{5}.$$

Reason for the Rule.

In each of the equivalent fractions, we have unity divided into 105 equal parts, and those fractions represent respectively 49, 50, and 48 of such parts; therefore the sum of the fractions must represent 49 + 50 + 48 or 147 parts, that is, must be $\frac{147}{105}$.

NOTE 1. If the sum of the fractions be a fraction which is not in its lowest terms, reduce it to its lowest terms; and if the result be an improper fraction, then reduce it to a whole or mixed number: thus $\frac{147}{105} = \frac{49}{35} = 1\frac{14}{35}$: the same remark applies to all results in Vulgar Fractions.

NOTE. 2. Before applying the rule, reduce all fractions to their lowest terms, improper fractions to whole or mixed numbers, and compound fractions to simple ones.

NOTE. 3. If any of the given numbers be whole or mixed numbers: the whole numbers may be added together as in Simple Addition, and the fractional parts by the Rule given above.

Ex. Find the sum of $\frac{3}{8}$, $3\frac{14}{15}$, $10\frac{2}{5}$, and $\frac{9}{22}$.

$$\frac{3}{8}+3\frac{14}{15}+10\frac{2}{5}+\frac{9}{22}=3+10+\frac{3}{8}+\frac{14}{15}+\frac{2}{5}+\frac{9}{22}$$

$$=13+\frac{3}{8}+\frac{14}{15}+\frac{2}{5}+\frac{9}{22},$$

and $\quad \frac{3}{8}+\frac{14}{15}+\frac{2}{5}+\frac{9}{22}=\frac{495}{1320}+\frac{1232}{1320}+\frac{528}{1320}+\frac{540}{1320}$

$$=\frac{495+1232+528+540}{1320}=2\frac{31}{264};$$

therefore the whole sum $\quad =13+2\frac{31}{264}=15\frac{31}{264}$.

Ex. XXVIII.

Add together,

(1) $\frac{2}{3}$ and $\frac{5}{7}$. (2) $\frac{2}{9}$ and $\frac{4}{5}$. (3) $\frac{6}{8}$ and $\frac{7}{8}$.
(4) $\frac{1}{7}$ and $\frac{4}{21}$. (5) $\frac{9}{16}$ and $\frac{5}{64}$. (6) $\frac{9}{11}$ and $1\frac{3}{21}$.
(7) $\frac{5}{12}$ and $\frac{7}{18}$. (8) $2\frac{1}{4}$ and $3\frac{1}{2}$. (9) $\frac{3}{5}$ and $2\frac{1}{6}$.
(10) $\frac{5}{12}$ and $\frac{7}{13}$. (11) $3\frac{5}{8}$ and $7\frac{2}{8}$. (12) $4\frac{7}{8}$ and $9\frac{11}{12}$.
(13) $\frac{3}{6}$, $\frac{5}{8}$, and $\frac{7}{12}$. (14) $\frac{3}{5}$, $\frac{7}{9}$, and $\frac{1}{5}$. (15) $\frac{1}{2}$, $\frac{2}{3}$, and $\frac{1}{16}$.
(16) $\frac{5}{12}$, $\frac{1}{8}$, and $\frac{2}{24}$. (17) $\frac{2}{5}$, $\frac{3}{8}$, and $\frac{7}{12}$. (18) $\frac{3}{7}$, $\frac{4}{7}$, and $\frac{7}{8}$.
(19) $\frac{1}{8}$, $\frac{5}{8}$, and $\frac{3}{21}$. (20) $\frac{1}{2}$, $\frac{1}{3}$, $\frac{1}{4}$, and $\frac{1}{5}$. (21) $\frac{5}{7}$, $\frac{2}{15}$, and $9\frac{9}{35}$.
(22) $\frac{3}{6}$, $2\frac{1}{7}$, and $13\frac{3}{10}$. (23) $\frac{3}{8}$, $\frac{4}{5}$ of $\frac{1}{3}$, and $9\frac{3}{20}$.
(24) $\frac{1}{3}$ of $\frac{3}{4}$ of $\frac{4}{5}$, $5\frac{1}{2}$, and $\frac{2}{19}$. (25) $\frac{7}{8}$, $\frac{4}{3}$, $\frac{1}{6}$, and $\frac{11}{30}$.
(26) $\frac{10}{14}$, $\frac{2}{15}$, and $\frac{18}{70}$. (27) $\frac{4}{5}$, $\frac{2}{16}$, $\frac{4}{7}$, and $\frac{2}{21}$.
(28) $\frac{1}{3}$, $\frac{7}{12}$, $\frac{5}{9}$, and $\frac{2}{26}$. (29) $\frac{1}{5}$, $6\frac{4}{8}$, and $\frac{4}{7}$ of $\frac{1}{3}$.
(30) $100\frac{2}{3}$, $64\frac{5}{8}$, $\frac{3}{6}$ of 701. (31) $261\frac{1}{2}$, $174\frac{3}{4}$, and $\frac{5}{6}$ of $10\frac{1}{2}$.
(32) $387\frac{1}{2}$, $285\frac{1}{4}$, $394\frac{1}{3}$, and $\frac{2}{5}$ of 3704.

Find the value of

(33) $\frac{11}{10}+\frac{11}{100}+\frac{11}{1000}+\frac{11}{10000}$. (34) $\frac{11}{12}+\frac{17}{18}+\frac{29}{30}+\frac{47}{48}+\frac{59}{60}$.
(35) $\frac{1}{2}+\frac{1}{5}+\frac{7}{17}+\frac{29}{28}+\frac{23}{34}$.
(36) $2\frac{11}{4}+6\frac{21}{34}+\frac{16}{67}+\frac{1}{3}$ of $1\frac{9}{7}+\frac{29}{28}+\frac{1}{4}$ of $2\frac{3}{4}$.
(37) $2\frac{2}{3}+3\frac{1}{4}+4\frac{4}{5}+5\frac{3}{7}+6\frac{5}{9}$. (38) $1\frac{4}{7}+3\frac{1}{16}+\frac{13}{15}+7\frac{5}{16}+\frac{9}{28}+\frac{2}{7}$ of $\frac{1}{5}$.
(39) $5\frac{1}{8}+\frac{5}{6}$ of $\frac{6}{9}$ of $3\frac{1}{2}+9\frac{3}{19}+\frac{3}{5}$ of $\frac{6}{8}$ of 4.
(40) $\frac{5}{8}$ of $12+\frac{3}{5}$ of $\frac{8}{9}+3\frac{3}{8}$ of $1\frac{6}{8}$ of $\frac{14}{155}+\frac{13}{16}$ of $3\frac{1}{3}$ of $\frac{1}{81}$ of $1\frac{1}{26}$.
(41) $270\frac{3}{4}+650\frac{3}{20}+5000\frac{1}{4}+53\frac{1}{4}+1\frac{1}{15}$.
(42) $\frac{1}{2}$ of $\frac{3}{4}+\frac{7}{21}$ of $(1+\frac{13}{15})+3\frac{1}{2}+\frac{29}{67}$ of $(1+\frac{1}{4})$.

SUBTRACTION.

118. Rule. Reduce the fractions to equivalent ones with the least common denominator, take the difference of the new numerators, and place the common denominator underneath.

Ex. Subtract $\frac{1}{2}$ from $\frac{7}{8}$.

Proceeding by the Rule given above, since 8 is clearly the least common multiple of the denominators, the equivalent fractions will be $\frac{4}{8}$ and $\frac{7}{8}$,

$$\text{and their difference} = \frac{7-4}{8} = \frac{3}{8}.$$

Reason for the Rule.

The unit in each of the equivalent fractions is divided into 8 equal parts, and there are 7 and 4 parts respectively taken, and therefore the difference must be 3 of such parts, or, in other words, the difference of the two fractions is $\frac{3}{8}$.

Note 1. Remember always, before applying the above Rule, to reduce fractions to their lowest terms, improper fractions to whole or mixed numbers, and compound fractions to simple ones.

Note 2. If either of the given fractions be a whole or mixed number, it is most convenient to take separately the difference of the integral parts and that of the fractional parts, and then add the two results together, as in the following examples.

Ex. 1. From $4\frac{3}{8}$ subtract $2\frac{1}{4}$.

Here $4-2=2$, and $\frac{3}{8}-\frac{1}{4}=\frac{3}{8}-\frac{2}{8}=\frac{1}{8}$;

therefore the difference of $4\frac{3}{8}$ and $2\frac{1}{4}=2\frac{1}{8}$.

Ex. 2. Take $2\frac{3}{8}$ from $4\frac{1}{4}$.

Now $\frac{3}{8}$ cannot be taken from $\frac{1}{4}$, since it is the greater of the two; we therefor add 1 to $\frac{1}{4}$, and take $\frac{3}{8}$ from $1+\frac{1}{4}$ or $\frac{5}{4}$; and then, in order that the difference may not be altered, we add 1 to the 2.

Now $\frac{5}{4}-\frac{3}{8}=\frac{10}{8}-\frac{3}{8}=\frac{7}{8},$

$4-3=1;$

therefore the difference of $4\frac{1}{4}$ and $2\frac{3}{8}=1\frac{7}{8}.$

ARITHMETIC.

For the process expressed at length is

$$4 + \tfrac{1}{4} - (2 + \tfrac{3}{8})$$

which $= 4 + 1 + \tfrac{1}{4} - (2 + 1 + \tfrac{3}{8})$ adding and subtracting 1),

$$= 4 + \tfrac{5}{4} - (3 + \tfrac{3}{8}) = 4 - 3 + \tfrac{5}{4} - \tfrac{3}{8}$$

$$= 1 + \tfrac{10}{8} - \tfrac{3}{8} = 1 + \tfrac{7}{8} = 1\tfrac{7}{8}.$$

Ex. XXIX.

Find the difference between

(1) $\tfrac{3}{4}$ and $\tfrac{1}{2}$. (2) $\tfrac{6}{8}$ and $\tfrac{1}{4}$. (3) $\tfrac{4}{9}$ and $\tfrac{3}{11}$.

(4) $\tfrac{7}{12}$ and $\tfrac{8}{15}$. (5) $\tfrac{17}{18}$ and $1\tfrac{1}{12}$. (6) $\tfrac{8}{15}$ and $\tfrac{9}{20}$.

(7) $2\tfrac{3}{4}$ and $1\tfrac{1}{3}$. (8) $37\tfrac{4}{15}$ and $33\tfrac{5}{24}$. (9) $6\tfrac{3}{8}$ and $4\tfrac{1}{2}$.

(10) $13\tfrac{5}{12}$ and $9\tfrac{7}{15}$. (11) $50\tfrac{1}{16}$ and $47\tfrac{1}{24}$. (12) 42 and $30\tfrac{6}{12}$.

(13) $15\tfrac{32}{125}$ and $1\tfrac{96}{243}$. (14) $90\tfrac{10}{11}$ and $25\tfrac{12}{125}$.

(15) 21 and $1\tfrac{173}{208}$. (16) 125 and $\tfrac{2}{7}$ of 14.

(17) $46\tfrac{5}{8}$ and $15\tfrac{1}{4}$. (18) $\tfrac{13}{58}$ and $\tfrac{2}{5}$ of $1\tfrac{1}{5}$.

(19) $\tfrac{1}{2}$ of $\tfrac{4}{5}$ of $\tfrac{3}{4}$ and $\tfrac{2}{5}$ of $\tfrac{3}{4}$.

(20) $\tfrac{2}{7}$ of $\tfrac{2}{3}$ of $\tfrac{4}{5}$ of $8\tfrac{3}{4}$ and $\tfrac{1}{8}$ of $\tfrac{1}{2}$ of $1\tfrac{1}{8}$ of $1\tfrac{7}{11}$.

(21) By how much does $\tfrac{5}{8}$ of $\tfrac{4}{15} - \tfrac{1}{3}$ of $\tfrac{1}{21}$ exceed $\tfrac{6}{8}$ of $\tfrac{2}{16} - \tfrac{3}{8}$ of $\tfrac{1}{18}$?

(22) Add $\tfrac{8}{15}$ of $\tfrac{6}{8}$ to $2\tfrac{3}{7}$ and subtract $\tfrac{6}{8}$ from the result.

(23) From the sum of $11\tfrac{2}{3}$ and $8\tfrac{7}{8}$ subtract $9\tfrac{13}{22}$.

(24) By how much does the difference of $5\tfrac{23}{31}$ and $2\tfrac{3}{7}$ exceed the sum of $\tfrac{5}{8} + \tfrac{5}{14} + \tfrac{1}{6}$?

(25) By how much does the sum of the fractions $1\tfrac{1}{28}$ and $\tfrac{8}{18}$ exceed their difference?

MULTIPLICATION.

114. Rule. Multiply all the numerators together for a new numerator, and all the denominators together for a new denominator.

Note 1. The word *MULTIPLICATION* has hitherto been used to mean "*the repeating* another *number* a NUMBER *of times;*" here the meaning is extended to include also PART or PARTS *of a time.*

Ex. Multiply $\tfrac{3}{7}$ by $\tfrac{5}{8}$.

MULTIPLICATION OF FRACTIONS.

Proceeding by the Rule given above,

$$\frac{3 \times 5}{7 \times 8} = \frac{15}{56}$$

Reason for the Rule.

If $\frac{3}{7}$ be multiplied by 5, the result is $\frac{15}{7}$, Art (97).

But this result must be 8 times too large, since, instead of multiplying by 5, we have only to multiply by $\frac{5}{8}$, which is 8 times smaller than 5, or in other words, is one-eight part of 5. Consequently the product above, viz., $\frac{15}{7}$ must be divided by 8, and $\frac{15}{7} \div 8 = \frac{15}{56}$, Art. (98).

NOTE 1. The same reasoning will apply, whatever be the number of fractions which have to be multiplied together.

NOTE 2. Before applying the above Rule, mixed numbers must be reduced to improper fractions.

NOTE 3. It has been shown that a fraction is reduced to its lowest terms by dividing its numerator and denominator by their greatest common measure, or, in others, by the product of those factors which are common to both: hence in all cases of multiplication of fractions, it will be well to split up the numerators and denominators as much as possible into the factors which compose them; and then, after putting the several fractions under the form of one fraction, the sign of × being placed between each of the factors in the numerator and denominator, to cancel those factors which are common to both before carrying into effect the final multiplication. Thus, in the following Examples:

Ex. 1. Muliply $\frac{3}{4}$ and $\frac{4}{5}$ together.

Prod^t. $= \frac{3 \times 4}{4 \times 5} = \frac{3}{5}$ dividing num^r. and den^r. by 4.

Ex. 2 Multiply $\frac{8}{9}$, $\frac{16}{24}$, $\frac{27}{30}$, and $\frac{45}{60}$ together.

Prod^t. $= \frac{8 \times 16 \times 27 \times 45}{9 \times 24 \times 30 \times 60}$

$= \frac{(2 \times 2 \times 2) \times (2 \times 2 \times 2 \times 2) \times (3 \times 3 \times 3) \times (3 \times 3 \times 5)}{(3 \times 3) \times (2 \times 2 \times 2 \times 3) \times (2 \times 5 \times 3) \times (3 \times 2 \times 3 \times 5)}$

$= \frac{2}{5}$ dividing by $2 \times 2 \times 2 \times 2 \times 2 \times 2 \times 3 \times 3 \times 3 \times 3 \times 5$.

ARITHMETIC.

Ex. 3. Multiply $2\tfrac{1}{2}$, $3\tfrac{3}{8}$, $10\tfrac{1}{8}$, $20\tfrac{4}{9}$, and $5\tfrac{9}{23}$, together.

$$\text{Prod}^{t.} = \frac{5}{2} \times \frac{27}{8} \times \frac{81}{8} \times \frac{184}{9} \times \frac{124}{23}$$

$$= \frac{5 \times (9 \times 3) \times (9 \times 9) \times (8 \times 23) \times (4 \times 31)}{2 \times (2 \times 4) \times 8 \times 9 \times 23}$$

$$= \frac{5 \times 3 \times 9 \times 9 \times 31}{2 \times 2} = \frac{37665}{4} = 9416\tfrac{1}{4}.$$

Ex. 4. Simplify $(\tfrac{6}{7} \text{ of } 1\tfrac{1}{4} \text{ of } \tfrac{14}{15} + 3\tfrac{1}{2} \text{ of } 2\tfrac{10}{21} - 2\tfrac{2}{3}) \times 3\tfrac{6}{7}$.

$$\text{Value} = \left(\frac{6}{7} \text{ of } \frac{5}{4} \text{ of } \frac{14}{15} + \frac{7}{2} \text{ of } \frac{52}{21} - \frac{3}{9} \right) \times \frac{27}{7}$$

$$= \left(\frac{3 \times 2 \times 5 \times 2 \times 7}{7 \times 2 \times 2 \times 3 \times 5} + \frac{7 \times 2 \times 26}{2 \times 3 \times 7} - \frac{8}{3} \right) \times \frac{27}{7}$$

$$= \left(1 + \frac{26}{3} - \frac{8}{3} \right) \times \frac{27}{7} = \frac{3 + 26 - 8}{3} \times \frac{27}{7} = \frac{21}{3} \times \frac{27}{7} = 27.$$

Ex. XXX.

Multiply

(1) $\tfrac{5}{7}$ by $\tfrac{3}{5}$. (2) $\tfrac{4}{5}$ by $1\tfrac{2}{3}$. (3) $\tfrac{3}{4}$ by $\tfrac{8}{9}$. (4) $\tfrac{5}{12}$ by $\tfrac{6}{15}$.

(5) $8\tfrac{1}{3}$ by $1\tfrac{9}{7}$. (6) $7\tfrac{1}{2}$ by $\tfrac{1}{4}$. (7) $3\tfrac{7}{8}$ by $2\tfrac{2}{3}$. (8) $7\tfrac{1}{4}$ by $\tfrac{1}{5}$ of $\tfrac{4}{9}$.

(9) 12 by $\tfrac{2}{3}$ of 5. (10) $\tfrac{1}{2}$ of $\tfrac{2}{3}$ by $5\tfrac{1}{2}$ of 3.

(11) $1\tfrac{2}{3}$ of $3\tfrac{2}{3}$ by $1\tfrac{1}{13}$ of $\tfrac{29}{21}$ of $\tfrac{2}{3}$.

(12) $1\tfrac{1}{18}$ of $1\tfrac{3}{22}$ of $\tfrac{7}{10}$ by $\tfrac{7}{12}$ of $37\tfrac{1}{35}$ of $3\tfrac{1}{4}$ of $\tfrac{1}{11}$.

(13) $\tfrac{3}{8}$ of $2\tfrac{1}{16}$ of $1\tfrac{1}{63}$ by $3\tfrac{5}{87}$ of $\tfrac{3}{11}$ of $\tfrac{19}{18}$.

(14) $5\tfrac{2}{19}$ of $3\tfrac{1}{4}$ of $\tfrac{8}{117}$ of 34 by $\tfrac{3}{187}$ of $\tfrac{9}{23}$ of $1\tfrac{1}{4}$ of 19.

Find the continued product of

(15) $\tfrac{1}{2}$, $\tfrac{2}{3}$, $\tfrac{3}{4}$, $\tfrac{4}{5}$, and $\tfrac{5}{6}$. (16) $\tfrac{15}{8}$, $\tfrac{16}{27}$, $1\tfrac{1}{5}$, $\tfrac{572}{385}$, and $2\tfrac{1}{55}$.

(17) $1\tfrac{17}{15}$, $2\tfrac{2}{3}$ of $1\tfrac{7}{57}$, $\tfrac{33}{42}$, $\tfrac{19}{328}$, $5\tfrac{1}{22}$ of 49, and $\tfrac{7}{72}$.

(18) $\tfrac{8}{9}$, $2\tfrac{3}{4}$, $3\tfrac{5}{11}$, $5\tfrac{2}{15}$, and $6\tfrac{1}{102}$. (19) $\tfrac{99}{116}$, $2\tfrac{1}{11}$, $\tfrac{87}{103}$, $\tfrac{128}{193}$, and $\tfrac{4}{9}$ of $1\tfrac{2}{3}$.

(20) $1\tfrac{47}{56}$, $1\tfrac{764}{861}$, $\tfrac{858}{323}$, $\tfrac{287}{357}$, and $1\tfrac{140}{153}$.

DIVISION.

115. Rule. Invert the divisor, i. e. take its numerator as a denominator and its denominator as a numerator, and proceed as in Multiplication.

DIVISION OF FRACTIONS

Ex. Divide $\frac{2}{11}$ by $\frac{3}{5}$.

Proceeding by the Rule given above,

$$\frac{2}{11} \div \frac{3}{5} = \frac{2}{11} \times \frac{5}{3} = \frac{10}{33}.$$

Reason for the Rule.

If $\frac{2}{11}$ be divided by 3, the result is $\frac{2}{11 \times 3}$ or $\frac{2}{33}$ (Art. 102).

This result is **5 times too small**, or, in other words, is only one-fifth part of the required quotient, **since**, instead of dividing by 3, we have to divide by $\frac{3}{5}$, which is only one-fifth part of 3; and the quotient of $\frac{2}{11}$ divided by $\frac{3}{5}$ must therefore be 5 times greater than if the divisor were 3. Hence the above result $\frac{2}{33}$ must be multiplied by 5, in order to give the true quotient.

Therefore, the quotient $= \frac{2}{33} \times 5 = \frac{2 \times 5}{33} = \frac{10}{33}$.

Note 1. Before applying this Rule, mixed numbers must be reduced to improper fractions, and compound fractions to simple ones, as in the following Examples:

Ex. 1. Divide $4\frac{1}{3}$ by $2\frac{3}{4}$.

$$4\tfrac{1}{3} \div 2\tfrac{3}{4} = \frac{13}{3} \div \frac{11}{4} = \frac{13}{3} \times \frac{4}{11} = \frac{52}{33} = 1\tfrac{19}{33}.$$

Ex. 2. Divide $\frac{3}{4}$ of $\frac{7}{8}$ by $\frac{15}{16}$ of 7.

$$\frac{3}{4} \text{ of } \frac{7}{8} \div \frac{15}{16} \text{ of } 7 = \frac{3 \times 7}{4 \times 8} \div \frac{15 \times 7}{16 \times 1}$$

$$= \frac{3 \times 7}{4 \times 8} \times \frac{16 \times 1}{15 \times 7} = \frac{3 \times 7 \times 16}{4 \times 8 \times 15 \times 7}$$

$$= \frac{3 \times 7 \times 4 \times 4}{4 \times 2 \times 4 \times 3 \times 5 \times 7} = \frac{1}{10}.$$

Note 2. COMPLEX FRACTIONS may by this Rule be reduced to simple ones.

Thus $\dfrac{1\frac{3}{4}}{2\frac{1}{2}}=\dfrac{\frac{7}{4}}{\frac{5}{2}}=\dfrac{7}{4}\div\dfrac{5}{2}$ (Art. 96) $=\dfrac{7}{4}\times\dfrac{2}{5}=\dfrac{7}{10}$

Again, $\dfrac{4\frac{1}{2}}{30}=\dfrac{\frac{9}{2}}{30}=\dfrac{\frac{9}{2}}{\frac{30}{1}}=\dfrac{9}{2}\div\dfrac{30}{1}=\dfrac{9}{2}\times\dfrac{1}{30}=\dfrac{3\times 3}{2\times 3\times 10}=\dfrac{3}{20}$.

Again, $\dfrac{30}{4\frac{1}{2}}=\dfrac{30}{\frac{9}{2}}=\dfrac{\frac{30}{1}}{\frac{9}{2}}=\dfrac{30}{1}\div\dfrac{9}{2}=\dfrac{30}{1}\times\dfrac{2}{9}=\dfrac{20}{3}=6\frac{2}{3}$.

NOTE 1. The words *DIVISION* and *QUOTIENT* are used in a more general sense than hitherto; Division *now* denotes the finding a quantity which, when multiplied by the divisor will equal the Dividend; *multiplied* being understood in the sense explained in note 1, p. 102. Quotient *now* signifies how many times or how many times and *part* or *parts* of a time.

Ex. XXXI.

Divide

(1) 3 by $\frac{3}{4}$. (2) $\frac{5}{8}$ by $\frac{1}{4}$. (3) $\frac{8}{9}$ by $1\frac{3}{18}$. (4) $\frac{13}{20}$ by $\frac{3}{5}$.
(5) $\frac{51}{64}$ by $3\frac{3}{8}$. (6) $\frac{16}{45}$ by $\frac{14}{15}$. (7) $2\frac{8}{12}$ by $4\frac{1}{6}$, (8) $\frac{27}{80}$ by $\frac{5}{8}$ of $3\frac{2}{7}$.
(9) $2\frac{11}{12}$ by $6\frac{2}{7}$ of $2\frac{1}{2}$. (10) $3\frac{3}{4}$ of $3\frac{1}{3}$ of $\frac{1}{2}$ by 75.
(11) $3\frac{1}{5}$ of $5\frac{5}{8}$ of $3\frac{5}{9}$ by $9\frac{3}{4}$ of $\frac{8}{33}$ of $7\frac{1}{3}$. (12) 119 by $\frac{7}{8}$.
(13) $\frac{3}{8}$ of $\frac{5}{7}$ of $80\frac{1}{3}$ of 9 by $\frac{3}{7}$ of $\frac{2}{9}$ of $\frac{4}{5}$ of $8\frac{3}{4}$.
(14) $\frac{5}{8}$ of $\frac{4}{5}$ of $\frac{11}{18}$ of $1\frac{7}{11}$ by $\frac{6}{17}$ of $\frac{35}{114}$ of $\frac{33}{44}$ of $1\frac{6}{11}$.
(15) Compare the product and quotient of $2\frac{1}{2}$ by $3\frac{7}{8}$.

Reduce to simple fractions the following complex fractions:

(16) $\dfrac{\frac{3}{4}}{1\frac{7}{8}}$. (17) $\dfrac{\frac{14}{27}}{2\frac{1}{3}}$. (18) $\dfrac{2\frac{11}{12}}{\frac{5}{9}}$. (22) $\dfrac{13\frac{9}{13}}{1\frac{11}{14}}$.
(19) $\dfrac{\frac{6}{11}}{\frac{18}{55}}$. (20) $\dfrac{13\frac{1}{3}}{20}$. (21) $\dfrac{56}{1\frac{5}{9}}$.

REDUCTION OF FRACTIONS.

116. *To find the value of a fractional part of a number of one or several denominations in terms of the same or lower denominations.*

RULE. Multiply the given number by the numerator, as in Compound Multiplication; and divide the product by the denominator, as in Compound Division.

REDUCTION OF FRACTIONS.

Ex. 1. Find the value of $\frac{7}{8}$ of £1.

Proceeding by the Rule given above,

$$\frac{7}{8} \text{ of } £1 = \frac{7 \times 20}{8} s. = \frac{7 \times 5}{2} s.$$

$$= \frac{35}{2} s. = 17\tfrac{1}{2} s.$$

and $\frac{1}{2}$ of 1s. $= \frac{1 \times 12}{2} d. = 6d.$;

therefore the value required $= 17s.\ 6d.$

Reason for the above process.

$\frac{7}{8}$ of £1 is the same as 7 times $\frac{1}{8}$ of £1,

and $\frac{1}{8}$ of £1. $= \frac{20s.}{8} = \frac{5s.}{2}$;

therefore 7 times $\frac{1}{8}$ of £1 = 7 times $\frac{5s.}{2} = \frac{35s.}{2} = 17\tfrac{1}{2}s.$

$= 17s.\ 6d.$

Ex. 2. Find the value of $\frac{5}{8}$ of 3 tons.

$\frac{5}{8}$ of 3 tons $= \frac{15}{8}$ of a ton $= 1\tfrac{7}{8}$ tons.

$\frac{7}{8}$ of a ton $= \frac{7 \times 20}{8}$ cwt. $= \frac{7 \times 5}{2}$ cwt. $= 17\tfrac{1}{2}$ cwt.,

$\frac{1}{2}$ cwt. $= \frac{1 \times 4}{2}$ qrs. $= 2$ qrs.;

therefore required value $= 1$ ton, 17 cwt. 2 qrs.

Ex. 3. Find the value of $\frac{5}{9}$ of a bushel $- \frac{5}{7}$ of a peck.

$\frac{5}{9}$ of a bush. $= \frac{5 \times 4}{9}$ pks. $= \frac{20}{9}$ pks. $= 2\tfrac{2}{9}$ pks.,

$\frac{2}{9}$ pk. $= \frac{2 \times 8}{9}$ qts. $= \frac{16}{9}$ qts. $= 1\tfrac{7}{9}$ qts.

therefore $\frac{5}{9}$ of a bush. $= 2$ pks., $1\tfrac{7}{9}$ qts.

$\frac{5}{7}$ of a pk. $= \frac{5 \times 8}{7}$ qts. $= \frac{40}{7}$ qts. $= 5\tfrac{5}{7}$ qts.;

therefore required value $= 2$ pks., $1\tfrac{7}{9}$ qts. $- 5\tfrac{5}{7}$ qts.

$= 1$ pk., $4\tfrac{4}{63}$ qts.

ARITHMETIC.

Ex. XXXII.

Find the respective values of

(1) $\frac{3}{8}$ of $2; $\frac{2}{3}$ of $3; $\frac{5}{6}$ of $4; $\frac{3}{4}$ of a cwt.; $\frac{5}{8}$ of $9.

(2) $\frac{3}{4}$ of £1. 10s.; $\frac{2}{3}$ of £2; $\frac{3}{8}$ of half-a-crown; $\frac{3}{5}$ of 13s. 4d.

(3) $2\frac{3}{7}$ of $1.68; $\frac{5}{7}$ of £2. 3s. 9d.; $\frac{8}{21}$ of 9 tons; $\frac{2}{3}$ of $45.

(4) $\frac{2}{3}$ of $1.55; $1\frac{3}{5}$ of £1. 2s. 9d.; $\frac{4}{11} \times 1\frac{1}{18}$ of $21; $\frac{1}{3}$ of $\frac{3}{5}$ of 9s. 10¼d.

(5) $3\frac{1}{12}$ of 2s. 6d.; $\frac{1}{11}$ of £4. 14s. 5d.; $\frac{3}{5}$ of $\frac{5}{21}$ of 10s. 6d.

(6) $\frac{5}{16}$ of a cwt.; $\frac{3}{4}$ of a lb. Avoird.; $\frac{4}{5}$ of a mile; $\frac{5}{8}$ of an acre.

(7) $\frac{5}{11}$ of a mile; $\frac{7}{16}$ of a day; $\frac{2}{3}$ of a yard, $\frac{3}{7}$ of 3 cwt., 1 qr., 11 lb.

(8) $7\frac{3}{8}$ of a lb. Avoird.; $1\frac{3}{4}$ of a lb. Troy; $2\frac{2}{3}$ of a gal.; $4\frac{5}{15}$ of an acre.

(9) $3\frac{5}{12}$ of a hhd. of beer; $2\frac{3}{8}$ of a tun of wine; $6\frac{26}{29}$ of a bus.

(10) $\frac{3}{8}$ of $\frac{3}{5}$ of $10\frac{3}{5}$ hrs.; £$\frac{15\frac{3}{7}}{7\frac{4}{5}}$; $\frac{7\frac{1}{8}}{8\frac{1}{7}}$ of $\frac{5\frac{3}{7}}{7\frac{1}{5}}$ of $5.

(11) $\frac{3\frac{7}{9}}{1\frac{5}{7} \text{ of } 1\frac{1}{3}}$ of £16. 8s. 1½d.; $\frac{2}{7}$ of $1\frac{2}{3}$ of $12\frac{1}{2}$ of $\frac{3}{14}$ of $2 \times \frac{3}{14}$.

(12) $\frac{5}{7}$ of $1 \times 5\frac{2}{3}$; $\frac{3}{4}$ of $\frac{2}{3}$ of $1 \div \frac{1}{4}$.

(13) $19\frac{3}{4}$ of 5 lbs., 8 oz., 6 dwt.; $2\frac{2}{8}$ of 8 mls., 14 po., $2\frac{7}{12}$ ft. $+8\frac{1}{3}$.

Find the values of

(14) $\frac{3}{8}$ of $4.80 $+ \frac{3}{8}$ of 60 cts. $+ \frac{5}{8}$ of 24 cts.

(15) $\frac{1}{7}$ of £1 $+ \frac{2}{3}$ of 1s. $+ \frac{5}{12}d$.

(16) $\frac{5}{28}$ of $1\frac{1}{3}$ of $2.52 $+ \frac{5}{8}$s. $+ \frac{1}{12}$ of 60 cts.

(17) £$3\frac{2}{3} + 7\frac{1}{4}s. + 4\frac{3}{4}d$.

(18) $\frac{\frac{3}{4}+\frac{2}{3}}{1\frac{7}{12}-\frac{2}{3}}$ of 6 tons $+ \frac{7}{18}$ 4 cwts. $+ \frac{4}{5}$ of a qr.

(19) $\frac{3}{8}$ of a ton $+ \frac{5}{8}$ of a cwt. $+ \frac{2}{3}$ lb.

(20) $\frac{3}{5}$ lb. Troy $+ \frac{2}{3}$ lb. Troy $- \frac{3}{6}$ oz. Troy.

(21) $\frac{7}{15}$ of a mile $- \frac{5}{8}$ of a fur. $+ \frac{4}{11}$ po.

(22) $\frac{8}{111}$ cub. yds. $+ 2\frac{3}{4}$ cub. ft.

(23) $\frac{3}{5}$ of a qr. $+ \frac{2}{3}$ of a bus. $- \frac{1}{5}$ of a qr.

(24) $\frac{3}{8}$ of 7 fur., 29 po., $2\frac{1}{4}$ yds. $+ \frac{5}{7}$ of 5 mi., 3 fur., 37 po., $4\frac{1}{2}$ yds.

(25) $7\frac{2}{5}$ of $365\frac{1}{4}$d. $+ 3\frac{5}{16}$ of $\frac{5}{8}$ wks. $+ \frac{3}{4}$ of $5\frac{3}{9}$ hrs.

(26) $\frac{1}{11}$ of 91 ac., 3 ro., 36 po., $2\frac{3}{4}$ yds $- \frac{3}{5}$ of 6 ac., 2 ro., 17 po., $25\frac{1}{4}$ yds.

117. *To reduce a number, or a fraction, of any denomination, to a fraction of another denomination.*

RULE. Reduce the given number, or fraction, and also the number or fraction to the fraction of which it is to be reduced, to their respective

REDUCTION OF FRACTIONS.

equivalent values in terms of some one and the same denomination: then the fraction of which the former is made the numerator, and the latter the denominator, will be the fraction required.

Ex. 1. Reduce $\frac{5}{8}$ of £1 to the fraction of 27s.

Proceeding by the above Rule,

$$\frac{5}{8} \text{ of } £1 = 20 \text{ times } \frac{5}{8} \text{ of } 1s.$$

$$= \frac{5 \times 20}{8} s. = \frac{5 \times 5}{2} s.$$

$$27s. = 27s.$$

therefore fraction required $= \dfrac{\frac{5 \times 5}{2}}{27} = \frac{5 \times 5}{2} \div \frac{27}{1}$

$$= \frac{5 \times 5}{2} \times \frac{1}{27} = \frac{25}{54}.$$

For 27s. is divided into 27 equal parts; and $\frac{5}{8}$ of £1 is divided into $\frac{25}{2}$ of such parts; therefore the part of unity, or 27s., which the latter represents, is $\dfrac{\frac{25}{2}}{27} = \dfrac{25}{54}.$

Ex. 2. What part of $\frac{1}{3}$ of a ton is $2\frac{2}{3}$ of $1\frac{1}{3}$ of $\frac{1}{7}$ of a cwt.?

$2\frac{2}{3}$ of $1\frac{1}{3}$ of $\frac{1}{7}$ of a cwt. $= \frac{8}{3}$ of $\frac{4}{3}$ of $\frac{1}{7}$ of a cwt.

$$= \frac{8}{3} \times \frac{4}{3} \times \frac{1}{7} \text{ cwt.}$$

$\frac{1}{3}$ of a ton $= \frac{20}{3}$ cwt.

Therefore fraction required $= \dfrac{\frac{8}{3} \times \frac{4}{3} \times \frac{1}{7}}{\frac{20}{3}}$

$$= \frac{8}{3} \times \frac{4}{3} \times \frac{1}{7} \times \frac{3}{20} = \frac{8}{105}.$$

Ex. XXXIII.

Reduce

(1) $2.25 to the fraction of $4; and $3.15 to the fraction of $5.

(2) 5d. to the fraction of 1s.; and 3s. 4½d. to the fraction of £1.

(3) £18. 7s. 6d. to the fraction of £2; and 6s. 7⅔d. to the fraction of 7s. 9d.

(4) 3 qrs., 19 lbs. to the fraction of a ton; and 61¼lbs. to the fraction of 4 oz.

(5) 3 qrs., 4 lbs. to the fraction of 2 cwt.; and 5 oz., $2\frac{3}{5}$ drs. to the fraction of a grain.

(6) 3 ro., $27\frac{1}{2}$ po. to the fraction of an acre; and $26\frac{2}{7}$ sq. yds. to the fraction of 2 acres.

(7) 126 yds., 2 ft., 6 in. to the fraction of a mile; and 6 cub. ft., 100 cub. in. to the fraction of a cubic yard.

(8) 2 qrs., $2\frac{2}{3}$ na. to the fraction of an Eng. ell; and 8 h., 3 m. to the fraction of a day.

(9) 2 ac., 1 ro. to the fraction of 9 ac. 2 ro.; and 1540 yds., 2 ft., 9 in. to the fraction of 2 miles.

(10) 1 ft. $\frac{7}{8}$ in. to the fraction of a sq. yd.; and 2 qts., $1\frac{1}{2}$ pt. to the fraction of a barrel.

(11) 2 wks. 5 days, 7 h., 27 m. to the fraction of a day; and 1 ro., 20 po. to the fraction of an acre.

(12) 4 bush., $2\frac{2}{5}$ qts. to the fraction of a load; and 3 quires 7 sheets to the fraction of a ream.

(13) 6 ft., $3\frac{3}{8}$ in. to the fraction of 13 ft., $8\frac{9}{80}$ in.; and $1\frac{1}{2}$ yds. to the fraction of $1\frac{1}{2}$ in.

(14) $\frac{3}{4}$ of \$1.20 to the fraction of \$4.80; and $\frac{4}{5}$ of a farthing to the fraction of 1*s*.

(15) $\frac{1}{6}$ of £74. 13*s*. 4*d*. to the fraction of £28; and $\frac{1}{6}$ of \$16 to the fraction of \$19.

(16) $\frac{3}{5}$ of a dwt. to the fraction of 1 lb.; and $\frac{5}{6}$ of 2 lbs. to the fraction of $2\frac{1}{4}$ tons.

(17) $\frac{3}{8}$ of a lb. to the fraction of a cwt.; and $\frac{5}{8}$ of a yd. to the fraction of a mile.

(18) 1 oz. Troy to the fraction of 1 oz. Avoirdupois; and $\frac{3}{9680}$ of a mile to the fraction of a yard.

(19) $\frac{3}{7}$ of a pole to the fraction of a league; and $3\frac{7}{8}$ furlongs to the fraction of $2\frac{7}{8}$ miles.

(20) $\frac{2}{5}$ of $7\frac{1}{2}$ of $16\frac{1}{2}$ yards to the fraction of a furlong; and $\frac{7}{316}$ of a lb. Troy to the fraction of a pennyweight.

(21) $\frac{3}{8}$ of a lb. Avoird. to the fraction of 2 lbs. Troy; and $\frac{5}{9}$ of a French ell to the fraction of a yd.

(22) $\frac{13}{23}$ of a sq. in. to the fraction of a sq. yd.; and $\frac{1}{5}$ of a yd. to the fraction of an English ell.

(23) What part of £9 is $\frac{1}{4}$ of $\frac{8}{15}$ of 2*s*. 6*d*.?

(24) What part of a second is $\frac{1}{100000}$ of a day?

MISCELLANEOUS EXAMPLES IN VULGAR FRACTIONS. 111

(25) What part of $\frac{3}{4}$ of a league is $\frac{7}{8}$ of a mile?
(26) What part of 3 weeks, 4 days, is $\frac{1}{2}$ of $5\frac{3}{5}$ sec.?
(27) What part of $\frac{1}{3}$ of an acre is $25\frac{9}{11}$ po.?
(28) What part of $\frac{7}{30}$ of a min. is $\frac{7}{183}$ of a month of 28 days?
(29) What part of $\frac{1}{4}$ of 4 tuns of wine is $3\frac{1}{4}$ hhds?
(30) What part of 3 fathoms is $\frac{3}{14}$ of $\frac{7}{9}$ of a pole?
(31) What fraction of $2\frac{1}{2}$ cwt. together with 3 qrs., 14 lbs. will give a ton and a half? (cwt.=112 lbs.).

118. *Miscellaneous Examples in fractions worked out.*

Ex. 1. What number added to $\frac{7}{8} + \frac{5}{12}$ will give $2\frac{1}{8}$?

This question in other words is the following: "What number will remain after $\frac{7}{8} + \frac{5}{12}$ has been subtracted from $2\frac{1}{8}$?

Now $2\frac{1}{8} - (\frac{7}{8} + \frac{5}{12}) = 2\frac{1}{8} - \frac{7}{8} - \frac{5}{12} = \frac{17}{8} - \frac{7}{8} - \frac{5}{12}$
$= \frac{10}{8} - \frac{5}{12} = \frac{30}{24} - \frac{10}{24} = \frac{20}{24} = \frac{5}{6}.$

Therefore the number required $= \frac{5}{6}$.

NOTE. It will be remembered, that all quantities within a vinculum are equally affected by any sign placed before the vinculum.

Thus in the above expression, $-(\frac{7}{8} + \frac{5}{12})$ means that the sum of $\frac{7}{8}$ and $\frac{5}{12}$ has to be subtracted from $2\frac{1}{8}$; whereas $-\frac{7}{8} + \frac{5}{12}$ would mean that $\frac{7}{8}$ had to be subtracted from $2\frac{1}{8}$; and then $\frac{5}{12}$ had to be added to the result.

Ex. 2. What number subtracted from $14\frac{3}{8}$, will leave $1\frac{3}{4}$ for a remainder?

Number required

$= 14\frac{3}{8} - 1\frac{3}{4} = (14 + 1 + \frac{3}{8}) - (1 + 1 + \frac{3}{4})$
$= (14 + 1\frac{3}{8}) - (2 + \frac{3}{4}) \quad 14 - 2 + 1\frac{3}{8} - \frac{3}{4}) = 12\frac{5}{8}.$

Or thus. $14\frac{3}{8} - 1\frac{3}{4} = 13\frac{3}{8} - \frac{3}{4} = \frac{107}{8} - \frac{6}{8} = \frac{101}{8} \quad 12\frac{5}{8}.$

Ex. 3. What number multiplied by $1\frac{3}{8}$ will produce $14\frac{3}{4}$?

This question in other words is the following: "If $14\frac{3}{4}$ be divided by $1\frac{3}{8}$, what will the quotient be?"

But $\dfrac{14\frac{3}{4}}{1\frac{3}{8}} = \dfrac{\frac{59}{4}}{\frac{11}{8}} = \frac{59}{4} \times \frac{8}{11} = \frac{118}{11} = 10\frac{8}{11}.$

Therefore the number required $= 10\frac{8}{11}$.

112   ARITHMETIC.

Ex. 4. What number divided by $1\frac{3}{8}$ will produce $10\frac{9}{11}$?

This question in other words is the following: "What is the product of $1\frac{3}{8}$ and $10\frac{9}{11}$?"

The product of $1\frac{3}{8}$ and $10\frac{9}{11}$
$$= 1\tfrac{3}{8} \times 10\tfrac{9}{11} = \tfrac{11}{8} \times \tfrac{119}{11} = \tfrac{119}{8} = \tfrac{59}{4} = 14\tfrac{3}{4}.$$

Ex. 5. Reduce the expression
$$\left(\frac{3\tfrac{1}{3}}{7} + \frac{2}{10\tfrac{1}{2}} - \frac{5}{18} \text{ of } \frac{4}{7} \right) \times 1\tfrac{3}{4}$$
to its simplest form.

$$\left(\frac{3\tfrac{1}{3}}{7} + \frac{2}{10\tfrac{1}{2}} - \frac{5}{18} \text{ of } \frac{4}{7} \right) \times 1\tfrac{3}{4} = \left(\frac{\tfrac{10}{3}}{7} + \frac{\tfrac{2}{1}}{2\tfrac{1}{2}} - \frac{5 \times 4}{18 \times 7} \right) \times \tfrac{7}{4}$$
$$= \left(\tfrac{10}{21} + \tfrac{4}{21} - \tfrac{10}{63} \right) \times \tfrac{7}{4} = \left(\tfrac{14}{21} - \tfrac{10}{63} \right) \times \tfrac{7}{4} = \tfrac{42-10}{63} \times \tfrac{7}{4} = \tfrac{32}{63} \times \tfrac{7}{4} = \tfrac{8}{9}.$$

Ex. 6. Simplify the expression $\dfrac{\tfrac{1}{2} + \tfrac{1}{3} + \tfrac{1}{4}}{\dfrac{1}{2\tfrac{1}{2}} + \dfrac{1}{3\tfrac{1}{2}} + \dfrac{1}{4\tfrac{1}{2}}}$.

$$\frac{\tfrac{1}{2}+\tfrac{1}{3}+\tfrac{1}{4}}{\dfrac{1}{2\tfrac{1}{2}}+\dfrac{1}{3\tfrac{1}{2}}+\dfrac{1}{4\tfrac{1}{2}}} = \frac{\tfrac{6+4+3}{12}}{\dfrac{1}{\tfrac{5}{2}}+\dfrac{1}{\tfrac{7}{2}}\times\dfrac{1}{\tfrac{9}{2}}} = \frac{\tfrac{13}{12}}{\tfrac{2}{5}+\tfrac{2}{7}+\tfrac{2}{9}} = \frac{\tfrac{13}{12}}{\dfrac{126+90+70}{315}}$$

$$= \frac{\tfrac{13}{12}}{\tfrac{286}{315}} = \tfrac{13}{12} \times \tfrac{315}{286} = \tfrac{105}{88} = 1\tfrac{17}{88}.$$

Ex. 7. Simplify the expression
$$\frac{1}{13} \text{ of } \cfrac{1}{1+\cfrac{1}{3+\tfrac{1}{4}}}.$$

Now, $\cfrac{1}{1+\cfrac{1}{3+\tfrac{1}{4}}} = \cfrac{1}{1+\tfrac{1}{\tfrac{13}{4}}} = \cfrac{1}{1+\tfrac{4}{13}} = \cfrac{1}{\tfrac{39}{13}} \text{ or } \cfrac{39}{43};$

therefore $\tfrac{1}{13} \text{ of } \cfrac{1}{1+\cfrac{1}{3+\tfrac{1}{4}}} = \tfrac{1}{13} \text{ of } \tfrac{39}{43} = \tfrac{3}{43}.$

Ex. 8. Simplify $\left\{ 2\tfrac{3}{4} + \tfrac{5}{2} \text{ of } \tfrac{7}{3\tfrac{2}{5}} - \tfrac{1\tfrac{3}{4}}{2\tfrac{1}{2}} \right\} \div 1\tfrac{77}{225}.$

QUESTIONS AND EXAMPLES IN FRACTIONS. 113

The expression

$$= \left\{ \frac{11}{4} + \frac{5}{2} \text{ of } \frac{7\frac{3}{5}}{1\frac{2}{6}} - \frac{\frac{5}{3}}{\frac{5}{5}} \right\} \div \frac{305}{228}$$

$$= \left\{ \frac{11}{4} + \frac{5}{2} \times \frac{7}{1} \times \frac{5}{19} - \frac{5}{3} \times \frac{2}{5} \right\} \frac{228}{305}$$

$$= \left\{ \frac{11}{4} + \frac{175}{38} - \frac{2}{3} \right\} \times \frac{228}{305}$$

(the least common multiple of 4, 38, and 3, = 38 × 2 × 3)

$$= \left\{ \frac{11 \times 19 \times 3 + 175 \times 2 \times 3 - 2 \times 38 \times 2}{38 \times 2 \times 3} \right\} \times \frac{228}{305}$$

$$= \left\{ \frac{627 + 1050 - 152}{228} \right\} \times \frac{228}{305}$$

$$= \frac{1677 - 152}{228} \times \frac{228}{305} = \frac{1525}{305} = 5.$$

Ex. XXXIV.

Miscellaneous Questions and Examples on Arts. (94–118).

I.

1. Define a fraction; what is the distinction between a Vulgar and Decimal fraction? How many different kinds of Vulgar fractions are there? Give an example of each kind.

2. Find the sum and difference of $\frac{2\frac{1}{3}}{\frac{5}{6}}$ of $7\frac{1}{3}$, and $1\frac{3}{4}$ divided by $2\frac{1}{8}$; and the sum of $5\frac{1}{3}$, $\frac{2}{3}$ of $3\frac{1}{2}$, and $\frac{3}{4} \div \frac{9}{7}$.

3. Simplify

  (1) $\{\frac{3}{4} + \frac{7}{8} \text{ of } 5\frac{1}{4}\} \times \{\frac{5}{8} + \frac{2}{3} + 3\frac{3}{4}\}$.   (2) $3\frac{1}{15}$ of $3\frac{4}{7} \div 3\frac{47}{15}$ of 9.

  (3) $\frac{3\frac{3}{4}}{4\frac{2}{7}} - \frac{3\frac{2}{3}}{4\frac{1}{4}} + \frac{\frac{4}{5}}{2\frac{1}{3}}$.   (4) $\frac{4\frac{1}{2} \times 4\frac{1}{3} \times 4\frac{1}{4} - 1}{4\frac{1}{3} \times 4\frac{1}{4} - 1}$.   (5) $3 + \frac{1}{7 + \frac{1}{13}}$.

4. Show that the fraction $\frac{2 + 4 + 6}{3 + 5 + 7}$ lies between the greatest and least of the fractions, $\frac{2}{3}$, $\frac{4}{5}$, and $\frac{6}{7}$.

5. The difference of two numbers is $15\frac{4}{35}$; the greater number is $20\frac{11}{15}$: find the smaller number.

II.

1. If the numerator and denominator of a fraction be both multiplied or both divided by the same number, the value of the fraction is not altered: prove this by means of an example.

2. What number subtracted from $41\frac{1}{4}$ leaves $19\frac{1}{2}$? and what number multiplied by $2\frac{4}{10}$ of $\frac{4}{13}$ produce $3\frac{1}{2}$ of $\frac{4}{5}$?

3. When is a fraction said to be in its *lowest* terms?

Reduce the fractions $\frac{63208}{13786}$ and $\frac{65462}{356784}$ to their lowest terms.

Simplify

(1) $\dfrac{2\frac{1}{2}}{3\frac{1}{4}} + \dfrac{1\frac{1}{2} - \frac{5}{6}}{1\frac{1}{4} + \frac{5}{6}} - 1\frac{2}{39}.$ (2) $3\frac{2}{6}$ of $5\frac{1}{2}$ of $\frac{7}{9} - \frac{1}{3}$ of $\frac{5}{12}.$

(3) $(\frac{2}{16} + \frac{1}{13}) \div (3 - \frac{1}{3}) \times (\frac{1}{3} + \frac{1}{5}).$ (4) $\frac{3}{14}$ of $\dfrac{4\frac{5}{8}}{6\frac{1}{6}}$ of $\dfrac{6\frac{8}{11}}{11\frac{5}{7}}.$

5. Divide the product of $2\frac{2}{16}$ and $2\frac{5}{8}$ by the difference of $2\frac{2}{5}$ and $2\frac{1}{4}$. Explain why it is necessary in the addition and subtraction of fractions to reduce the fractions to a common denominator.

III.

1. Show by an example that multiplying the numerator of a fraction by any number, is the same in effect as dividing the denominator by that number, and conversely.

2. Simplify

(1) $275\frac{1}{3} + 62\frac{11}{120} + 1031\frac{1}{5} + \frac{7}{8}$ of $41\frac{1}{5}0\frac{1}{7}.$ (2) $\frac{32}{61} \div 1\frac{12}{85} \times \frac{189}{207} \div 1\frac{13}{28}.$

(3) $\dfrac{1}{3\frac{1}{6}} - \dfrac{2\frac{1}{4}}{9} + \dfrac{3\frac{5}{6}}{2} + \dfrac{\frac{4}{7}}{4\frac{4}{7}}.$ (4) $\dfrac{4\frac{1}{4} - 3\frac{3}{6}}{4\frac{1}{4} + 3\frac{3}{6}} + \dfrac{3 - 2\frac{1}{3}}{4 - 3\frac{1}{3}}.$

3. Which is the greater, $\frac{1}{3}$ of 4 or $\frac{1}{4}$ of 5? and by how much?

4. Divide the sum of the fractions $\frac{3}{9}$ and $\frac{4}{13}$ by the product of $\frac{8}{11}$ and $1\frac{3}{4}$; and reduce the result to its lowest terms.

5. What number is that, from which if you deduct $\frac{8}{9} - \frac{3}{7}$, and to the remainder add the quotient of $\frac{2}{15}$ divided by $2\frac{1}{4}$, the sum will be $1\frac{8}{63}$?

IV.

1. Define a Vulgar fraction; an improper fraction; and the terms numerator and denominator of a fraction.

QUESTIONS AND EXAMPLES IN FRACTIONS.   115

Prove by means of an example the rule for the multiplication of fractions; and multiply the sum of $\frac{1}{7}$ of $\frac{1}{4}$ and $1\frac{1}{6}$ by the difference of $1\frac{4}{1}$ and $\frac{1}{5}$.

2. Reduce to their most simple forms the following expressions:

(1) $\frac{3}{4} \times 1\frac{7}{1}$ $8\frac{1}{5} \div \frac{2}{5}$ths of $(7\frac{3}{4} + \frac{5}{2})$.   (2) $\frac{1}{8} - 1\frac{1}{2} + 1\frac{1}{5} - \frac{1}{20}$.   (3) $\frac{27243}{37378}$.

(4) $\frac{1}{13}$ of $(1 + 5\frac{1}{2}) + \frac{5}{8}$ of $2\frac{1}{27}$ of $(7 - 2\frac{2}{3}) - \frac{1}{3}$.   (5) $\dfrac{\frac{10}{3} + \frac{5}{8} - 2\frac{9}{21}}{\frac{5}{6} - \frac{4}{7}}$.

3. What number added to $\frac{3}{4}$ of $(\frac{1}{5} + \frac{1}{8} - \frac{4}{15} + \frac{1}{8})$ makes $3\frac{1}{4}$? and what number divided by $\frac{1}{4}$ of $\frac{1}{6}$ of $\frac{1}{8}$ will give $\frac{64}{315}$?

4. If I pay away $\frac{1}{3}$ of my money, then $\frac{1}{4}$ of what remains, and then $\frac{1}{4}$ of what still remains; what fraction of the whole will be left?

5. Explain the method of 'comparing' fractions.

Compare the product and quotient of the sum and difference of $5\frac{1}{2}$ and $5\frac{1}{4}$.

V.

1. State the rules for multiplying and dividing one fraction by another; and prove them by means of an example.

Divide $\dfrac{2+3}{4+5}$ by $\dfrac{4+3\frac{1}{2}}{5+5\frac{1}{2}}$; and multiply the sum of $\frac{1}{2}, 1\frac{3}{3}$, and $\frac{5}{3}$, by the difference of $1\frac{4}{5}$ and $2\frac{3}{20}$, and divide the product by $1\frac{1}{3}$ of $1\frac{1}{1}\frac{3}{8}$.

2. Reduce to their simplest forms

(1) $(\frac{5}{2} - \frac{2}{5}) \div (\frac{4}{3} - \frac{3}{4})$.   (2) $\dfrac{1\frac{41}{82} - \frac{9}{40} - \frac{3}{54}}{\frac{4}{9} + \frac{1}{4} - 1\frac{3}{14}}$.

(3) $\frac{3}{8}$ of $1\frac{3}{10} - \dfrac{1\frac{2}{3}}{6\frac{2}{3}}$ of $1\frac{9}{20} + \frac{3}{7}$ of $\dfrac{6\frac{5}{12}}{3\frac{2}{3}}$.

(4) $\dfrac{\frac{3}{5}}{\frac{5}{8}} + \dfrac{\frac{5}{6}}{2\frac{1}{2} \times 1\frac{1}{3}} \times \frac{1}{50}$.   (5) $2\frac{1}{2} \times \dfrac{1}{3\frac{1}{3} + \dfrac{1}{4\frac{1}{4}}}$

(6) $\dfrac{\frac{5}{7} \text{ of } \frac{3}{10} + \frac{1}{4} \text{ of } \frac{8}{21}}{\frac{2}{3} \text{ of } \frac{9}{14} - \frac{5}{8} \text{ of } \frac{2}{15}}$.   (7) $\dfrac{11\frac{5}{7} - 7\frac{5}{11}}{3\frac{1}{2} + 5\frac{6}{22}}$.

3. What is meant by the symbol $\frac{a}{b}$?

Find the least fraction which added to the sum of $\frac{3}{4}, \frac{7}{8}$, and $\frac{29}{12}$, shall make the result an integer.

4. Find the sum of the greatest and least of the fractions $\frac{3}{8}, \frac{5}{12}, \frac{1}{4}$, and $\frac{7}{20}$; the sum of the other two; and the difference of these sums.

5. A man has $\frac{3}{8}$ of an estate, he gives his son $\frac{1}{4}$ of his share; what portion of the estate has he then left?

VI.

1. State the rules for addition and subtraction of vulgar fractions; and prove them by means of an example.

2. Simplify

(1) $\frac{4}{5}$ of $\frac{1}{2} - \frac{2}{3}$ of $\frac{9}{17} + \frac{3}{5}$ of $1\frac{10}{17}$.

(2) $\dfrac{2\frac{1}{3} + 3\frac{2}{5}}{4\frac{1}{8} + 5\frac{1}{4}} + \dfrac{3\frac{1}{3}}{10\frac{1}{3}}$.

(3) $\{\frac{5}{7} \times \frac{2}{9} \times 13\frac{1}{2}\} \div \{\frac{1}{3} \times \frac{3}{7} + 40\}$.

(4) $\dfrac{2\frac{4}{11}}{2\frac{2}{5}} \div \dfrac{2\frac{7}{11}}{8\frac{7}{10}}$.

3. Define a *proper*, *mixed*, and *compound* fraction. Explain the method of reducing a compound fraction to a simple one.

Ex. $\frac{2}{3}$ of $\frac{5}{9}$ of $\frac{31}{111}$ of $1\frac{1}{5}$.

4. Show by means of an example how a fraction is affected if the same number be added to its numerator and denominator.

5. Multiply $3\frac{1}{5}$ by $3\frac{1}{15}$, and divide $\dfrac{20\frac{1}{4}}{3}$ by $\dfrac{41\frac{1}{4}}{4}$, and find the difference between the sum and difference of these results.

6. What number added to $3\frac{1}{35} + \frac{10}{21}$ will produce $8\frac{27}{30}$? and what number divided by $2\frac{1}{15}$ will produce $\frac{1}{15}$?

VII.

1. Show from the nature of fractions that $\frac{2}{3} + \frac{5}{7} = \frac{29}{21}$; that $\frac{2}{3}$ of $\frac{5}{7} = \frac{10}{21}$; and that $\frac{2}{3} \div \frac{5}{7} = \frac{14}{15}$.

2. Simplify

(1) $\dfrac{\frac{3}{4}}{\frac{5}{4}} \cdot \dfrac{2\frac{1}{3} - \frac{3}{7}}{5\frac{1}{4} + \frac{1}{14}} - \dfrac{2}{3\frac{4}{7}}$.

(2) $2\frac{1}{6} + 3\frac{3}{8} + \frac{9}{14} + \frac{1}{15} + 6\frac{17}{20}$.

(3) $\{(3\frac{1}{3} \text{ of } 4\frac{1}{2})\} \div \{(2\frac{1}{2} - \frac{1}{3}) \text{ of } (3\frac{1}{2} - \frac{1}{3})\}$.

(4) $\{(2\frac{1}{6} \text{ of } 3\frac{1}{4}) + (\frac{3}{8} + \frac{27}{34})\} - \left\{\left(\dfrac{1}{1\frac{1}{7}} - \dfrac{1\frac{1}{3}}{8}\right) \div (2 - \frac{5}{9})\right\}$.

3. Simplify $\dfrac{\frac{2}{3} \text{ of } \frac{5}{7} \text{ of } \frac{35}{9}}{\frac{4}{7} \text{ of } \frac{2}{3} \text{ of } \frac{21}{31}}$, and take the result from the sum of $10\frac{1}{4}$, $3\frac{9}{10}$, $7\frac{21}{25}$.

4. Add together $\frac{1}{2}$, $\frac{1}{3}$, $\frac{1}{4}$, and $\frac{1}{5}$, subtract the sum from 2, multiply the result by $\frac{2}{3}$ of $\frac{27}{17}$ of 8, and find what fraction this is of 99.

5. In a match of cricket, a side of 11 men made a certain number of runs, one obtained $\frac{1}{4}$th of the number, each of two others $\frac{1}{10}$th, and each of three others $\frac{1}{20}$th, the rest made up between them 126; which was the remainder of the score, and 4 of these last scored 5 times as many as the other. What was the whole number of runs, and the score of each man?

DECIMALS

119. Figures in the units' place of any number express their *simple values*, while those to the *left* of the units' place increase in value *tenfold* each step from the units' place; therefore, according to the same notation, as we proceed from the units' place to the *right* every successive figure would decrease in value *tenfold*. We can thus represent whole numbers or **integers** and certain fractions under a uniform notation by means of figures in the units' place and on each side of it; for instance, in the number 5673·211, the figures on the left of the *dot* represent *integers*, while those on the right of the dot denote *fractions*. The number written at length would stand thus:

$$5 \times 1000 + 6 \times 100 + 7 \times 10 + 3 + \frac{2}{10} + \frac{4}{100} + \frac{1}{1000}.$$

The dot is termed **the** decimal point, and all figures to the right of it are called DECIMALS, or DECIMAL FRACTIONS, because they are fractions with either 10, **100** or 10×10, 1000 or 10×10×10, &c., as their respective denominators.

The *extended Numeration Table* will be represented thus:

&c.	Millions.	Hundreds of Thousands.	Tens of Thousands.	Thousands.	Hundreds.	Tens.	Units.	.	Tenths.	Hundredths.	Thousandths.	Ten Thousandths.	Hundred Thousandths.	Millionths.	&c.
	7	6	5	4	3	2	1	.	2	8	4	5	6	7	

120. **10, called** the *first* POWER of **10,** is written thus, 10^1.
10×10, or 100, called the *second* POWER of 10, is written thus, 10^2.
10×10×10, or 1000, called the *third* POWER of 10, is written thus, 10^3, and so on; similarly of other numbers: thus the fifth power of 4 is 4×4×4×4×4, and is written thus, 4^5.

The small figures 1, 2, 3, &c., at the right of the number, a little above the line, are called INDICES.

121. From the preceding it appears that

First, $\quad \cdot 2345 = \dfrac{2}{10} + \dfrac{3}{100} + \dfrac{4}{1000} + \dfrac{5}{10000}$

Now the least common multiple of the denominators of the fractions is 10000: therefore, reducing the several fractions to equivalent ones with their least common denominator, we get

$$\cdot 2345 = \dfrac{2}{10} \times \dfrac{1000}{1000} + \dfrac{3}{100} \times \dfrac{100}{100} + \dfrac{4}{1000} \times \dfrac{10}{10} + \dfrac{5}{10000}.$$

$$= \dfrac{2000 + 300 + 40 + 5}{10000} = \dfrac{2345}{10000}.$$

Secondly, $\quad \cdot 00324 = \dfrac{0}{10} + \dfrac{0}{100} + \dfrac{3}{1000} + \dfrac{2}{10000} + \dfrac{4}{100000}$

(the least common multiple of the denominators is 100000)

$$= \dfrac{0}{10} \times \dfrac{10000}{10000} + \dfrac{0}{100} \times \dfrac{1000}{1000} + \dfrac{3}{1000} \times \dfrac{100}{100} + \dfrac{2}{10000} \times \dfrac{10}{10} + \dfrac{4}{100000}$$

$$= \dfrac{300 + 20 + 4}{100000} = \dfrac{324}{100000}.$$

Thirdly, $\quad 56 \cdot 816 = 5 \times 10 + 6 + \dfrac{8}{10} + \dfrac{1}{100} + \dfrac{6}{1000}$

(the least common multiple of the denominators is 1000)

$$= \dfrac{5 \times 10}{1} \times \dfrac{1000}{1000} + \dfrac{6}{1} \times \dfrac{1000}{1000} + \dfrac{8}{10} \times \dfrac{100}{100} + \dfrac{1}{100} \times \dfrac{10}{10} + \dfrac{6}{1000}$$

$$= \dfrac{50000 + 6000 + 800 + 10 + 6}{1000} = \dfrac{56816}{1000}.$$

Hence, we infer that every decimal, and every number composed of integers and decimals, can be put down in the form of a vulgar fraction, with the figures comprising the decimal or those composing the integer and decimal part (the dot being in either case omitted) as a numerator, and with 1 followed by as many zeros as there are decimal places in the given number for the denominator.

122. Conversely, any fraction having 10 or any power of 10 for its denominator, as $\dfrac{56816}{1000}$, may be represented in the form 56·816.

DECIMALS.

For $\dfrac{56816}{1000} = \dfrac{5 \times 10000 + 6 \times 1000 + 8 \times 100 + 1 \times 10 + 6}{1000}$

$= \dfrac{5 \times 10000}{1000} + \dfrac{6 \times 1000}{1000} + \dfrac{8 \times 100}{1000} + \dfrac{1 \times 10}{1000} + \dfrac{6}{1000}$

$= 5 \times 10 + 6 + \dfrac{8}{10} + \dfrac{1}{100} + \dfrac{6}{1000}$

$= 56\cdot816$ (by the notation we have assumed).

123. Again, by what has been said above, it appears that

$$\cdot 327 = \dfrac{327}{1000}, \quad \cdot 0327 = \dfrac{327}{10000}, \quad \cdot 3270 = \dfrac{3270}{10000} = \dfrac{327}{1000}.$$

We see that $\cdot 327$, $\cdot 0327$, and $\cdot 3270$ are respectively equivalent to fractions which have the same numerator, and the first and third of which have also the same denominator, while the denominator of the second is greater.

Consequently, $\cdot 327$ is equal to $\cdot 3270$, but $\cdot 0327$ is less than either.

The value of a decimal is therefore not affected by *affixing* cyphers to the right of it; but its value is decreased by *prefixing* cyphers: which effect is exactly opposite to that which is produced by affixing and prefixing ciphers to integers.

124. Hence it appears that a decimal is *multiplied* by 10, if the decimal point be removed *one* place towards the *right* hand; by 100, i. e. *two* places; by 1000, if *three* places; and so on: and conversely, a decimal is *divided* by 10, if the point be removed *one* place to the *left* hand; by 100, if *two* places; by 1000, if *three* places; and so on.

Thus $\quad 5\cdot6 \times 10 = \dfrac{56}{10} \times 10 = 56$.

$\quad\quad\quad 5\cdot6 \times 1000 = \dfrac{56}{10} \times 1000 = 5600$.

$\quad\quad\quad 5\cdot6 \div 10 = \dfrac{56}{10} \times \dfrac{1}{10} = \dfrac{56}{100} = \cdot 56$.

$\quad\quad\quad 5\cdot6 \div 1000 = \dfrac{56}{10} \times \dfrac{1}{1000} = \dfrac{56}{10000} = \cdot 0056$.

125. The advantage arising from the use of decimals consists in this; viz., that the addition, subtraction, multiplication, and division of *decimal* fractions are much more easily performed than those of *vulgar* fractions; and although all vulgar fractions cannot be reduced to finite decimals, yet we can find decimals so near their true value, that the error arising from using the *decimal* instead of the *vulgar* fraction is not perceptible.

Ex. XXXV.

1. Express as vulgar fractions in their lowest terms:

·075; ·818; 3·02; 3·434; 343·4; ·03434; ·050005; 230·409; 2·30409; 2137·2; 91300·0008; 24·000625; 8213·7169125; ·00083276; 1·0000009; ·000000001.

2. Express as decimals,

$\frac{1}{10}$; $\frac{3}{10}$; $\frac{7}{10}$; $\frac{53}{100}$; $\frac{3}{100}$; $\frac{7}{1000}$; $\frac{0173}{10000}$; $\frac{0178}{100}$; $\frac{0178}{100000}$; $\frac{21}{10000}$; $\frac{9}{100000}$; $\frac{5203}{10}$; $\frac{90}{100}$; $\frac{30142}{10000}$; $\frac{672312}{100000}$; $\frac{672312}{1000000000}$; $\frac{47281202}{10000}$.

3. Multiply

·7 separately by 10, 100, 1000, and by 100000;
·006 separately by 100, 10000, and by 10000000;
·0431 separately by 100, and by 1000000;
16·201 separately by 10, 1000, and by a million;
9·0016 by ten hundred thousand, and by 100.

4. Divide

·51 separately by 10, 1000, and by 100000;
·008 separately by 100, and by a million;
5·016 separately by 1000, and by 100000;
378.0186 separately by 1000, and by a million.

5. Express according to the decimal notation, five-tenths; seven-tenths; nineteen hundredths; twenty-eight hundredths; five thousandths; ninety-seven tenths; one millionth; fourteen and four tenths; two hundred and eighty, and four ten-thousandths; seven and seven thousandths; one hundred and one hundred-thousandths; one one-thousandth and one ten-millionth; five billionths.

6. Express the following decimals in words:

·4; ·25; ·75; ·745; ·1; ·001; ·00001; 23·75; 2·375; ·2375; ·00002375; 1·000001; ·1000001; ·00000001.

ADDITION OF DECIMALS.

126. RULE. Place the numbers under each other, units under units, tens under tens, &c., tenths under tenths, &c.; so that the decimals be all under each other; add as in whole numbers, and place the decimal point in the sum under the decimal point above.

ADDITION OF DECIMALS

Ex. Add together 27·5037, ·042, 342, and 2·1.
Proceeding by the Rule given above,

$$27·5037$$
$$·042$$
$$342·$$
$$2.1$$
$$\overline{371·6457}$$

NOTE. The same method of explanation holds for the fundamental rules of decimals, which has been given at length in explaining the Rules for Simple Addition, Simple Subtraction, and the other fundamental rules in whole numbers.

Reason for the above process.

If we convert the decimals into fractions, and add them together as such, we obtain

$$27·5037 + ·042 + 342 + 2·1$$
$$= \frac{275037}{10000} + \frac{42}{1000} + \frac{342}{1} + \frac{21}{10};$$

(or reducing the fractions to a common denominator),

$$= \frac{275037}{10000} + \frac{420}{10000} + \frac{3420000}{10000} + \frac{21000}{10000}$$

$$= \frac{3716457}{10000} = 371·6457, \text{(Art. 122)}.$$

Ex. XXXVI.

Add together:

(1) ·234, 14·3812, ·01, 32·47, and ·00075.
(2) 232·15, 3·225, 21, ·0001, 34·005, and ·001304.
(3) 14·94, ·00857, 1·5, 5607·25, 530, and ·0057.

Express in one sum:

(4) ·08 + 165 + 1·327 + ·0003 + 2760·1 + 9.
(5) 346 + ·0027 + ·25 + ·186 + 72·505 + ·0014 + ·00004.
(6) 6·3084 + ·006 + 36·207 + ·0001 + 364 + ·008022.
(7) 725·1201 + 34·00076 + ·04 + 50·9 + 143·713.
(8) 67·8125 + 27·105 + 17·5 + ·000375 + 255 + 3·0125.

Add together:

(9) 2·0068, ·04137, ·987641, 1·0000009, 57, and 1·5; and prove the result.

6

(10) ·0003025, 29·99987, 143·2, 5·000025, 9000, and 3·4073; and verify the result.

(11) 21·74, ·075, 103·00375, ·0005495, and 4957·5; and verify the result.

(12) Five hundred and nine-hundredths; three hundred and seventy-five; twenty thousand and eighty-four, and seventy-eight hundred-thousandths; eleven millions, two thousand, and two hundred and nine millionths; eleven millionths: one billion, and one billionth.

SUBTRACTION OF DECIMALS.

(127) RULE. Place the less number under the greater, units under units, tens under tens, &c., tenths under tenths, &c.; suppose cypher to be supplied if necessary in the upper line to the right of the decimals then proceed as in Simple Subtraction of whole numbers, and place the decimal point in the remainder under the decimal point above.

Ex. Subtract 5·473 from 6·23.

Proceeding by the Rule given above,

$$\begin{array}{r} 6·23 \\ 5·473 \\ \hline ·757 \end{array}$$

Reason for the above process.

If we convert decimals into fractions, and subtract the one from the other as such, we obtain

$$6·23 - 5·473 = \frac{623}{100} - \frac{5473}{1000} = \frac{6230}{1000} - \frac{5473}{1000}$$

$$= \frac{757}{1000} = ·757, \text{ (Art. 122)}.$$

Ex. XXXVII

(1) Find the difference between 2·1354 and 1·0436; 7·835 and 2·0005; 15·67 and 156·7; ·001 and ·0009; ·305 and ·000683.

Find the value of

(2) 213·5 − 1·8125.
(3) ·0516 − ·0094187.
(4) 603 − ·6584003.
(5) 17·5 − 13·0046.
(6) ·582 − ·09647.
(7) 9·233 − ·0536.

MULTIPLICATION OF DECIMALS.

(8) **Take** ·01 from ·1 ; 57·704 from **713·00683** : 35·009876 **from 56·078** ; 27·148 from 9816 ; and prove the truth of each result.

(9) Required the difference between seven and seven-tenths ; **also** between seven tenths and seven millionths ; also between seventy-four + three hundred and four thousandths and one hundred and seventy-four + **one hundredths; and verify** each result.

MULTIPLICATION OF DECIMALS.

128. RULE. Multiply **the** numbers **together as if they were whole numbers,** and point off in the product **as many decimal places as there are** decimal places in both the multiplicand **and the multiplier ; if there are** not figures enough, supply the deficiency by **prefixing cyphers.**

Ex. Multiply 5·34 by ·0021.

Proceeding by the Rule given above,

$$\begin{array}{r} 5·34 \\ ·0021 \\ \hline 534 \\ 1068 \\ \hline 11214 \end{array}$$

The number of decimal places in the multiplicand + the number of those in the multiplier = 2 + 4 = 6 ; but there **are only** 5 figures in the product ; therefore we must prefix **one zero,** and place a point before it, thus ·011214.

Reason for the above process.

$$5·34 \times ·0021 = \frac{534}{100} \times \frac{21}{10000} = \frac{11214}{1000000} = ·011214.$$

Ex. XXXVIII.

Multiply together :

(1) 3·8 **and** 42 ; ·38 and ·42 ; 3·8 and 4·2 ; ·038 and ·0042.

(2) 417 and **·417** ; ·417 and ·417 ; 71956 and ·000025.

(3) 2·052 **and** ·0031 ; 4·07 and ·916 ; 476 and ·00026.

Multiply (proving the **truth of** the result in each case).

(4) 81·4632 by ·0378. (5) 27·35 by 7·70071. (6) ·04375 by ·0754,

(7) ·0046 by 7·85   (8) ·00846 by ·00324.   (9) ·314 by ·0021.
(10) ·009 by 00846.   (11) ·009207 by 6·056.   (12) ·00948 by 29.
(13) Find the continued product of 1, ·01, ·001, and 100; also of ·12, 1·2, ·012, and 120; and prove the truth of the results.

Find the value of

(14) $7·6 \times ·071 \times 2·1 \times 29$.

(15) $·007 \times 700 \times 760·3 \times ·00416 \times 100000$.

DIVISION OF DECIMALS.

129. *First. When the number of decimal places in the dividend exceeds the number of decimal places in the divisor.*

RULE. Divide as in whole numbers, and mark off in the quotient a number of decimal places equal to the excess of the number of decimal places in the dividend over the number of decimal places in the divisor; if there are not figures sufficient, prefix cyphers as in Multiplication.

Ex. 1. Divide 1·1214 by 5·34.

Proceeding by the Rule given above,

$$5·34) 1·1214 (21$$
$$1068$$
$$\overline{534}$$
$$534$$

Now the number of decimal places in the dividend − the number of decimal places in the divisor = 4 − 2 = 2;

therefore the quotient = ·21.

Ex. 2. Divide ·011214 by 53·4.

$$53·4) ·011214 (21$$
$$1068$$
$$\overline{534}$$
$$534$$

Now the number of decimal places in the dividend − the number of decimal places in the divisor

= 6 − 1 = 5;

therefore we prefix three cyphers, and the quotient is ·00021.

DIVISION OF DECIMALS. 125

Reason for the above process.

$$1{\cdot}1214 \div 5{\cdot}34 = \frac{11214}{10000} \div \frac{534}{100} = \frac{11214}{10000} \times \frac{100}{534} = \frac{11214}{534} \times \frac{1}{1000}$$

$$= \frac{21}{1} \times \frac{1}{100}; \left(\text{since } \frac{11214}{534} = 21, \text{ and } \frac{100}{10000} = \frac{1}{100}\right)$$

$$= \frac{21}{100} = {\cdot}21.$$

Again,
$${\cdot}011214 \div 53{\cdot}4 = \frac{11214}{1000000} \div \frac{534}{10} = \frac{11214}{1000000} \times \frac{10}{534}$$

$$= \frac{11214}{534} \times \frac{10}{1000000} = \frac{21}{1} \times \frac{1}{100000} = \frac{21}{100000} = {\cdot}00021.$$

180. *Secondly.* When the number of decimal places in the dividend *is less* than the number of decimal places in the divisor.

RULE. Affix cyphers to the dividend until the number of decimal places in the dividend equals the number of decimal places in the divisor; the quotient up to this point of the division will be a whole number; if there be a remainder, and the division be carried on further, the figures in the quotient after this point will be decimals.

Ex. Divide 1121·4 by ·534.

Proceeding by the Rule given above,

```
·534) 1121·400 (2100
      1068
       534
       534
```

Reason for the above process.

$$1121{\cdot}4 \div {\cdot}534 = \frac{11214}{10} \div \frac{534}{1000} = \frac{11214}{10} \times \frac{1000}{534}$$

$$= \frac{11214}{534} \times \frac{1000}{10} = 21 \times 100 = 2100.$$

NOTE. In order to prevent mistakes in the proof of examples in Division of Decimals, always contrive in the process to separate 10, 100, &c. in the two fractions from the other figures, as in the above examples; and be sure never to effect the multiplication if there be tens left in the denominator; nor, if there be tens left in the numerator, to effect it until the last step of the operation.

Ex. Divide 172·9 by ·142 to three places of decimals.

·142) 172·900000 (1217·605
142

309
284

250
142

1080
994

860
852

800
710

90

Here we must affix 5 cyphers to 172·9; for if we affix two according to the Rule, the division up to that point will give the integral part of the quotient only, and therefore as the quotient is to be obtained to three places of decimals, we must affix three cyphers more, that is, we must affix five altogether.

Reason for the above process.

$$172\cdot 9 \div \cdot 142 = \frac{1729}{10} \div \frac{142}{1000} = \frac{1729}{142} \times \frac{1000}{10}$$

$$= \frac{1729}{142} \times \frac{100000}{1000} = \frac{172900000}{142} \times \frac{1}{1000}$$

Now $\quad \dfrac{172900000}{142} = 1217605 \ldots$ from above;

therefore the result $= \dfrac{1217605\ldots}{1000} = 1217\cdot 605$.

Ex. XXXIX.

Divide, (proving the truth of each result by Fractions):

(1) 10·836 by 5·16, and 34·96818 by ·381.

(2) ·025075 by 1·003, and ·02916 by ·0012.

(3) ·00081 by 27, and 1·77089 by 4·785.

DIVISION OF DECIMALS. 127

(4) 1 by ·1, by ·01, and by ·0001.
(5) 31·5 by ·126, and 5·2 by ·32.
(6) 3217 by ·0625, and ·03217 by 6250.
(7) 4·63638 by 81·34, and 15·4546 by ·019.
(8) ·429408 by 59·64, and 2147·04 by ·036.
(9) 12·6 by ·0012, and ·065341 by ·000475.
(10) 3·012 by ·0006, and 293916·669 by 541·283.
(11) 130·4 by ·0004 and by 4, and 46·634205 by 4807·65.
(12) 1·69 by 1·3, by ·13, by 13, and also by ·013.
(13) ·00281 by 1·405, by 1405, and by ·001405.
(14) 72·36 by 36 by ·0036, and ·003 by 1·6.
(15) 6725402·3544 by 7089, and by ·7089.
(16) 10363284·75 by 396·25, and ·09844 by ·0046.
(17) 816 by ·0004, and ·0019610652875 by 2·38645.
(18) 18368830·5 by 2315, by 231·5, and by ·2315.
(19) ·00005 by 2·5, by 25, and by ·0000025.
(20) 684·1197 by 1200·21, and also by ·0120021.

Divide to four places of decimals each of the following, and prove the truth of the results by Fractions:

(21) 32·5 by 8·7; ·02 by 1·7; 1 by ·013.
(22) ·009384 by ·0063; 51846·734 by 1·02.
(23) 7380·964 by ·023; 6·5 by 3·42; 25 by **19**.
(24) 176432·76 by ·01257; 7457·1345 by 6535496·2.
(25) 37·24 by 2·9; ·0719 by 27·53.

Find the quotient (verifying each result) of
(26) ·0029202 by 157, and by 1·57.
(27) 5005 by 1953125; of 50·05 by 195·3125; of ·05005 by ·001953125.
(28) ($7\frac{1}{2}$ of $\frac{1}{8} \times \frac{1}{25}$) by ·0005; of 31·008 by $\frac{128}{125}$ of $1\frac{1}{4}$ of $\frac{525}{1573}$; ·7575 by $16\frac{3}{4}$.

131. *Certain* **Vulgar Fractions** *can be expressed accurately as Decimals.*

RULE. Reduce the fraction to its lowest terms; then place a dot after the numerator and affix cyphers for decimals; divide by the denominator, as in division of decimals, and the quotient will be the decimal required.

Ex. 1. Convert $\frac{3}{5}$ into a decimal.

$$5 \,|\, \underline{3\cdot 0}$$
$$\cdot 6$$

There is one decimal place in the dividend and none in the divisor; therefore there is one decimal place in the quotient.

NOTE. In reducing any such fraction as $\frac{3}{50}$ or $\frac{3}{500}$ to a decimal, we may proceed in the same way as if we were reducing $\frac{3}{5}$; taking care however, in the result to move the decimal point one place further to the left for each cypher cut off.

Thus, $\quad \dfrac{3}{5} = \cdot 6, \quad \dfrac{3}{50} = \cdot 06, \quad \dfrac{3}{500} = \cdot 006.$

for in fact, we divide by 5, and then by 10, 100, &c., according as the divisor is 50, 500, &c.

Ex. 2. Reduce $\dfrac{5}{16}$ to a decimal.

```
16) 5·0000 (·3125
    48
    ――
    20
    16
    ――
    40
    32
    ――
    80
    80
    ――
```

or thus, $16 \begin{cases} 4 \,|\, 5\cdot 00 \\ 4 \,|\, \overline{1\cdot 2500} \\ \overline{\cdot 3125} \end{cases}$

$\therefore \tfrac{5}{16} = \cdot 3125$

Ex. 3. Convert $\dfrac{3}{512}$ and $\dfrac{3}{51200}$ into decimals.

Now $512 = 8 \times 64 = 8 \times 8 \times 8$

$$8 \,|\, \underline{3\cdot 000}$$
$$8 \,|\, \underline{\cdot 375000}$$
$$8 \,|\, \underline{\cdot 046875000}$$
$$ \cdot 005859375$$

or $\dfrac{3}{512}$ is equivalent to $\cdot 005859375$

and $\dfrac{3}{51200}$ is equivalent to $\cdot 00005859375.$

VULGAR FRACTIONS EXPRESSED AS DECIMALS.

Ex. 4. Convert $\frac{3}{5} + 3\frac{1}{8} + 2\frac{9}{40} + 6\frac{11}{125}$ into a decimal.

$$\frac{3}{5} + 3\frac{1}{8} + 2\frac{9}{40} + 6\frac{11}{125} = 11 + \frac{3}{5} + \frac{1}{8} + \frac{9}{40} + \frac{11}{125}.$$

$$8 \,|\, 1\cdot 000$$
$$\overline{\cdot 125}$$

$$5 \,|\, 3\cdot 0 \qquad 5 \,|\, 11 \qquad\qquad 4 \,|\, 9\cdot 00$$
$$\overline{\cdot 6} \qquad\quad 5 \,|\, \overline{2\cdot 20} \qquad\quad \overline{2\cdot 25}$$
$$\qquad\qquad\quad 5 \,|\, \overline{\cdot 440} \qquad \therefore \tfrac{9}{40} = \cdot 225$$
$$\qquad\qquad\qquad \overline{\cdot 088}$$

therefore $\quad \tfrac{3}{5} = \cdot 6, \; \tfrac{1}{8} = \cdot 125, \; \tfrac{9}{40} = \cdot 225, \; \tfrac{11}{125} = \cdot 088 \,;$

therefore the whole expression

$$= 11 + \cdot 6 + \cdot 125 + \cdot 225 + \cdot 088$$
$$= 12\cdot 038.$$

Ex. XL.

Reduce to decimals:

(1) $\tfrac{1}{4}\,;\; \tfrac{3}{4}\,;\; \tfrac{5}{8}\,;\; \tfrac{9}{25}\,;\; \tfrac{6}{16}\,;\; \tfrac{19}{20}.$

(2) $\tfrac{68}{125}\,;\; \tfrac{54}{125}\,;\; \tfrac{570}{200}\,;\; \tfrac{170}{125}\,;\; \tfrac{1}{160}.$

(3) $6\tfrac{11}{64}\,;\; \tfrac{57}{240}\,;\; \tfrac{13}{256}\,;\; \tfrac{3}{512}\,;\; 15\tfrac{589}{78125}.$

(4) $3\tfrac{3}{8}$ of $\tfrac{1}{512}.$ (5) $\tfrac{1}{2} + \tfrac{1}{4} + \tfrac{1}{16} + \tfrac{3}{32}.$ (6) $\tfrac{1}{64} \times \cdot 0064.$

(7) $\tfrac{3}{5} + \cdot 061.$ (8) $\tfrac{1}{2} + \tfrac{1}{5} - \tfrac{1}{8}.$ (9) $\dfrac{47\tfrac{5}{8}}{94}$ of $\dfrac{11\tfrac{1}{4}}{7\cdot 5}.$

(10) $\dfrac{7\cdot 75}{9}$ of $\dfrac{2\tfrac{1}{7}}{2\tfrac{7}{9}}$ of $\dfrac{20}{31}.$ (11) $5\tfrac{5}{540} + \cdot 75$ of $\tfrac{8}{5}$ of $7\tfrac{1}{2}.$

(12) $3\tfrac{4}{25} + 1\tfrac{3}{110} + 81\tfrac{27}{1000} + \dfrac{7\tfrac{1}{2}}{3\tfrac{1}{8}}.$ (13) $\dfrac{247}{5} + 1\tfrac{513}{108} + \dfrac{17}{7\tfrac{5}{9}} + 200\tfrac{7}{10} + \dfrac{11}{62\cdot 5}.$

132. We have seen that, in order to convert a vulgar fraction into a decimal, we have in fact, after reducing the fraction to its lowest terms and affixing ciphers to the numerator, to divide 10, or some multiple of 10 or of its powers, by the denominator: now $10 = 2 \times 5$, and these are the only factors into which 10 can be broken up; therefore, when the fraction is in its lowest terms, if the denominator be not composed solely of the factors 2 and 5, or one of them, or of powers of 2 and 5, or one of them, then the division of the numerator by the denominator will never terminate. Decimals of this kind are called indeterminate decimals, and they are also called CIRCULATING, REPEAT-

130 ARITHMETIC.

ing, or Recurring Decimals, from the fact that, when a decimal does not terminate, the same figures must come round again, or recur, or be repeated: for since we always affix a cipher to the dividend, whenever any former remainder recurs, the quotient will also recur. Now, when we divide by any number, the remainder must always be less than that number, and therefore some remainder must recur before we have obtained a number of remainders equal to the number of units in the divisor.

133. Pure Circulating Decimals are those which recur from the beginning: thus, ·333.., ·2727.., are pure circulating decimals.

Mixed Circulating Decimals are those which do not begin to recur till after a certain number of figures. Thus, ·128888.., ·0113636.., are mixed circulating decimals.

Pure and mixed circulating decimals are generally written down only to the end of the first period, a dot being placed over the first and last figures of that period.

Thus ·3̇ represents the pure circulating decimal ·333..

·3̇6̇.. ·3636..

·6̇39̇.. ·639639..

·13̇8̇................mixed........................ 1388..

·013̇6̇.. ·0113636..

134. *Pure Circulating Decimals may be converted into their equivalent Vulgar Fractions by the following Rule.*

Rule. Make the period or repetend the numerator of the fraction, and for the denominator put down as many *nines* as there are figures in the period or repetend.

This fraction, reduced to its lowest terms, will be the fraction required in its simplest form.

Ex. Reduce the following pure circulating decimals, ·3̇, ·2̇7̇, ·8̇5714̇2̇, to their respective equivalent vulgar fractions.

Proceeding by the Rule given above,

$$\dot{3} = \frac{3}{9} = \frac{1}{3}\ ;\quad \dot{2}\dot{7} = \frac{27}{99} = \frac{3}{11}.$$

CIRCULATING DECIMALS.

The truth of these results will appear from the following considerations.

$$\frac{1}{9} = \cdot 111111 \text{ &c., hence } \frac{4}{9} = \cdot 4444 \text{ &c.}, \frac{7}{9} = \cdot 7777 \text{ &c.}$$

$$\text{therefore } \cdot \dot{3} = \frac{3}{9} = \frac{1}{3}.$$

Again,

$$\frac{1}{99} = \frac{1}{9} \div 11 = \cdot 111111 \text{ &c.} \div 11 = \cdot 010101 \text{ &c.;}$$

hence $\frac{7}{99} = \cdot 070707$ &c., $\frac{17}{99} = \cdot 171717$ &c.

$$\text{therefore } \cdot \dot{2}\dot{7} = \frac{27}{99} = \frac{3}{11},$$

In like manner,

$$\frac{1}{999} = \frac{1}{9} \div 111 = \cdot 111111 \text{ &c.} \div 111 = \cdot 001001 \text{ &c.}$$

and $\frac{1}{9999} = \frac{1}{9} \div 1111 = \cdot 111111$ &c. $\div 1111 = \cdot 00010001$ &c.;

hence $\frac{1\,2}{999} = \cdot 206206$ &c., and $\frac{3214}{9999} = \cdot 32143214$ &c.;

$$\text{therefore } \cdot \dot{8}5714\dot{2} = \frac{857142}{999999} = \frac{142857 \times 6}{142857 \times 7} = \frac{6}{7}.$$

135. *Mixed Circulating Decimals may be converted into their equivalent Vulgar Fractions by the following Rule.*

RULE. Subtract the figures which do **not circulate from** the figures taken to the end of the first period, *as if both were whole numbers*.

Make the result the numerator, and write down as many *nines* as there are figures in the circulating part, followed by as many *zeros* as there are figures in the non-circulating part, for the denominator.

Ex. Reduce the following mixed circulating decimals, $\cdot 1\dot{4}$, $\cdot 01\dot{3}\dot{8}$, $\cdot 241\dot{8}$, to their respective equivalent vulgar fractions.

Proceeding by the Rule given above,

$$\cdot 1\dot{4} = \frac{14-1}{90} = \frac{13}{90}; \quad \cdot 01\dot{3}\dot{8} = \frac{138-13}{9000} = \frac{125}{9000} = \frac{1}{72};$$

$$\cdot 241\dot{8} = \frac{2418-2}{9990} = \frac{2416}{9990} = \frac{1208}{4995}.$$

The reason of the rule will appear from the following considerations.

Let $\cdot 27\dot{8}3\dot{6}$ be the mixed circulating decimal, we have $27 \cdot \dot{8}3\dot{6}$ by multiplying in this case, the given decimal by 100

$$= 27\tfrac{836}{999} \quad \text{Art. (134).}$$

But this value is 100 times too great;

therefore, $\quad = \dfrac{27}{100} + \dfrac{836}{99900} \quad$ true value

$$= \dfrac{27 \times 999 + 836}{99900} = \dfrac{27 \times (1000-1) + 836}{99900}$$

$$= \dfrac{27000 - 27 + 836}{99900} = \dfrac{27836 - 27}{99900} = \dfrac{27809}{99900}.$$

NOTE 1. Always multiply by such a number as will make the non-circulating part a whole number.

NOTE 2. Sometimes a decimal of very long period may be carried out easily to many places, as in the following example:

Reduce $\dfrac{1}{17}$ to a decimal.

$$
\begin{array}{r}
17)\,1\cdot 00 \\
85 \\ \hline
150 \\
136 \\ \hline
140 \\
136 \\ \hline
4
\end{array}
\quad (\cdot 0588\tfrac{4}{17}.
$$

hence $\dfrac{1}{17} = \cdot 0588\tfrac{4}{17}$; $\therefore \dfrac{4}{17} = \cdot 2352\tfrac{4}{17}$,

hence $\dfrac{1}{17} = \cdot 05882352\tfrac{4}{17}$ (by substitution);

$\therefore \dfrac{16}{17} = \cdot 94117632\tfrac{4}{17}$,

hence $\dfrac{1}{17} = \cdot 05882352941176322\tfrac{4}{17}$ (by substit$^\text{n}$.).

$= \cdot 05882352941176470\tfrac{1}{17}$.

By the above process, we *double* at every step the number of figures previously obtained.

Ex. XLI.

1. Reduce the following vulgar fractions and mixed numbers to circulating decimals:

(1) $\tfrac{5}{4}$; $\tfrac{2}{11}$; $\tfrac{1}{37}$; $\tfrac{3}{7}$.   (2) $\tfrac{17}{30}$; $\tfrac{308}{405}$; $\tfrac{14}{81}$; $15\tfrac{53}{335}$.

(3) $\tfrac{3231}{3520}$; $7\tfrac{962}{3367}$; $\tfrac{17}{55055}$.   (4) $24\tfrac{88}{575}$; $17\tfrac{13}{700}$; $2\tfrac{13086}{53532}$.

(5) $\tfrac{1}{19}$; $\tfrac{1}{23}$; $\tfrac{1}{29}$; $\tfrac{1}{31}$.

CIRCULATING DECIMALS.

Find the vulgar fractions equivalent to the recurring decimals:

(6) $\cdot\dot{7}$; $\cdot 0\dot{7}$; $\cdot 2\dot{2}\dot{7}$.   (7) $\cdot 5\dot{8}\dot{3}$; $\cdot\dot{1}3\dot{5}$; $\cdot 26\dot{3}$.

(8) $\cdot 0018\dot{5}$; $3\cdot 02\dot{4}$; $\cdot 012\dot{3}\dot{6}$.   (9) $\cdot\dot{1}4285\dot{7}$; $\cdot\dot{3}9791\dot{6}$; 382142857.

(10) $\cdot\dot{3}0769\dot{2}$; $\cdot 6\dot{3}0769\dot{2}$; $2\cdot 785714\dot{2}$. (11) $\cdot 3\dot{4}275\dot{3}$; $\cdot 03\dot{1}3\dot{2}$; $8\cdot 0208\dot{3}$.

(12) $85\cdot 6080\dot{6}$; $3\cdot\dot{6}42857\dot{1}$; $127\cdot 0002209\dot{5}$.

136. The value of the **circulating decimal** $\cdot\dot{9}99\ldots$ is found by Art. (134) to be $\frac{9}{9}$ or 1; but since the difference between 1 and $\cdot 9 = \cdot 1$, between 1 and $\cdot 99 = \cdot 01$, between 1 and $\cdot 999 = \cdot 001$, &c., it appears that however far we continue the recurring decimals, **it can never** at any stage be *actually* $=1$. But the recurring decimal is considered $=1$, **because** the difference between **1 and** $\cdot 99\ldots$ becomes less and less, the more figures we take in the decimal, which **thus, in fact, approaches nearer to 1 than** by any difference that can be assigned.

In like manner, it is in this sense that any vulgar fraction **can be said to be the value of a** circulating **decimal**; **because there is no assignable difference between** their values.

137. In arithmetical operations, where circulating decimals are concerned, and the result is only required to be true to a certain number of decimal places, it will be sufficient to carry on the circulating part to two or three decimal places more than the number required: taking care that the last figure retained be increased by 1, if the succeeding figure be 5, or greater than 5; because, for instance, if we have the mixed decimal $\cdot 6\dot{2}8\dot{8}$, and stop at $\cdot 628$, it is clear that $\cdot 628$ is less, and $\cdot 629$ is greater than the **true value** of the decimal: but $\cdot 628$ is less than the true **value by** $\cdot 000111\ldots$ and $\cdot 629$ is greater than the true value by $\cdot 000111\ldots$

Now $\cdot 000111\ldots$ is less than $\cdot 000888\ldots$;

Therefore $\cdot 629$ is nearer the true value than $\cdot 628$.

Ex. 1. Add together $\cdot\dot{3}\dot{3}$, $\cdot 0\dot{4}3\dot{2}$, $2\cdot\dot{3}4\dot{5}$, so **as to be correct to 5 places of decimals.**

$\cdot 3333333$
$\cdot 0432432$
$2\cdot 3454546$

$2\cdot 7220311$ *Ans.* $2\cdot 72203$.

ARITHMETIC.

Ex. 2. Subtract $\cdot 291\dot{6}$ from $\cdot 98958\dot{3}$, so as to be correct to 5 places of decimals.

$$\cdot 9895833$$
$$\cdot 2916667$$
$$\overline{\cdot 6979166}$$

Ans. $\cdot 69791$.

NOTE. This method may be advantageously applied in the Addition and Subtraction of circulating decimals. In the Multiplication and Division, however, of circulating decimals, it is always preferable to reduce the circulating decimals to Vulgar Fractions, and having found the product or quotient as a Vulgar Fraction, then, if necessary, to reduce the result to a decimal.

Ex. XLII.

Find the value (correct to 6 places of decimals) of

(1) $2\cdot 41\dot{8} + 1\cdot 1\dot{6} + 3\cdot 00\dot{9} + \cdot 735\dot{4} + 24\cdot 042.$

(2) $234\cdot \dot{6} + 9\cdot 92\dot{8} + \cdot 0\dot{1}2345678\dot{9} + \cdot \dot{0}04\dot{4} + 456.$

(3) $6\cdot 4\dot{5} - \cdot \dot{3}$; and $7\cdot 7\dot{2} - 6\cdot 04\dot{5}$; and $309 - \cdot 9472\dot{4}.$

(4) Express the sum of $\frac{4\,9}{5\,2}$, $\frac{2\,7\,8}{3\,9\,3}$, and $\frac{7}{1\,2}$, and the difference of $18\frac{1}{1\,5}$ and $4\frac{5}{1\,4}$, as recurring decimals.

Multiply

(5) $2\cdot \dot{3}$ by $5\cdot \dot{6}$; $\cdot 757\dot{5}$ by $\cdot 36\dot{6}.$ (6) $\cdot 40\dot{6}$ by 62; 825 by $\cdot 3\dot{6}.$

(7) $7\cdot 5\dot{2}$ by $48\cdot \dot{3}$; 368 by $\dot{6}.$ (8) $3\cdot 14\dot{5}$ by $\cdot 429\dot{7}$; $20\frac{4}{7}$ by $\cdot 8\dot{4}.$

Divide

(9) $195\cdot 0\dot{2}$ by 4; $\cdot 3759\dot{2}$ by $\cdot 0\dot{5}.$ (10) 54 by $\cdot 1\dot{7}$; $13\cdot \dot{2}$ by $5\cdot \dot{6}.$

(11) $411\cdot 351\dot{9}$ by $58\cdot 76 4\dot{5}$; $2\cdot 1659\dot{5}$ by $\cdot 0\dot{4}$; $\cdot 655990\dot{3}$ by $48\cdot 7\dot{6}.$

REDUCTION OF DECIMALS.

138. *To reduce a decimal of any denomination to its proper value.*

RULE. Multiply the decimal by the number of units connecting the next lower denomination with the given one, and point off for decimals as many figures in the product, beginning from the right hand, as there are figures in the given decimal. The figures on the left of the

REDUCTION OF DECIMALS.

decimal point will represent the whole numbers in the next denomination. Proceed in the same way with the decimal part for that denomination, and so on.

Ex. 1. Find the value of ·0484 of £1.

Proceeding by the Rule given above,

£.
·0484
 20
─────
·9680s.
 12
─────
11·6160d.
  4
─────
2·4640q.

For, £·0484 of £1 $= \frac{484}{10000}$ of £1.

$= \frac{9680}{10000}$ s. $\qquad = \frac{116160}{10000}$ d.

$= 11\frac{616}{1000} = 11d. + \frac{616 \times 4}{1000}$ q.

$= 11d. + \frac{2464}{1000}$ q.

$= 11d. + 2\frac{484}{1000}$ q.

$= 11\frac{1}{2}\frac{29}{250}d.$

therefore the value of ·0484 of £1 $= 11\frac{1}{2}\frac{29}{250}d$.

Ex. 2. Find the value of 13·3375 acres.

Acres.
13.3375
   4
─────
1·3500 ro.
   40
─────
14·0000 po.

therefore the value is 13 ac., 1 ro., 14 po.

Ex. Find the value of ·972916̇ of £1.

1st method.

£.
·972917
  20
─────
19·458340s.
  12
─────
5·500080d.
   4
─────
2·000320q.

2d method.

·972916̇ of £1 $= \frac{972916 - 97291}{900000}$ of £1 Art. (185).

$= \frac{875625}{900000}$ of £1 $\quad \left(\frac{467}{180} \times 20\right) s.$

$= \frac{467}{24}$ s. $= 19s. \ 5\frac{1}{4}d.$

therefore the value is 19s. 5¼d. nearly.

Note. The 2nd method is generally the better one to adopt.

ARITHMETIC.

Ex. Find the value of $\frac{133}{400}$ of $3\frac{3}{4}$ tons $-\cdot 3\dot{4}0\dot{5}$ of $1\frac{2}{3}$ qrs. $+\frac{\cdot 21334\dot{8}}{\cdot 32\dot{6}}$ of 1 cwt., 63 lbs.

$$\frac{133}{400} \text{ of } 3\frac{3}{4} \text{ tons} = \left(\frac{113}{400} \times \frac{15}{4}\right) \text{ tons} = \frac{113 \times 3}{80 \times 4} \text{ tons,}$$

$$= \left(\frac{133 \times 3}{80 \times 4} \times 20\right) \text{cwt.} = \frac{399}{16} \text{ cwt.}$$

$$= 24 \text{ cwt., 3 qrs., } 18\frac{3}{4} \text{ lbs.}$$

$$\cdot 3\dot{4}0\dot{5} \text{ of } 1\frac{2}{3} \text{ qrs.} = \left(\frac{3405-3}{9990} \text{ of } \frac{5}{3}\right) \text{ qrs.}$$

$$= \left(\frac{3402}{9990} \times \frac{5}{3} \times 25\right) \text{ lbs.}$$

$$= \left(\frac{21 \times 25}{37}\right) \text{ lbs.} = 14\frac{7}{37} \text{ lbs.}$$

$$\frac{\cdot 21334\dot{8}}{\cdot 32\dot{6}} \text{ of 1 cwt., 63 lbs.} = \left(\frac{213348-21334}{900000} \times \frac{1000}{326} \text{ of } 163\right) \text{ lbs.}$$

$$= \frac{96007}{900} \text{ lbs.}$$

$$= 106\tfrac{307}{900} \text{ lbs.}$$

therefore the value of the expression

$$= 24 \text{ cwt., 3 qrs., } 18\tfrac{3}{4} \text{ lbs.} - 14\tfrac{7}{37} \text{ lbs.} + 106\tfrac{307}{900} \text{ lbs.}$$
$$= 24 \text{ cwt., 3 qrs., } 4\tfrac{83}{148} \text{ lbs.} + 1 \text{ cwt., } 6\tfrac{307}{900} \text{ lbs.}$$
$$= 1 \text{ ton, 5 cwt., 3 qrs., } 11\tfrac{4}{1387} \text{ lbs.}$$

XLIII.

Find the respective values of

(1) ·45 of \$1 ; ·16875 of \$4 ; ·87708 of \$5.
(2) ·28125 of £1 ; ·7962 of £1 ; ·359375 of £2.
(3) ·086 of \$5 ; ·5783 of \$10 ; ·075 of \$16.
(4) ·875 of a lea.; 2·5384375 of a day; ·6 of 1 lb Troy.
(5) ·85076 of a cwt,; ·07325 of a cwt.; ·045 of a mile.

REDUCTION OF DECIMALS.

(6) 4·16525 of a ton ; ·3625 of a cwt. ; ·05 of an acre.

(7) 3·8343 of a lb. Troy; 2·46875 of a qr.; 4·106 of 3 cwt., 1 qr., 21 lbs.

(8) 3·8375 of an acre ; 3·5 of 18 gallons.

(9) ·925 of a furlong ; ·34375 of a lunar month.

(10) 5·06325 of $100; 3·8 of an Eng. ell.

(11) 2·25 of 3½ acres ; 2·0396 of 1 m., 530 yds.

(12) 4·751 of 2 sq. yds., 7 sq. ft. ; 2·009943 of 2 miles.

(13) ·38$\dot{3}$ of $1 ; ·4708$\dot{3}$ of $4 ; ·469$\dot{4}$ of 1 lb. Troy.

(14) ·574$\dot{0}$ of 27s. ; ·13$\dot{8}$ of 10s. 6d. ; 2·$\dot{6}$ of 5s.

(15) 4·05 of 1½ sq. yds.; ·16$\dot{3}$ of 2½ miles ; 4·9$\dot{0}$ of 4d., 8 hrs.

(16) 3·24$\dot{2}$ of 2¼ acres ; $\dfrac{·0931\dot{8}}{·568\dot{1}}$ of 2 $\tfrac{1}{12}$ of 2·5 days.

(17) Find the difference between ·77777 of a pound and 8s. 6·6648d.; and between ·70323 of $4.80 and 3·5646 of 24 cents.

(18) ·26$\dot{8}$ cwt. + ·056$\dot{2}$ ton − ·578$\dot{6}$ qr.

(19) £·634375 + ·025 of 25s. + ·31$\dot{6}$ of 30s.

(20) 2·$\dot{8}\dot{1}$ of 365¼ days + 5·75 of a week − ⅜ of 5⅝ hours.

(21) ⅞ of $\tfrac{3}{11}$ of 3 acres − 2·00875 square yards + ·02$\dot{2}\dot{7}$ of 3¼ square feet.

139. *To reduce a number or fraction of one or more denominations, to the decimal of another denomination of the same kind.*

RULE. Reduce the given number or fraction to a fraction of the proposed denomination ; and then reduce this fraction to its equivalent decimal.

Ex. 1. Reduce 13s. 6¼d. to the decimal of £1.

$$13s.\ 6¼d. = 162¼d. = \tfrac{649}{4}d.$$
$$£1 = 240d. ;$$

therefore the fraction $= \dfrac{\tfrac{649}{4}}{240} = \dfrac{649}{960} = ·67\dot{6}$.

or thus,
```
  4 | 1·00
 12 | 6·25
 2,0| 13·52083
     ·6760416
```

We first reduce $\frac{1}{4}d.$ to the fraction of a penny, which is ·25 ; next 6·25$d.$ to the decimal of a shilling by dividing by 12, which is ·52083s ; then 13·52083 to the decimal of a £1 by dividing by 20, which ·6760416.

Ex. 2. Reduce 3 bus., 1 pk. to the decimal of a load : and verify the result.

$$40 \begin{cases} 8 \\ 5 \end{cases} \begin{array}{c|l} 4 & 1\cdot 00 \\ \hline 8 & 3\cdot 25 \\ \hline 5 & \cdot 40625 \\ \hline & \cdot 08125 \end{array}$$

therefore ·08125 is the decimal required.

·08125 ld.
5
·40625 qrs.
8
8·25000 bush.
4
1·00000 pk.

therefore 0·8125 of a load = 3 bus., 1 pk.

Ex. 3. Express the sum of ·428571 of $72, $\frac{1}{4}$ of $\frac{1}{2\frac{1}{6}}$ of $\frac{1}{6}$ of \$7.68, and $\frac{5}{9}$ of 12 cts., as the decimal of \$48.

·428571 of \$72 = $\frac{428571}{999999}$ of \$72.
     = $\frac{3}{7}$ of \$72 = \$21$\frac{3}{7}$
     = \$30.85$\frac{5}{7}$

$\frac{1}{4}$ of $\frac{1}{2\frac{1}{6}}$ of $\frac{1}{6}$ of \$7.68 = $\frac{1}{4}$ of $\frac{5}{13}$ of $\frac{1}{6}$ of \$7.68
     = 54$\frac{6}{7}$ cts.

$\frac{5}{9}$ of 12 cts. = 6$\frac{6}{9}$ cts. :

therefore the sum = \$30.85$\frac{5}{7}$ + 54$\frac{6}{7}$ cts. + 6$\frac{6}{9}$ cts.
     = \$31.47$\frac{3}{7}$

therefore the decimal required = $\frac{31.47\frac{3}{7}}{48}$ = ·655714

REDUCTION OF DECIMALS.

Ex. 4. Convert £17. 9s. 6d. into pounds, florins, &c., and verify the result.

First reduce 9s. 6d. to the decimal of £1.

$$\begin{array}{r|l} 12 & 6\cdot 0 \\ 2{,}0 & 9\cdot 5 \\ \hline & \cdot 475 \end{array}$$

∴ £17. 9s. 6d. = £17·475
= £17. 4 fl. 7 c. 5 m.

Again, £17. 4 fl. 7 c. 5 m.
= £17·475

$$\begin{array}{r} 20 \\ \hline 9\cdot 500s. \\ 12 \\ \hline 6\cdot 000d. \end{array}$$

∴ £17. 4 fl. 7 c. 5 m. = £17. 9s. 6d.

Ex. 5. Reduce the difference between a cent (New Coinage) and a penny to the decimal of 3s. 4d.

$1d. = £\frac{1}{240}$; $1 c. = £\frac{1}{100}$;

∴ difference $= £(\frac{1}{100} - \frac{1}{240}) = £\frac{140}{1000} = £\frac{7}{1200}$
$= (\frac{7}{1200} \times 20 \times 12)d. = \frac{7}{5}d.$

3s. 4d. = 40d.

∴ fraction $= \frac{\frac{7}{5}}{40} = \frac{7}{200} = \frac{35}{1000}$;

∴ decimal = ·035.

XLIV.

Reduce

(1) $1.25 to the decimal of $2; and $3.75 to the decimal of $4.
(2) 4s. 7½d. to the dec¹. of £1; and 15s. 11¼d. to the dec¹. of £1.
(3) 10s. 0¾. to the dec¹. of £1; and 5s. 8¾d. to the dec¹. of £5.
(4) 2 oz., 13 dwts. to the dec¹. of 1 lb.; and 4 lbs., 2 sc. to the dec¹. of 1 oz.
(5) 2 qrs., 21 lbs. to the dec¹. of 1 ton; and 3 cwt., 3 oz. to the dec¹. of 10 cwt.

(6) 2 fur., 41 yrds. to the dec¹. of a mile; and 1 fur., 30 po. to the dec¹. of a league.

(7) 2 sq. ft., 73 sq. in to the dec¹. of a sq. yd.; and 3 ro., 20 po. to the dec¹. of an acre.

(8) 4 days, 18 hrs. to the decl. of a wk.; and 11 sec. to the dec¹, of 5 days.

(9) 1 lb. Troy to the dec¹. of 1 lb. Avoird.; and $2\frac{1}{4}$ in. to the dec¹. of $2\frac{1}{2}$ mls.

(10) $3\frac{3}{4}$ pks. to the dec¹. of $3\frac{1}{2}$ qrs.; and $27\frac{1}{2}$ gals. to the dec¹. of $1\frac{1}{2}$ qts.

(11) $5\frac{3}{4}$ yds. to the dec¹. of 2 Fr. ells; and 1 ton, $2\frac{1}{4}$ cwt. to the dec¹. of 1 cwt., $2\frac{1}{4}$ qrs.

(12) 3 wks., $5\frac{1}{4}$ d. to the decl. of $5\frac{1}{2}$ hrs.; and 1 min., $2\frac{1}{4}$ sec. to the decl. of $\frac{2}{25}$ of a lunar month.

(13) 3 reams to the dec¹. of 19 sheets; and $3\frac{1}{2}$ ac. to the dec¹. of $3\frac{1}{4}$ sq. yds.

(14) 33 yds. to the dec¹. of a mile; 3s. $5\frac{13}{200}d.$ to the dec¹. of a dollar (a dollar being 4s. 3d.); and 7s. $8\frac{103}{10000}d.$ to the dec¹. of 10s. 6d.

(15) $\frac{3}{7}$ of $7 to the decl. of $10; and $63\frac{3}{4}$ cwt. to the dec¹. of a ton.

(16) $\frac{15}{26}$ of $8 to the dec¹. of $7; and $\frac{7}{8}$ pk. to the dec¹. of 2 qrs.

(17) $\frac{3}{7}$ of a guinea to the dec¹. of £2; and $\frac{21}{15000}$ of a year to the dec¹. of a day.

(18) $\frac{3}{7}$ of $\frac{1}{10}$ of 40 yds. to the dec¹. of $\frac{1}{8}$ of 2 mls.; and $\frac{1}{4}$ of $3\frac{1}{2}$ sq. yds. to the dec¹. of 2 ac., 1 ro.

(19) $\frac{3}{8}$ of $4\frac{1}{2}$ hrs. to the dec¹; of $365\frac{1}{4}$ days; and $9\frac{6}{11}$ of $\frac{11}{15}$ pks. to the dec¹. of $3\frac{1}{2}$ of 8 bush.

(20) 8 lbs., 6 oz. Troy to the dec¹. of 10 lbs. Avoird.; and $\frac{1}{8}$ oz. Avoird. to the dec¹. of $\frac{1}{3}$ oz. Troy.

(21) Add together $\frac{3}{5}$ of a day, $\frac{2}{3}$ of an hour, and $\frac{4}{5}$ of 6 hours; and express the result as the decimal of a week.

(22) Express the value of ·83 of $1.92 + ·05 of $5.04 + 1·8 of $1.20 as the dec¹. of $2.52.

(23) Add $5\frac{1}{4}$ cwt. to 3·125 qrs.; and reduce the sum to the decimal of a ton.

(24) Convert the following sums of money into the New Decimal Coinage of pounds, florins, &c., and verify each result:

1. 6d.	2. 10d.	3. $4\frac{1}{2}d.$	4. 5s.
5. 10s. 6d.	6. 16s.		7. £5. 12s. 6d.
8. £54. 7s. 4d.	9. £20. 19s. $7\frac{1}{2}d.$		10. 15s. $4\frac{3}{4}d.$
11. 14s. 8·16d.	12. £2. 15s. 11·088d.		13. £3. 0s. 11d. 3·04q.

XLV.

Miscellaneous Questions and Examples on Arts. (119-139).

I.

(1) Define a decimal; and show how its value is affected by affixing and prefixing ciphers. Reduce ·0625, 3·14159 to fractions; and express the difference between $20\frac{5}{13}$ and $17\frac{1}{13}$ as a decimal.

(2) Find the value of $10\frac{3}{8} + 1\frac{5}{16} + \frac{7}{5} + 1\frac{3}{16}$ both by vulgar fractions and by decimals; and show that the results coincide.

(3) Find the sum, difference, product, and quotient of 573·005 and ·000754; and of 1·015 and ·01015, and prove the truth of each result.

(4) If a vulgar fraction, being converted into a decimal, do not terminate, prove that it must recur. What must be the limit to the number of figures in the recurring part? Is $\frac{2}{5144}$ convertible into a terminating decimal.

(5) Simplify 1. $2\frac{1}{4} + 72\frac{2}{3} + 316\frac{1}{2} + 2\cdot 875$. 2. $\cdot 026649 \div 2\frac{1}{8}\frac{5}{8}$.

3. $\dfrac{1-\cdot 05}{5+\cdot 5} \times \dfrac{3-\cdot 8}{3\cdot 8} \div \dfrac{1}{10}$. 4. $\{\cdot 18 + \cdot 009\} \div \cdot 01\dot{6}$.

(6) Divide $\dfrac{48\frac{1}{7}}{1085\frac{7}{15}}$ by $\dfrac{7\frac{3}{11}}{174\frac{9}{17}}$; reduce the quotient to the form $1\cdot 071428\dot{5}$. Divide $91\cdot 86\dot{3}$ by $87\cdot 5\dot{6}$.

II.

(1) Write down in a decimal form seven hundred thousand four hundred and nine billionths. Express 12·1345 as a fraction, and $\frac{32548}{1000000}$ as a decimal.

(2) State the effect as regards the decimal point of multiplying and dividing a decimal by any given power of 10. Write down in words the meaning of 397008·405009; multiply it by 1000, and also divide it by 1000; and write down the meaning of each result in words.

(3) What decimal multiplied by 125 will give the sum of $\frac{2}{5}, \frac{7}{16}, \frac{2}{5}$, ·09375 and 2·46?

(4) Multiply 1·05 by 10·5; and reduce the result to a fraction in its lowest terms. Divide ·8727588 by 1620; find the value of $\dfrac{\cdot 0003 \times \cdot 004}{\cdot 006}$; reduce $\frac{1}{15} + \frac{8}{165} - \frac{12}{825}$ to a decimal.

(5) Simplify, expressing each result in a decimal form

1. $\frac{1}{10000}$ of $\frac{2\frac{1}{4}}{2\frac{1}{3}}$.   2. $(2\frac{1}{2}+6) \div (3\frac{1}{4} - \frac{1}{11})$.

3. $\dfrac{4 \cdot 4 + \frac{3}{5}}{7 \cdot 375 + \frac{3}{4} - \frac{1}{3}}$.   4. $2\frac{1}{3000} + 1\frac{5}{1000} + 5\frac{1}{6000} + 2 \cdot 000875$.

(6) Find a number which multiplied into 3132·458 will give a product which differs only in the 7th decimal place from 7823·6572.

III.

(1) Divide 684·1197 by 1200·21, and also by ·0120021; and 594·27 by ·047 to three places of decimals, and explain fully how the position of the decimal point is determined in each of the quotients.

(2) Simplify, expressing each result in a fractional and decimal form,

1. $\dfrac{\cdot 015 \times 2 \cdot 1}{\cdot 035}$.   2. $\dfrac{3\frac{1}{2} - \cdot 04}{5 - \cdot 0625}$.

3. $\frac{3}{8} + \cdot 14 + \frac{3}{4}$ of $1 \cdot 0784$.   4. $(\frac{1}{4} - \frac{1}{5}) \times (\frac{3}{5} + 1\frac{1}{2})$.

(3) What is meant by a "Recurring Decimal"? What kind of vulgar fractions produce such decimals? State the rules for reducing any recurring decimal to a vulgar fraction. Multiply 5·81 by ·4583, and divide 1·13 by ·000132. Is $\frac{23}{1000}$ reducible to a recurring decimal?

(4) Show that if $1\frac{1}{12}$, $2\frac{2}{15}$, $3\frac{3}{20}$, $4\frac{4}{27}$ be added together, (1) as fractions, and (2) as decimals, the results coincide.

(5) If a man walk in 4 days 60 miles; in each of the three first days he walked an equal distance, in the fourth day he walked 13.95 miles; find the amount of his daily walking.

(6) A person has ·1875 part of a mine, he sells ·17 part of his share; what fractional part of the mine has he still left?

IV.

(1) State the Rules for the Addition and Subtraction of decimals. Add together 1·23, ·123, ·0123, ·00123, and 123; and find the vulgar fraction corresponding to the result. Find the fraction equivalent to 31·457457, and subtract it from the fraction $\frac{424}{125}$.

(2) Write down in figures the number, three millions six thousand and five. Also write down in words the signification of the same figures when the last is marked off as a decimal.

(3) Compare the values of $5 \times \cdot 05$, $1 \cdot 5 \times \cdot 75$, and $2 \cdot 625 \div 5$.

MISCELLANEOUS QUESTIONS.

(4) Find the product of $\cdot 014714\dot{7}$ by $\cdot 3\dot{3}\dot{3}$; and the quotients of $\cdot 1269\dot{3}$ by $19\cdot 3\dot{9}$; 132790 by $\cdot 245$; of $\cdot 014904$ by $3\frac{4}{25}$; of 61061 by $3\cdot 05$; and of $6106\cdot 1$ by 305000.

(5) Shew that the decimal $\cdot 90437532$ is more nearly represented by $\cdot 90438$ than by $\cdot 90437$; and find the value of

$$16 \times \left\{ \frac{1}{5} - \frac{1}{3 \times 5^3} + \frac{1}{5 \times 5^5} - \frac{1}{7 \times 5^7} + \&c. \right\} - \frac{4}{239}$$

accurately to 5 places of decimals.

(6) A person sold $\cdot 15$ of an estate to one person, and then $\frac{5}{17}$ of the remainder to another person. What part of the estate did he still retain?

V.

(1) Express $\frac{1}{2}(6\frac{1}{2} + 2\frac{2}{3} - 3)$, $\frac{8073}{8125}$, and also the product of $3\frac{5}{8}$ and $(3\frac{1}{4} - \frac{5}{7})$ of $\frac{6}{7}$ as decimals.

(2) Simplify

1. $\dfrac{4\cdot 255 \times \cdot 032}{\cdot 00016}$.   2. $(\frac{1}{2} + \frac{1}{4} + \frac{1}{8} + \frac{1}{4} + \frac{1}{10}) \div (\frac{1}{3} + \frac{1}{5} + \frac{1}{6} \pm \frac{1}{15}.)$

3. $(\frac{9}{11} \text{ of } 35\frac{1}{3} - 3\frac{1}{4}) + (2\cdot 5625 + 7\frac{1}{4})$.   4. $\cdot 59\dot{3} \div 1\cdot 78 \times \cdot 3\dot{6} \div \cdot 072$.

(3) State at length the advantages which decimals possess over vulgar fractions; what disadvantages have they?

Shew whether $\frac{22}{7}$ or $\frac{333}{106}$ is nearer to the number $3\cdot 14159$.

(4) Find the value of $1 + \dfrac{1}{1} + \dfrac{1}{1 \times 2} + \dfrac{1}{1 \times 2 \times 3} + \&c.$, to 7 places of decimals; and also of

$$\frac{1}{10^3} \times \left(1 - \frac{3}{10^2} + \frac{3 \times 4}{1 \times 2} \times \frac{1}{10^4} + \frac{3 \times 4 \times 5}{1 \times 2 \times 3} \times \frac{1}{10^6} \right)$$

expressing at (1) as a decimal, and (2) as a fraction.

(5) Find the Earth's equatorial diameter in miles, supposing the Sun's diameter, which is $111\cdot 454$ times as great as the equatorial diameter of the Earth, to be 883345 miles.

(6) In what sense is a vulgar fraction said to be the value of a recurring decimal? Explain how a sufficient degree of accuracy may be obtained in the addition and subtraction of circulating decimals to any given number of decimal places, without converting the decimals into fractions.

Ex. Find the sum of $\cdot 12\dot{5}$, $4\cdot 1\dot{6}\dot{3}$, and $9\cdot 45\dot{7}$, correct to 5 places of decimals.

VI.

(1) Prove the Rule for Multiplication of decimals by means of the example 404·04 multiplied by ·030303. Multiply ·345 by $\frac{·111}{4·3}$; and divide ·04813489963 by ·6593, and ·006593.

(2) Explain the meaning of 7^2, and 7^3; and find what vulgar fraction is equivalent to the sum of 20·5 and 2·05 divided by the difference.

(3) Reduce to their lowest terms $\frac{123·48}{1033·2}$, and $\frac{36·595}{5·7980}$,

(4) Shew that $\dfrac{·375 \times ·375 - ·025 \times ·025}{·375 - ·025} = \dfrac{2}{5}$, and that

$$8 + \dfrac{1}{7} + \dfrac{1}{\frac{1}{16}} = 3·14159 \text{ nearly.}$$

Reduce ·1298i3i to its equivalent vulgar fraction.

(5) What decimal added to the sum of $1\frac{7}{14}$, $\frac{3}{8}$, and $1\frac{3}{8}$ will make the sum total equal to 3?

(6) The quotient being $2\frac{11}{12}$ and the divisor ·15, find the dividend.

VII.

(1) A and B can finish a piece of work in $1\frac{1}{2}$ days, A and C in 2 days, and B and C in 3 days. If $1.44 be paid for the piece of work, what are a day's wages of each workman?

(2) A tax of $2544 is to be raised from 3 towns, the number of inhabitants of which are respectively 2500, 3000, and 4200. How much should each town pay, and each person in it?

(3) The wages of 25 men amount to £76. 13s. 4d. in 16 days; how many boys must work 24 days to receive £103. 10s., the daily wages of the latter being one-half those of the former?

(4) A person rows a distance of $1\frac{1}{2}$ miles *down* a stream in 20 minutes, but without the aid of the stream it would take him half an hour; what is the rate of the stream per hour? and how long would it take him to return against it?

(5) A and B engage to do a piece of work for $7.20. A could do the work alone in 4 days, B in 5 days; with the help of a boy it is completed in 2 days; how should the money be divided?

(6) A person buys 3 lbs. of tea at 74 cents per lb., and mixes them with 5 lbs. at 56 cents per lb. What will 2 lbs. of his tea cost him?

PRACTICE.

140. *An aliquot part* of a number is such a part as, when taken a certain number of times, will exactly make up that number.

Thus, 3 is **an aliquot** part of 9 ; $6 of $18.

TABLE OF ALIQUOT PARTS.

Parts of a cwt. (100 lbs.)

50 lbs. or 2 qrs. =	$\frac{1}{2}$ cwt.
25 lbs. or 1 qr. =	$\frac{1}{4}$ "
20 lbs. =	$\frac{1}{5}$ "
10 lbs. =	$\frac{1}{10}$ "
5 lbs. =	$\frac{1}{20}$ "

NOTE. The parts of a $ are the same as of the cwt. (100 lbs.)

Parts of a cwt. (112 lbs.)

56 lbs. or 2 qrs. =	$\frac{1}{2}$ cwt.
28 lbs. or 1 qr. =	$\frac{1}{4}$ "
16 lbs. =	$\frac{1}{7}$ "
14 lbs. =	$\frac{1}{8}$ "
7 lbs. =	$\frac{1}{16}$ "
4 lbs. =	$\frac{1}{28}$ "
2 lbs. =	$\frac{1}{56}$ "

Parts of a £1.

10s. =	$\frac{1}{2}$ £1.
6s. 8d. =	$\frac{1}{3}$ "
5s. =	$\frac{1}{4}$ "
4s. =	$\frac{1}{5}$ "
3s. 4d. =	$\frac{1}{6}$ "
2s. 6d. =	$\frac{1}{8}$ "
2s. =	$\frac{1}{10}$ "
1s. 8d. =	$\frac{1}{12}$ "
1s. 4d. =	$\frac{1}{15}$ "
1s. 3d. =	$\frac{1}{16}$ "
1s. =	$\frac{1}{20}$ "

Parts of a shilling.

6d. =	$\frac{1}{2}$ of 1s.
4d. =	$\frac{1}{3}$ "
3d. =	$\frac{1}{4}$ "
2d. =	$\frac{1}{6}$ "
1½d. =	$\frac{1}{8}$ "
1d. =	$\frac{1}{12}$ "
¾d. =	$\frac{1}{16}$ "
½d. =	$\frac{1}{24}$ "
¼d. =	$\frac{1}{48}$ "

NOTE. In working examples in Practice, the above tables will often have **to be varied**; the knowledge which the scholar now has will render him expert in taking such aliquot parts as he may require in any particular example.

141. *Practice* is **a** short method of finding the value of any number of articles by means of *aliquot parts*, when the value of a unit of any denomination is given. Practice may be divided into two cases, SIMPLE and COMPOUND.

SIMPLE PRACTICE.

I. In this case the given number is expressed in the same denomination as the unit whose value is given; as, for instance, 27 bushels of wheat at $1:10 per bushel.

The Rule for Simple Practice will be easily shewn by the following examples.

Ex. 1. Find the value of 1296 things at 16s. 10½d. each.

The method of working such an example is the following:

If the cost of the things be £1 each;
then the total cost = £1296.

∴ cost at

		£	s.	d.
10s. 0d. each = ½ of the above sum................. =		648	0	0
5s. 0d. each = ½ the cost at 10s. each.............. =		324	0	0
1s. 3d. each = ¼ the cost at 5s. each............... =		81	0	0
0s. 7½d. each = ½ the cost at 1s. 3d. each.......... =		40	10	0

∴ by adding up the vertical columns,
cost at 16s. 10½d. each   = £1093 . 10 . 0

The operation is usually written thus:

	£	s.	d.	
10s. = ½ of £1.	1296	0	0	= cost at £1 each.
5s. = ½ of 10s.	648	0	0	= cost at 10s. each.
1s. 3d. = ¼ of 5s.	324	0	0	= cost at 5s. each.
7½d. = ½ of 1s. 3d.	81	0	0	= cost at 1s. 3d. each.
	40	10	0	= cost at 7½d. each.
	£1093	10	0	= cost at 1s. 10½d. each.

NOTE. The student must use his own judgment in selecting the most convenient 'aliquot' parts; taking care that the sum of those taken make up the *given price of the unit*.

Ex. 2. Find the value of 825 bushels of wheat at $1.30 per bus.

If 1 bus. cost $1, cost of 825 bus. = $825 at $1 each.

	$825.00 = value at $1 each.
20 cts. = ½ of $1.	165.00 = value at 20 cts. each.
10 cts. = ½ of 20 cts.	82.50 = value at 10 cts. each.
	$1072.50 = value at $1.30 each.

PRACTICE.   147

COMPOUND PRACTICE.

II. In this case the given number is not wholly expressed in the same denomination as the unit whose value is given; as for instance, 1 cwt. 2 qrs., 14 lbs., at $10.24 per cwt.

The Rule for Compound Practice will be easily shewn from the following examples.

Ex. 1. Find the value of 60 cwt., 3 qrs., 5 lbs. of sugar at $8.50 per cwt.

The method of working such an example is the following:
The value of 1 cwt. of sugar being $8.50;

$\therefore$ **value** of 60 cwt. $= \$(8.50 \times 60)$ $\qquad = \$510.00$

$\qquad$ 2 qrs. $= \frac{1}{2}$ (value of **1 cwt.**)
$\qquad\qquad = \frac{1}{2} (\$8.50)$ $\qquad = \$4.25$

$\qquad$ 1 qr. $= \frac{1}{2}$ (value of 2 qrs.)
$\qquad\qquad = \frac{1}{2} (\$4.25.)$ $\qquad = \$2.12\frac{1}{2}$

$\qquad$ 5 lbs. $= \frac{1}{5}$ (value of 1 (qr.)
$\qquad\qquad = \frac{1}{5} (\$2.12\frac{1}{2})$ $\qquad = \$0.42\frac{1}{2}$

Therefore adding up the **vertical** columns,
value of 60 cwt., 3 qrs., 5 lbs. $\qquad = \$516.80$.

The operation is usually written **thus**.

2 qrs. $= \frac{1}{2}$ cwt.	$8.50 = value of 1 cwt.
	10
	85.00 = value of 10 cwt.
	6
	510.00 = value of 60 cwt.
	4.25
1 qr. $= \frac{1}{2}$ of 2 qrs.	2.12$\frac{1}{2}$
5 lbs. $= \frac{1}{5}$ of 1 qr.	42$\frac{1}{2}$
	$516.80 = value of 60 cwt., 3 qrs., 5 lbs.

Ex. 2. Find the value of 319 cwt., 3 qrs., 16 lbs. at £2. 12s. 6d. per cwt.

148   ARITHMETIC.

		£.	s.	d.	
2 qrs. = ½ cwt.		2 .	12 .	6	= value of 1 cwt.
			10		
		26 .	5 .	0	= value of 10 cwt.
			4		
		105 .	0 .	0	= value of 40 cwt.
			8		
		840 .	0 .	0	= value of 320 cwt.
subtracting		2 .	12 .	6	= value of 1 cwt.
		837 .	7 .	6	= value of 319 cwt.
		1 .	6 .	3	= value of 2 qrs.
1 qr. = ½ of 2 qrs.		0 .	13 .	1½	= value of 1 qr.
14 lbs. = ½ of 1 qr.		0 .	6 .	6¾	= value of 14 lbs.
2 lbs. = ⅐ of 14 lbs.		0 .	0 .	11¼	= value of 2 lbs.
	£	839 .	14 .	4½	= value of 319 cwt., 3 qrs., 16 lbs.

XLVI.

Find the value of

(1) 275 articles at 25 cents each ; 125 articles at 30 cents each.
(2) 92 articles at 45 cents each : 80 articles at 50 cents each.
(3) 120 articles at 75 cents each ; 215 articles at 85 cents each.
(4) 225 articles at $1.10 each ; 350 articles at $1.25 each.
(5) 128 bus. oats at 53 cts. each ; 75 bus. wheat at $1.10 each.
(6) 318 yds. cloth at 72 cts. a yd. ; 48 bus. pease at 63 cts. a bus.
(7) 7 tons, 2 cwt., 3 qrs., 10 lbs. of sugar at $10 per ton.
(8) 87 ac., 2 ro., 22 per., at $8 an acre.
(9) 210 lbs. tea at 42 cts. a lb. ; 812 lbs. sugar at 10 cts. a pound.
(10) 626 lbs., 7 oz., 19 dwts. at $1.27 per dwt. 78 things at $2.36 each.
(11) Find the value of 282 ac., 17 per. at $0.60 per perch.
(12) Find rent of 100 acres at 87½ cents a rood.

In the following examples the cwt. = 112 lbs.

(13) Find the value of 5 cwt., 2 qrs., 14 lbs. at £2. 5s. 6d. per cwt.
(14) Find the value of 36 cwt., 3 qrs., 7 lbs. at £6. 7s. 8d. per cwt.
(15) Find the value of 72 cwt., 3 qrs., 17 lbs. of sugar at £1 4s. 6d. per cwt.

PRACTICE. 149

(16) Find the value of 60 cwt., 3 qrs., 12 lbs. at £7. 13s. 6d. per cwt.
(17) Find the value of 3 cwt., 2 qrs., 16 lbs. at £3. 7s. 6d. per cwt.
(18) Find the value of 9 yds., 2 ft., 10 in. at 5s. 7¼d. per yard.
(19) Find the value of 39 cwt., 10 lbs. at £3. 15s. 7¾d. per cwt.
(20) Find the cost of 30 cwt., 2 qrs., 14 lbs. at £1. 17s 8¼d. per qr.
(21) Find the value of 15 oz., 6 dwt., 17 grs. at 5s 10d. per oz.
(22) What will 2789 lbs. of pork, cost at $8.50 per 100 lbs.
(23) Find value of 28800 ft. fire-wood, at $6 per cord.

Find the amount of the following account:

(24) 24 lbs. crushed sugar at 12 cts. a lb.; 7¾ lb. tea at 75 cts. a lb. 4¾ lbs. coffee at 32 cts. a lb.; 5 lbs. rice 7 cts. a lb.; 20¼ lbs. cheese at 11½ cts. per lb.; 17½ lbs. ham at 19 cts. a lb.

142. Examples which are usually classed under particular Rules, such as the Rule of Three, &c., can nevertheless be readily solved independently by means of the foregoing principles.

The following examples, which are worked out, are intended to exemplify various methods of reasoning. In the examples for practice which follow them, questions will be found the solution of which may be easily arrived at in a similar way: the number of such questions in this place must necessarily be very limited, and therefore the student is strongly recommended to apply to all questions which are hereafter classed under particular Rules, an independent method of solution, as well as the one denoted by the Rule to which they are respectively affixed.

Ex. 1. Express a degree (69½ m.) in metres, 32 metres being = 35 yards.

$$35 \text{ yards} = 32 \text{ metres.}$$

$$\therefore 1 \text{ yard} = \frac{32}{35} \text{ metres.}$$

$$\therefore 1 \text{ degree} = (69\tfrac{1}{2} \times 1760) \text{ yards} = (139 \times 880) \text{ yards.}$$

$$= \left(\frac{139 \times 880 \times 32}{35}\right) \text{ metres} = 111835\tfrac{3}{7} \text{ metres.}$$

Ex. 2. If ⅔ds of a lottery ticket be worth $880, what is the value of $\tfrac{3}{11}$ ths of the same?

$$\therefore \tfrac{2}{3}\text{rds of the ticket} = \$880.$$

$$\therefore \tfrac{1}{3}\text{rd of the ticket} = \$440.$$

$$\therefore \text{whole ticket} = \$(440 \times 3) = \$1320.$$

$$\therefore \tfrac{3}{11}\text{ths of the ticket} = \tfrac{3}{11} \text{ of } \$1320 = \$\frac{1320 \times 3}{11} = \$360.$$

Ex. 3. A person has $\frac{3}{7}$ths of an estate of 4000 acres left him; he sells $\frac{2}{3}$rds of his share: how many acres has he remaining, and what fraction of the whole estate will they be?

He sells $\frac{2}{3}$ of $\frac{3}{7}$ of 4000 acres, or $\frac{2}{7}$ of 4000 acres.

$\therefore$ he has remaining $\left(\frac{3}{7} \text{ of } 4000 - \frac{2}{7} \text{ of } 4000\right)$ acres.

$= \frac{1}{7}$ of 4000 acres $= 571\frac{3}{7}$ acres.

Ex. 4. The sum of $1000 is to be raised in a school section, the assessment of which is $100000; what is the rate in the dollar?

$100000 produce $1000,

$\therefore$ $1 produce $ $\left(1000 \times \frac{1}{100000}\right)$, or $\left(1000 \times \frac{1}{100000} \times 100\right)$cts.

or $\frac{100000}{100000}$ cts. or 1 ct.

Ex. 5. After taking from my purse $\frac{1}{3}$ of my money, I find that $\frac{3}{4}$ of what is then left amounts to 7s. 6d.; what money had I in my purse at first?

Let unity, or 1, denote the sum in the purse at first. After taking away $\frac{1}{3}$, $\frac{2}{3}$ remains. Now by the question

$\frac{2}{3}$ of $\frac{3}{4}$ of unity, or $\frac{2}{3}$ of $\frac{3}{4}$ of the sum in the purse at first $= 7s.\ 6d.$

or $\frac{1}{2}$ of the sum in the purse at first $= 7s.\ 6d.$

$\therefore$ sum in the purse at first $= 15s.$

Ex. 6. A met two beggars, B and C, and having $\frac{3\cdot\frac{7}{1}}{4\frac{2}{7}}$ of $\frac{10\frac{1}{2}}{7\frac{1}{2}}$ of $\frac{77}{540}$ of a moidore in his pocket, gave B $\frac{1}{7}$ of $\frac{3}{4}$ of that sum, and C $\frac{3}{6}$ of the remainder; what did each receive?

A had at first $\dfrac{\frac{40}{11}}{\frac{30}{7}}$ of $\dfrac{\frac{75}{7}}{\frac{15}{2}}$ of $\frac{77}{540}$ of $27s.$, or $\frac{14}{3}$ s.

MISCELLANEOUS EXAMPLES WORKED OUT.   151

B received $\frac{1}{7}$ o $\frac{3}{4}$ of $\frac{14}{3}$ s., or $\frac{1}{2}$ s., or 6d.

A had left afterwards $\left(\frac{14}{3} - \frac{1}{2}\right)$ s. $= \frac{25}{6}$ s.,

$\therefore C$ received $\frac{3}{5}$ of $\frac{25}{6}$ s., or $\frac{5}{2}$ s., or 2s. 6d.

Ex. 7. A farmer pays a corn-rent of 5 quarters of wheat and 3 quarters of barley, Winchester measure: what is the money value of his rent, when wheat is at 60s., and barley at 54s. per quarter, imperial measure; 32 imperial gallons being =33 Winchester gallons?

Rent is 5 qrs. of Wheat Win. mea. + 3 qrs. of barley Win. mea.

But 1 Win. Gal. $= \frac{32}{33}$ imp. gal..

$\therefore$ 1 Win. qr. $= \frac{32}{33}$ imp. qr.

$\therefore$ rent is $5 \times \frac{32}{33}$ imp. qrs. of wheat $+ 3 \times \frac{32}{33}$ imp. qrs. of barley.

$\therefore$ money value of rent $= (5 \times \frac{32}{33} \times 60 + 3 \times \frac{32}{33} \times 54)$ s. $= £22.\ 8s.$

Ex. 8. If £1 sterling be worth 25 francs, 60 centimes; and also worth 6 thalers, 20 silbergroschen; how many francs and centimes is a thaler worth? (One thaler=30 silbergroschen; 1 franc=100 centimes.)

6 thalers, 20 silbergroschen=25 francs, 60 centimes,

or $6\frac{2}{3}$ thalers $= 25\frac{60}{100}$ francs,

1 thaler $= (25\frac{3}{5} \div 6\frac{2}{3})$ francs

$= \frac{384}{100}$ francs $=3$ francs, 84 centimes.

Ex. 9. Standard gold contains 11 parts of pure gold to one part of alloy and 20lb. Troy are coined into 934 sovereigns and a half-sovereign; find the weight of pure gold in a sovereign.

Number of parts $= 11 + 1 = 12$, of which $\frac{11}{12}$ is pure gold.

By the question

$$934\tfrac{1}{2} \text{ sovereigns weigh 20 lbs. Troy,}$$

$$\therefore 1 \text{ sov. weighs } \frac{20 \times 2}{1869} \text{ lbs Troy}$$

$$\therefore \text{ weight of pure gold in a sov.} = \left(\frac{11}{12} \times \frac{20 \times 2}{1869}\right) \text{ lb. Troy}$$

$$= 113\tfrac{1}{628} \text{ grs.}$$

Ex. 10. If a person, travelling $13\tfrac{3}{4}$ hours a day, perform a journey in $27\tfrac{5}{8}$ days, in what length of time will he perform the same if he travel $10\tfrac{2}{7}$ hours a day?

If he travel $13\tfrac{3}{4}$ hrs. a day, he does the journey in $27\tfrac{5}{8}$ days,

.... 1 hr .... $(27\tfrac{5}{8} \times 13\tfrac{3}{4})$ days,

.... $10\tfrac{2}{7}$ hrs .... $\dfrac{27\tfrac{5}{8} \times 13\tfrac{3}{4}}{10\tfrac{2}{7}}$ days,

which, worked out, gives $36\tfrac{214}{360}$ days.

Ex. 11. If 858 men in 6 months consume 234 quarters of wheat, how many quarters will be required for the consumption of 979 men for three months and a half?

858 men in 6 months consume 234 quarters,

$$\therefore 1 \text{ man in 1 month consumes } \frac{234}{858 \times 6} \text{ qrs.}$$

$$\therefore 979 \text{ men in 1 month consume } \frac{979 \times 234}{858 \times 6} \text{ qrs.}$$

$$\therefore 979 \text{ men in } 3\tfrac{1}{2} \text{ months consume } \left(\frac{979 \times 234}{858 \times 6} \times \frac{7}{2}\right) \text{ qrs. or } 155\tfrac{3}{4} \text{ qrs.}$$

Ex. 12. If 5 men or 7 women can do a piece of work in 37 days; in what time will 7 men and 5 women do a piece of work twice as great?

$$5 \text{ men} = 7 \text{ women,}$$

$$\therefore 1 \text{ man} = \frac{7}{5} \text{ woman,}$$

$$\therefore 7 \text{ men} = \frac{49}{5} \text{ women,}$$

$$\therefore 7 \text{ men and 5 women} = \left(\frac{49}{5} + 5\right) \text{ women} = \frac{74}{5} \text{ women}$$

Now by the question,

7 women in 37 days do the piece of work,

∴ 1 woman in (37 × 7) days does............

∴ 74 women in $\dfrac{37 \times 7}{74}$ days do...................

∴ $\dfrac{74}{5}$ women in $\dfrac{37 \times 7 \times 5}{74}$ days do...............

∴ $\dfrac{74}{5}$ women in $\dfrac{37 \times 7 \times 5 \times 2}{74}$ or in 35 days do twice as much.

Ex. 13. A bankrupt owes three creditors, A, B, and C, $250, $330, and $420 respectively, and his property is worth $125; how much will each creditor receive, and how many cents in the dollar?

Debts amount to $(250 + 330 + 420), or $1000.

If the bankrupt has $1, he pays $\dfrac{1}{1000}$ part of debt.

.... $125 .... $\dfrac{125}{1000}$ part of debt.

.... $\dfrac{1}{8}$ part of debt.

∴ A gets $31.25, B gets $41.25, and C gets $52.50. He pays ⅛ of $1, or 12½ cents in the dollar.

Ex. 14. Gunpowder being composed of nitre 15 parts, charcoal 3 parts, and sulphur 2 parts; find how much of each is required for 16 cwt. of powder.

The whole number of parts = (15 + 3 + 2) = 20

Of every 20 parts,

$\dfrac{15}{20}$ or $\dfrac{3}{4}$ is nitre, $\dfrac{3}{20}$ is charcoal, $\dfrac{2}{20}$ or $\dfrac{1}{10}$ is sulphur.

∴ $\dfrac{3}{4}$ of 16 cwt., or 12 cwt. = quantity of nitre required,

$\dfrac{3}{20}$ of 16 cwt., or 2⅖ cwt. = ... charcoal ....

$\dfrac{1}{10}$ of 16 cwt., or 1⅗ cwt. = ... sulphur ....

Ex. 15. The price of a work which comes out in parts is £2. 16s. 8d.

But if the price of each part were 13d. more than it is, the price of the work would be £3. 7s. 6d. How many parts are there?

$$£2.\ 16s.\ 8d. + (\text{number of parts} \times 13)d. = £3.\ 7s.\ 6d.$$
$$\therefore (\text{number of parts} \times 13)d. = 10s.\ 10.$$
$$= 130d.$$
$$\therefore \text{number of parts} = \tfrac{130}{13} = 10.$$

Ex. 16. Divide 1860 dollars between A, B, and C, so that as often as A gets \$5, B shall get \$4, and as often as B gets \$3, C shall get \$1.

It is clear that B's share = 3 times C's share,
4 times A's share = 5 times B's share,
or, A's share = $\tfrac{5}{4}$ times B's share,
$= \left(\tfrac{5}{4} \times 3\right)$ times C's share,

but A's share + B's share + C's share = 1860 dollars;

$\therefore \tfrac{15}{4} C$'s share + 3 C's share + C's share = 1860 dollars,

or $\left(\tfrac{15}{4} + 4\right) C$'s share = 1860 dollars,

or $\tfrac{31}{4} C$'s share = 1860 dollars.

$\therefore C$'s share = $\left(\tfrac{1860}{1} \times \tfrac{4}{31}\right)$ dollars = 240 dollars.

B's share = 720 dollars, and A's share = $\left(240 \times \tfrac{15}{4}\right)$ dollars = 900 dollars.

Ex. 17. Of a certain dynasty, $\tfrac{1}{3}$ of the kings are of the same name, $\tfrac{1}{4}$ of another, $\tfrac{1}{8}$ of a third, and $\tfrac{1}{12}$ of a fourth, and there are 5 besides; how many are there of each name?

Representing the whole dynasty by unity, or 1.

$\tfrac{1}{3}$ = number of kings of one name,

$\tfrac{1}{4}$ = ... of a second..,

$\tfrac{1}{8}$ = ... of a third....,

$\tfrac{1}{12}$ = ... of a fourth....

MISCELLANEOUS EXAMPLES WORKED OUT.   155

Now $\frac{1}{3} + \frac{1}{4} + \frac{1}{8} + \frac{1}{12} = \frac{19}{24}$.

∴ whole dynasty $- \frac{19}{24}$, or $1 - \frac{19}{24}$, or $\frac{5}{24}$ no. of remaining kings in it.

But by the question,

$\frac{5}{24}$ of unity, or $\frac{5}{24}$ of the whole dynasty $= 5$;

∴ 1, or the whole dynasty, $= 5 \times \frac{24}{5} = 24$;

∴ there are 8 kings of the 1st name, 6 of the 2nd, 3 of the 3rd, and 2 of the 4th.

Ex. 18. *A* can do a piece of work in 5 days, *B* can do it in 6 days, and *C* can do it in 7 days; in what time will *A*, *B*, and *C*, all working at it, finish the work? Find also in what time *A* and *B* working together, *A* and *C* together, and *B* and *C* together, could respectively finish it,

Representing the work by unity, or 1.

In one day *A* does $\frac{1}{5}$ part of the work,

In one day *B* does $\frac{1}{6}$ part of the work,

.... *C* does $\frac{1}{7}$ ....;

∴ .... $A + B + C$ do $\left(\frac{1}{5} + \frac{1}{6} + \frac{1}{7}\right)$ or $\frac{107}{210}$ part;

∴ time in which $A + B + C$ would finish the work

$= \frac{1}{\frac{107}{210}}$ days $= \frac{210}{107}$ days $= 1\frac{103}{107}$ days.

Again in one day $A + B$ do $\left(\frac{1}{5} + \frac{1}{6}\right)$, or $\frac{11}{30}$, of the work; therefore time in which they would finish it $= \frac{1}{\frac{11}{30}}$ or $2\frac{8}{11}$ days.

In like manner, it may be shown that *A* and *C* would finish the work in $2\frac{11}{12}$ days; and *B* and *C* in $3\frac{3}{13}$ days.

Ex. 19. It being given that A and B can do a piece of work in $2\frac{8}{11}$ days; and that A and C can do the same in $2\frac{11}{12}$ days; and that B and C can do it in $3\frac{3}{12}$ days: find the time in which A, B, and C would do the work: working, first, all together, secondly, separately.

In one day A and B do $\frac{11}{30}$ of the work,

.... A and C do $\frac{12}{35}$ ....,

.... B and C do $\frac{13}{42}$ ....

$\therefore$ by addition,

In one day $2A + 2B + 2C$ would do $\left(\frac{11}{30} + \frac{12}{35} + \frac{13}{42}\right)$; or $\frac{214}{210}$, of the work,

$\therefore$ In one day $A + B + C$ do $\frac{107}{210}$ ....

$\therefore$ time required $= \dfrac{1}{\frac{107}{210}} = \frac{210}{107}$ days $= 1\frac{103}{107}$ days.

Again,

work done by $A + B + C$ in one day $-$ work done by $B + C$ in one day,

or, work done by A in one day $= \frac{107}{210} - \frac{13}{42} = \frac{1}{5}$

therefore time required, in which A could do the work, $= 5$ days.

In like manner it may be shown that B would do the work in 6 days, and that C would do it in 7 days.

Ex. 20. A cistern is fed by a spout which can fill it in 2 hours, how long would it take to fill it if the cistern has a leak which would empty it in 10 hours?

In one hour spout fills $\frac{1}{2}$ of the cistern.

.... leak empties $\frac{1}{10}$ ....

Therefore in one hour, when the spout and leak are both open, the part of the cistern filled by what runs in $-$ what runs out,

$$= \left(\frac{1}{2} - \frac{1}{10}\right) = \frac{2}{5},$$

$\therefore$ time required for filling the cistern $= \dfrac{1}{\frac{2}{5}}$ hrs. $= \frac{5}{2}$ hrs. $= 2\frac{1}{2}$ hrs.

Ex. 21. A and B can do a piece of work in 15 and 18 days respectively; they work together at it for 3 days, when B leaves, but A continues, and after 3 days is joined by C, and they finish it together in 4 days; in what time would C do the piece of work by himself?

Representing the work by unity, or 1.

In one day $A + B$ do $\left(\dfrac{1}{15} + \dfrac{1}{18}\right)$ of the work,

In 3 days they do $\left(\dfrac{1}{15} + \dfrac{1}{18}\right) \times 3$ ....

or $\dfrac{11}{30}$ ....

$\therefore \dfrac{19}{30}$ of the work remains to be done.

In 3 days more A does $\dfrac{3}{15}$ or $\dfrac{1}{5}$ of the work;

$\therefore$ when A is joined by C,

$\dfrac{19}{30} - \dfrac{1}{5}$, or $\dfrac{13}{30}$ of the work remains to be done.

In 4 days A does $\dfrac{4}{15}$ of the work;

$\therefore$ work which has to be done by C in 4 days

$= \dfrac{13}{30} - \dfrac{4}{15} = \dfrac{5}{30} = \dfrac{1}{6}$;

$\therefore$ part of work to be done by C in one day $= \dfrac{1}{24}$,

$\therefore$ time in which C would do the whole work $= 24$ days.

Ex. XLVII.

Miscellaneous Questions and Examples on preceding Arts.

I.

(1) State the rules for the multiplication and division of decimals, and divide $34\cdot17$ by $3\frac{1}{4}$.

(2) What is the value in English money of $1556\cdot85$ francs, when the exchange is at $24\cdot25$ francs per £?

ARITHMETIC.

(3) Reduce $\frac{1}{3} + \frac{1}{6} + \frac{1}{14} + \frac{3}{56}$ to a decimal fraction. What decimal of a cwt. is 1 qr. 7 lbs.?

(4) If $\frac{5}{7}$ of an estate be worth $4818·50, what is the value of $\frac{2}{3}$ of it?

(5) If a bankrupt pay 17 cents in the dollar, what will be received on a debt of $17658.

(6) A person possessing $\frac{3}{11}$ of an estate, sold $\frac{2}{7}$ of $\frac{1}{3\frac{1}{5}}$ of his share for £120$\frac{3}{8}$; what would $\frac{1}{5}$ of $\frac{3}{11}$ of the estate sell for at the same rate?

(7) A man, his wife, and 3 children earn $24.75 a week; the wife earns twice as much as each child, and the man three times as much as his wife? required the man's weekly earnings.

(8) If £1. sterling be worth 12 florins, and also worth 25 francs, 56 centimes; how many francs and centimes is one florin worth? (100 centimes = 1 franc.)

(9) The wages of 5 men for six weeks being $405, how many weeks will 4 men work for $540.

II.

(1) What is meant by saying that one sum is a certain fraction (for example $\frac{2}{3}$) of another? If 26 francs are equivalent to a pound, what fraction of a shilling is a franc? Give the reasons for the process which you adopt in answering the question.

(2) Express $\frac{2}{3}$ of 1$\frac{5}{8}$ of a mile in terms of a metre, supposing 32 meters = 35 yards.

(3) A, B, and C rent a pasture for $192. A puts in 8 cattle, B, 9, and C, 11: how much should each pay for his share?

(4) Reduce 3$\frac{3}{4}$d. to the decimal of 10s., and divide the result by 12·5. Explain the process employed.

(5) If the property in a town be assessed at $288000, what must be the rate in the dollar in order that $12000 may be raised?

(6) If the circumference of a circle = Diameter × 3·14159; find the number of revolutions passed over by a carriage-wheel 5 ft. in diameter in 10 miles.

(7) A farmer has to pay yearly to his landlord the price of 7$\frac{1}{4}$ bushels of wheat at 4s. 9d. per bushel, and 9$\frac{1}{4}$ of malt at 5s. 3d., and 6$\frac{3}{4}$ of oats at 2s. 4d. What is the whole amount of his rent?

MISCELLANEOUS QUESTIONS AND EXAMPLES. 159

If there were a decimal coinage of pounds, florins, &c., how many of them would he have to pay?

(8) A alone can do a piece of work in 10 hours, and B can do it in 12 hours; find the time in which both working together can do it.

(9) Ten excavators dig 12 loads of earth in 16 hours, whilst 12 others can dig only 9 loads in 15 hours; in what time will they jointly dig 100 loads?

III.

(1) Divide 28 tons, 4 cwt., 3 qrs., into 36 equal portions; and find the value of one of them at $36.16 per cwt. (cwt. = 112 lbs.)

(2) Reduce 186 yds., 2 ft. $8\frac{1}{25}$ in. to the decimal of a chain. If one chain = 10 chainlets = 100 links = 1000 linklets; express the above in chains, chainlets, links, linklets.

(3) If £1 sterling be worth 45 Pauls, 9 Baiocchi (Roman), and be worth $25\frac{1}{2}$ francs (French); show that a Napoleon of 20 francs = 36 Pauls. (10 Baiocchi = 1 Paul.)

(4) If the rents of a parish amount to £2514. 7s. 6d. and a rate is granted of £83. 16s. 3d., how much is that in the £? and how much must be paid by an estate whose rental is £115. 12s. 6d.

(5) If a tradesman with a capital of $4800 gains $432 in 7 months, in what time will he gain $97.20 with a capital of $1512.

(6) In the civil year 97 days are intercalated in 400 years; what is the average length of the year?

(7) If 15 horses and 148 sheep can be kept 9 days for £75. 15s., what sum will keep 10 horses and 132 sheep for 8 days, supposing 5 horses to eat as much as 84 sheep?

(8) A, B, and C are three workmen: A can do a piece of work in 3 hours, being twice as much as B can do; and A, B, and C can together do the whole in $2\frac{1}{4}$ hours. Show that C can do in 5 hours as much as B can do in 9 hours.

(9) Two persons gained in trade $1800; one having put in $2400 and the other $4080; what part of the profit ought each person to receive?

IV.

(1) Explain how whole numbers are represented in the decimal or common system of notation. Multiply 729 by 37, and explain the process.

(2) Add together the fifth of a $0.24, two-sevenths of a $1·20, and four-ninths of a $5.04; and reduce the result to the decimal of $120.

(3) Taking the circumference of a circle at $3\tfrac{1}{7}$ times its diameter, find the cost of a marble column of two feet breadth, and 5 yards height, marble being at 15s. 6d. per cub. ft. (Area of circle = $\tfrac{1}{2}$ circumference × semi-diameter.)

(4) If a certain number of men can throw up an intrenchment in 12 days, when the day is 6 hours long, in what time will they do it when the day is 8 hours long?

(5) Find the entire cost of 10 lbs. of tea at 4s. 3d. per lb., 18 lbs. of coffee at 1s. $3\tfrac{1}{2}d.$ per lb., 23 lbs. of sugar at $4\tfrac{1}{4}d.$ per lb., and 16 lbs. of candles at $7\tfrac{3}{4}d.$ per lb., and divide the amount equally among 14 persons.

(6) Reduce $2375\tfrac{3}{4}$ Spanish dollars to English money, the exchange being at 3s. 4d. per dollar. And find the value of 1,000,000 rupees at 2s. $3\tfrac{7}{8}d.$ each.

(7) The roller used for rolling a bowling-green, being 6 ft. 6 in. in circumference, by 2 ft. 3 in. wide, is observed to make 12 revolutions as it rolls from one extremity of the green to the other; find the area rolled when the roller has passed 10 times the whole length of it.

(8) Divide $1400 among A, B, and C, in such a manner that as often as A gets $5, B shall get $4, and as often as B gets $3, C shall get $2.

(9) A fraudulent wine-merchant sells, as brandy, a mixture of brandy and rum at $5.40 a gallon, which is the proper price of his brandy, that of his rum being $2.52 a gallon. Supposing one-third of the whole mixture to be rum, ascertain how much a gallon he gains by his dishonesty.

V

(1) Divide 550974 by 1472; find the quotient and remainder. Explain the operation, and prove the result.

(2) Show that the value of a fraction is not altered by multiplying the numerator and denominator by the same number.

Express the fractions $\tfrac{6}{15}$, $\tfrac{5}{7}$, and $\tfrac{2}{35}$ by corresponding fractions having the same denominator, and find the sum.

(3) If 1 lb. Avoirdupois be equivalent to 7000 grains Troy, and 1869 sovereigns weigh 40 lbs. Troy, how many sovereigns will weigh 1 Avoirdupois ounce?

(4) A quarter of wheat is consumed annually by each person in England; if wheat be 45s. a quarter, and the population 27,500,000, what is the value of a quarter of a year's consumption.

(5) A certain number of men mow **4 acres in 3 hours**; and a certain number of others mow **8 acres in 5 hours**; how long will they be mowing 11 acres, if all work together?

(6) If a man can do a piece of work in $8\frac{1}{2}$ days by working 6 hours a day, how many hours a day must he work to finish it in 5 days?

(7) If 7 men or 11 women can finish a piece of work in 17 days, how many days will it take 11 men and 7 women to finish it?

(8) A bankrupt owes A $2475, B $1953.60, and C $1406.52; his estate is worth $4377.45; how much can be paid in the dollar, and what will A, B, and C each receive?

(9) How many francs must be transmitted from Paris to Berlin to discharge a debt of 420 thalers? a thaler being equivalent to 3 shillings, and 24 franks to one pound sterling.

VI.

(1) Can two concrete numbers be multiplied together? multiply £10. 17s. $6\frac{3}{4}d.$ by 8764.

(2) A bankrupt's assets amounted to $2603, and his creditors received 55 cents in the dollar: find the amount of his debts.

(3) A piece of cloth, when measured with a yard measure which is two-thirds of an inch too short, appears to be $10\frac{1}{4}$ yards long, what is its true length?

(4) Estimate the cost of a dish of almonds and raisins consisting of six ounces of almonds and three-quarters of a pound of raisins: supposing almonds to be ten cents, and raisins eleven cents a pound.

(5) If 5 cwt., 3 qrs., 14 lbs. cost $6 per cwt., what will be the cost per pound when the cost of the whole has been reduced by $11.78.

(6) A grocer buys 10 cwt. 3 qrs. 21 lbs. of sugar for $54.80, and pays $2.74 for expenses, at what rate must he sell it at per pound to clear $8.22 by his bargain?

(7) If a snail, on the average, creep 2 ft. 7 in. up a pole during 12 hrs. in the night, and slip down 16 in. during the 12 hrs. in the day, how many hours will he be in getting to the top of a pole 35 ft. high?

(8) The profits of a tradesman average £54. 6s. 5d. per week, out

of which he pays 3 foremen, 10 shopmen, and 5 assistants, at the rate of 2 guineas, 1 guinea, and 17s. 6d. per week respectively: his yearly outgoings for rent, &c., **amounted to £723. 11s. 8d.** Find his net annual profit.

(9) In an orchard of fruit trees, $\frac{1}{3}$ of them bear apples, $\frac{1}{4}$ pears, $\frac{1}{5}$ plums, and 50 cherries; how many trees are there in all.

VII.

(1) What is meant by a fraction? Find the value of $\frac{3}{4}$ of $\frac{1}{2}$ of $6: and then express the result as the fraction and decimal of $237.50.

(2) By what number must £5. 6s. 3¼d. be multiplied, in order to give as product £85. 0s. 4d.? Divide £34. 18s. into 3 parts, one of which shall be twice and the other 4 times as great as the third.

(3) If a year consist of 365·242264 days, in how many years will its defect from the civil year of 365¼ days amount to one day?

(4) If 15 men take 17 days to mow 300 acres of grass, how long will 27 men take to mow 167 acres?

(5) If 20 men can perform a piece of work in 12 days, how many men will accomplish another piece of work, which is six times as great, in a tenth part of the time?

(6) I am owner of $\frac{3}{4}$ of $\frac{2}{3}$ of $\frac{1}{2}$ of a ship worth $30,000, and sell $\frac{1}{5}$th of the ship; what part of her will then belong to me, and what will it be worth?

(7) A bankrupt owes $900 to his three creditors, and his whole property amounts to $675; the claims of two of his creditors are $125 and $375 respectively; what sum will the remaining creditor receive for his dividend?

(8) There are in a manufactory a certain number of workmen who receive $13 a week, twice as many who receive $10 a week, and eleven times as many who receive $8 a week, and the total amount of the workmen's wages for one week is $847; find the number of workmen.

(9) Reduce £405. 6s. 8d to francs and centimes, at the rate of 25¼ francs to £1, and 100 centimes to a franc.

VIII

(1) Find the value at $15.60 per oz. of 13 lbs. 9 oz. 3 dwt. of gold dust.

MISCELLANEOUS QUESTIONS AND EXAMPLES.

(2) If a florin be made the unit of money, what number will represent £1. 11s. 6¼d.?

(3) If £1. be worth 12 guldens, and one penny 8 kreutzers, what fraction of one gulden is 5 kreutzers.

(4) A creditor receives on a debt of $1420.80 a dividend of $61\frac{2}{3}$ cents in the dollar, and he receives a further dividend, upon the deficiency, of $18\frac{3}{4}$ cents in the dollar; what does the creditor receive in the whole?

(5) Reduce 12 ft. 4½ in. to the fraction of a mile, and find the corresponding decimal.

(6) A man has an income of £200 a year; an income-tax is established of 7d. in the pound, while a duty of $1\frac{1}{2}d$. per lb. is taken off sugar; what must be his yearly consumption of sugar that he may just save his income-tax?

(7) If A can do as much work in 5 hours as B can do in 6 hours, or as C can do in 9 hours, how long will it take C to complete a piece of work, one-half of which has been done by A working 12 hours and B working 24 hours?

(8) Find the number of shillings and pence which are equivalent to the recurring decimal ·3333 .... of a pound.

(9) The gross earnings of an undertaking average $14400, and the expenses $3723.40 per week, one-tenth of the remainder is put aside for wear and tear, and the annual charges amount to $115880.08. What is the net annual profit? (1 year = 52 weeks.)

IX.

(1) Explain the process of Long Division.
Reduce $\frac{9375}{371} + \frac{15242512}{715}$ to its equivalent whole number.

(2) Shew how to convert any proper fraction into a decimal.
Reduce $\frac{3}{8}$ and $\frac{75}{1875}$ to the decimal form.

(3) State what kind of vulgar fractions can be expressed in finite decimals. Can the quantity $\frac{1}{4} - \frac{1}{6} - \frac{1}{21}$ be so expressed?

How many cents should be given in exchange for $\dfrac{\frac{1}{3}+\frac{1}{4}}{\frac{1}{4}+\frac{1}{5}}$ of a dollar?

(4) If two-thirds of an academic term exceed one-half of it by $13\frac{1}{2}$ days, how many days are there in the whole term?

(5) In a decimal coinage of pounds, florins, &c., how many of these may be obtained for £19. 17s. 6¼d.? How much is lost by the exchange?

(6) A butler concocts a bowl of punch, of which the following are the ingredients: milk 2¼ quarts, the rind of one lemon, 2 eggs, 1 pint of rum, and half-a-pint of brandy. Compute the value of the punch, reckoning milk at 3d. a quart, lemons at 2s. a dozen, eggs at 16 a shilling, rum at 13s. per gallon, and brandy at £1. 4s. 8d. per gallon.

(7) A Cochin China hen eats a pint of barley and lays a dozen eggs, while an English hen eats half-a-pint of barley and lays 5 eggs. Supposing the eggs of the English hen to be half as large again as those of the Cochin China, which is the more economical layer?

(8) If 72 men dig a trench 20 yds. long, 1 ft. 6 in. broad, 4 feet deep, in 3 days of 10 hours each, how many men would be required to dig a trench 30 yards long, 2 ft. 3 in. broad, and 5 feet deep, in 15 days of 9 hours each?

(9) A crew consists of 420 men, and a certain number of boys; the men receive each $14.40 per month; and the amount of wages of the whole crew is $7200 per month; find the number of boys supposing each to receive $7.20 per month.

X.

(1) Explain the rule for the addition of decimals; add together ⅜ and ·061; subtract ·003 from ·02; and divide ·0072 by ·006.

(2) Subtract ½ of ⅝ from ⅔ of $\frac{7}{11}$, and multiply the result by ¾ of ⅘.

(3) If £1 sterling = 10 florins = 100 cents = 1000 mills, shew that £25. 10s. 7½d. = 255 florins, 3 cents, 1½ mills.

(4) If 6 men earn $90 in 7½ days, how much will 10 men earn in 11¾ days?

(5) A person expends $345.60 in the purchase of cloth, how much can he buy at the rate of 52 cents a yard.

(6) What is the cost per hour of lighting a room with ten burners, each consuming 4 cub. in. of gas per second; the price of gas being 6s. for a thousand cubic feet?

(7) What is the value of 8 qrs., 5 bushels, 3 pecks of wheat at $1.20 a bushel?

If 8 qrs., 6 bushels, 2 pecks of malt cost £21. 3s., what is the price per bushel?

(8) If 36 men, working 8 hours a day for 16 days, can dig a trench 72 yards long, 18 wide, and 12 deep, in how many days will 32 men

working 15 hours a day, dig a trench 64 yards long, 27 wide, and 18 deep?

(9) If a sheet of paper $5\frac{1}{2}$ feet long by $2\frac{1}{2}$ feet broad be cut into strips an inch broad; how many sheets would be required to form a strip that would reach round the earth (25,000 miles)?

XI.

(1) Express $\frac{11}{133}$ as a decimal; and thence find its value when unity represents $300.

(2) A person has city property yielding a rental of $3070; a rate of 2 cts. in the dollar being levied, what will he have to pay?

(3) Find the price of 2 tons, 16 cwt., 17 lbs. of sugar at 20 cts. for $2\frac{1}{2}$ lbs.

(4) If 1 cwt. of an article cost $33.60, at what price per lb. must it be sold to gain $\frac{1}{16}$ of the outlay?

(5) Find in inches and fractions of an inch the value of ·00003551136 of a mile. Explain the process employed.

(6) Express each silver coin now current in England by a decimal of $2\frac{3}{4}d$. If $\frac{1}{10}$th of $2\frac{3}{4}d$ be the unit of money, what decimal will express a halfpenny?

(7) A Canadian dollar is 4s. $3\frac{7}{8}d$., and is 5·45 francs; find the number of francs in £1 sterling, and express both a dollar and a franc in terms of the unit of money mentioned in the last question.

(8) A and B can do a piece of work in 6 days, B and C in 7 days, and A, B, and C can do it in 4 days; how long would A and C take to do it?

(9) A bag contains a certain number of sovereigns, three times as many shillings, and four times as many pence and the whole sum in the bag is £280; find how many sovereigns, shillings, and pence it contains respectively.

SECTION V.

RATIO AND PROPORTION.

143. We may ascertain the relation which one abstract number bears to another abstract number, or one concrete number to another concrete number of the same kind, in respect of magnitude, in two different ways; either by considering how much one is greater or less than the other; or by considering what multiple, part, or parts, one is of the other, that is, how many times or parts of a time, or both, one number is contained in the other. Thus if we compare the number 12 with the number 3, we observe, adopting the first mode of comparison, that 12 is greater than 3 by the number 9; or, adopting the second mode of comparison, that 12 contains 3 four times, and is thus $\frac{12}{3}$ or four times as great as 3. Again if we compare the number 7 with the number 13, we observe, according to the first mode of comparison, that 7 is less than 13 by the number 6; and, according to the second, that as 1 is one thirteenth part of 13, so 7 is seven thirteenth parts of 13, or $\frac{7}{13}$ths of 13.

144. The relation of one number to another in respect of magnitude is called Ratio; and when the relation is considered in the first of the above methods, that is, when it is estimated by the difference between the two numbers, it is called Arithmetical Ratio; but when it is considered according to the second method, that is, when it is estimated by considering what multiple, part, or parts, one number is of the other, or, which is seen from above to be the same thing, by the fraction which the first number is of the second, it is called Geometrical Ratio. Thus for instance, the arithmetical ratio of the numbers 12 and 3 is 9; while their geometrical ratio is $\frac{12}{3}$ or 4. In like manner the arithmetical ratio of 7 and 13 is 6, while their geometrical ratio is $\frac{7}{13}$.

145. It is more common, however, in comparing one number with another to estimate their relation to one another in respect of magnitude according to the second method, and to call that relation so estimated by the name of Ratio. According to this mode of treatment, which we shall adopt in what follows, "Ratio is the relation which one number has to another in respect of magnitude, the comparison being made by considering what multiple, part, or parts, the first num-

RATIO AND PROPORTION.

ber is of the second, or how many times or parts of a time, or both, the second is contained in the first."

146. It is plain that, for any two numbers, the fraction in which the first is numerator and the second denominator, will correctly express the multiple or part, or both, which the first number is of the second, or the number of times or parts of a time, or both, of a time, the second is contained in the first. Thus if we take the numbers 12 and 3, the fraction $\frac{12}{3}$, which is equivalent to the whole number 4, shews the multiple which 12 is of 3, or the number of times 3 is contained in 12. And again, if we take the numbers 7 and 13, the fraction $\frac{7}{13}$ will express the part or parts which the number 7 is of 13, or will express the part or parts of a time that 13 is contained in 7: for 1 is one thirteenth part of 13, so that 7 must be seven thirteenth parts of 13, that is, $\frac{7}{13}$ths of it: and 1 is contained 7 times in 7, so that 13 must be contained only $\frac{7}{13}$ths of a time in 7. We conclude therefore that the ratio of one number to another may be estimated and expressed by the fraction in which the former number is the numerator and the latter the denominator.

147. The ratio of one number to another is often denoted by placing a colon between them. Thus the ratio of 7 to 13 is denoted by 7 : 13. As we have shewn that the ratio of one number to another may be expressed by the fraction in which the former is the numerator and the latter the denominator, we see that 7 : 13 is $=\frac{7}{13}$. The two numbers which form a ratio are called its *terms*; the first number being called the ANTECEDENT, and the second number THE CONSEQUENT, of the ratio.

148. If the two numbers to be compared together be concrete, they must be of the *same kind*. We cannot compare together 7 days and 13 miles in respect of magnitude; but we can compare 7 days with 13 days; and it is clear that 7 days will have the same relation to 13 days in respect of magnitude, which the number 7 has to the number 13, so that the ratio of 7 days to 13 days will be the same as the ratio of the abstract number 7 to the abstract number 13, and may be expressed by the fraction $\frac{7}{13}$. Since 3d. reduced to the fraction of 12s.$=\frac{3}{144}$, it is clear that when we have two concrete numbers of the same kind, but of different denominations, we must in order to find their ratio, reduce them to one and the same denomination, and may then treat them as abstract numbers.

149. When two Ratios are equal, in other words, when they can be expressed by the same fraction, they are said to form a PROPORTION, and the four numbers are called PROPORTIONALS. Thus the ratio of 8 to 9 is equal to that of 24 to 27, for $8 : 9 = \frac{8}{9}$, and $24 : 27 = \frac{24}{27} = \frac{8}{9}$.

The Ratios being equal, Proportion exists among the numbers, 8, 9, 24, 27; and those numbers are Proportionals.

150. The existence of Proportion between the numbers 8, 9, 24, 27 is denoted thus, $8 : 9 = 24 : 27$, or $8 : 9 :: 24 : 27$, which is usually read thus 8 is to 9 as 24 is to 27.

151. It has been stated that proportion is the equality of two ratios, and we have explained that the two numbers constituting a ratio must either be both abstract, or (if concrete) both of the same kind. In a proportion if one of the ratios be formed by two abstract numbers, the other may arise from two concrete numbers. For it has been explained (Art. 148) that if a ratio consist of two concrete numbers, we may reduce them both to the same denominations, and then treat the resulting numbers as abstract, the ratio of those abstract numbers being the same as that of the two concrete numbers from which they have arisen. For the same reason, one of the two ratios constituting a proportion may be formed from concrete numbers of one kind, while the other is formed from concrete numbers of a different kind; for 7 days : 13 days :: 7 miles : 13 miles, each ratio being in fact that of 7 to 13. Indeed it appears by (Art. 148) that the ratio of two concrete numbers may always be expressed by a ratio of two abstract numbers. If both or either of the ratios in a proportion be formed from concrete numbers, we may thus replace each such ratio by one arising from abstract numbers, and in this way every term of the proportion will become an abstract number; so that, notwithstanding the remark in note (Art. 23), any one of the terms may then be multiplied or divided by any other.

152. In any Proportion, as $8 : 9 :: 24 : 27$, *the product of the 1st and 4th, i. e. the extreme terms = the product of the 2nd and 3rd, i. e. the mean terms;*

$$\frac{8}{9} = \frac{24}{27}\; ; \; \therefore \; \frac{8}{9} \times 9 \times 27 = \frac{24}{27} \times 9 \times 27, \text{ or } 8 \times 27 = 24 \times 9.$$

RULE OF THREE.

153. *If four numbers be proportionals when taken in a certain order, they will also be proportionals when taken in the contrary order.* For instance, 8, 9, 24, 27 are proportionals;

$$\therefore \frac{8}{9} = \frac{24}{27}; \therefore 1 \div \frac{8}{9} = 1 \div \frac{24}{27}; \text{ or } \frac{9}{8} = \frac{27}{24}, \text{ or } \frac{27}{24} = \frac{9}{8};$$

$$\therefore 27 : 24 :: 9 : 8.$$

154. *Of any three terms of a proportion be given, the remaining term may always be found.*

For since in any Proportion

$$\text{1st term} \times \text{4th term} = \text{2nd term} \times \text{3rd term};$$

$$\therefore \text{1st term} = \frac{\text{2nd} \times \text{3rd}}{\text{4th}}, \quad \text{2nd term} = \frac{\text{1st} \times \text{4th}}{\text{3rd}},$$

$$\text{3rd term} = \frac{\text{1st} \times \text{4th}}{\text{2nd}}, \quad \text{4th term} = \frac{\text{2nd} \times \text{3rd}}{\text{1st}}.$$

Ex. 1. Find the 4th term in the proportion 2, 3, 18.

$$2 : 3 :: 18 : \text{4th term}; \therefore \text{4th term} = \frac{3 \times 18}{2} = 27.$$

Ex. 2. Find the 2nd term in the proportion 8, 32, and 24,

$$8 : \text{2nd term} :: 32 : 24; \therefore \text{2nd term} = \frac{8 \times 24}{32} = 6.$$

Ex. XLVIII.

Find the 4th term in each of the following proportions:

(1) $18 : 48 :: 15 :$ (2) $\frac{5}{7} : \frac{3}{8} :: \frac{10}{21} :$

(3) $8 : 4 :: \frac{1}{2} :$ (4) $1 \cdot 2 : 3 \cdot 6 :: 1 \cdot 3 :$

Find the 2nd term in each of the following proportions:

(5) $\frac{1}{2} : \quad :: \frac{1}{4} : \frac{1}{5}.$ (6) $\cdot 05 : \quad :: \cdot 79 : 12 \cdot 63.$

Find the 1st term in each of the following proportions:

(7) $\quad : 10 :: 4\frac{1}{2} : 15.$ (8) $\quad : 1\frac{23}{40} :: 2 \cdot 94 : \cdot 072.$

RULE OF THREE.

155. THE RULE OF THREE is a method by which we are enabled, from three numbers which are given, to find a fourth which shall bear the same ratio to the third as the second to the first; in other words, it is a rule by which, when three terms of a proportion are given, we can determine the fourth.

156. Rule. Find out of the three quantities which are given, that which is of the same kind as the fourth or required quantity; or that which is distinguished from the other terms by the nature of the question: place this quantity as the third term of the proportion.

Now consider whether, from the nature of the question, the fourth term will be greater or less than the third; if greater, then put the larger of the other two quantities in the second term, and the smaller in the first term; but if less, put the larger in the first term and the smaller in the second term.

Then multiply the second and third terms together, and divide by the first, treating all three as abstract numbers. The quotient will be the answer to the question, in the denomination to which the third term was reduced.

Note 1. The first and second terms must be brought to one and the same denominations.

Note 2. Although we have said in the Rule, multiply the second and third terms together, and then divide their product by the first; it will be found in most cases advisable not to perform the actual multiplication until we have discovered, by putting the expression in the form of a fraction, whether there be any factor or factors common to the numerator and denominator, and if so, have rejected such factor or factors.

157. It may be proper to observe that the Rule of Three is applicable in two different kinds of cases, according to which it is called the Rule of Three Direct or the Rule of Three Inverse. The method just stated (Art. 156) is applicable to both kinds of cases.

The Rule of Three Direct is that in which more requires more, or less requires less; or, in other words, in which a greater number requires a greater answer, or a less number a less answer. Thus in the question, "If 4 acres of land cost $250, find the cost of 15 acres, at the same rate." The 15 acres being more than the 4 acres, will require a larger sum than $250 for their purchase, and so, in this case, more requires more. Again in the question, "If 15 acres of land cost $937.50, find the cost of 4 acres, at the same rate," the 4 acres being less than the 15 acres, will require a less sum than $937.50 for their purchase, and therefore, in this case, less requires less. Such cases belong to the Rule of Three Direct.

RULE OF THREE.   171

The Rule of Three Inverse is that in which more requires less, or less requires more: or, in other words, in which a greater **number requires a less answer, or a less number a greater answer.** Thus in the question, "If 4 men can mow a certain meadow in 3 days, find the time in which 6 men ought to mow it," the six men being more than the four, should perform the work in less time, and so, in this case, more requires less. Again, in the question, "If 6 men can mow a certain meadow in 2 days, find the time in which 4 men ought to mow it," the 4 men, being fewer than the 6, will require a longer time for performing the work, and therefore, in this case, less requires more. Such cases belong to the Rule of Three Inverse.

Ex. 1. Find the value of 37 yards of silk, when 25 yards cost $50.

There are here three given quantities, 25 yards, 37 yards, and $50, and we have to find a fourth which will be the price of 37 yards. It is manifest that the three given quantities, 25 yards, 37 yards, $50, and the required sum, must form a proportion, because the 25 yards must have the same relation in respect of magnitude to the 37 yards, which the $50 (cost of 25 yards) has to the required sum (cost of 37 yards). Proceeding then by Rule (Art. 156) we observe that $50 is of the same kind as the required term, viz. money; we make that the third term of the proportion; and since the required sum (cost of 37 yards) must necessarily be greater than $50 (cost of 25 yards), we make 37 the second term, and 25 the first. We have thus the first three terms arranged as follows:

$$25 \text{ yds.} : 37 \text{ yds.} :: \$50.$$

And the entire proportion will be as follows:

$$25 \text{ yards} :: 37 \text{ yds.} :: \$50 : \text{required cost}.$$

The first and second terms are in one and the same denomination, and require no **reduction.** And by previous reasoning we must now treat the numbers as abstract, therefore

$$\text{cost required} = \$ \frac{37 \times 50}{25} = \$74.$$

Reason for the above process.

We have the cost of 25 yards given, viz. $50, in order to enable us to find the cost of 37 yards.

ARITHMETIC.

It is manifest that the required sum must have the same relation in respect of magnitude to $50, which 37 yards have to 25 yards; that is, the ratio of the required sum to $50, must be equal to that of 37 yards to 25 yards.

Now the ratio of the number of dollars in the required sum to $50, is the same as that of the *abstract* number which indicates how many dollars the required sum contains to the abstract number 50, and may (if the former number be called the *required number*) be expressed by the fraction $\dfrac{\text{required number}}{50}$.

And the ratio of 37 yards to 25 yards is the same as that of the abstract number 37 to the abstract number 25, and may therefore, in like manner be expressed by the fraction $\dfrac{37}{25}$.

$$\therefore \frac{\text{required number}}{50} = \frac{37}{25};$$

$$\therefore \frac{\text{required number}}{50} \times 50 = \frac{37}{25} \times 50.$$

$$\text{or } \frac{\text{required number} \times 50}{50} = \frac{37 \times 50}{25},$$

$$\text{or required number} = \frac{37 \times 50}{25}, \text{ (Art. 103),}$$

$$\text{or } = \frac{50 \times 37}{25}.$$

This result shews that if we arrange the three given terms, 25 yards, 37 yards, and $50 in the following manner,

yds. yds
25 : 37 :: $50,

and then consider the numbers to be abstract, as if they had been written

25 : 37 :: 50,

we shall obtain the abstract number which will shew us how many dollars there are in the required sum by multiplying the second and third terms together and dividing the product by the first; and then

RULE OF THREE. 173

by treating this number as concrete, that is, as so many dollars, we have the required answer in dollars.

The student is strongly advised to work each example *independently*, thus:

$$25 \text{ yds. cost } \$50.$$

$$\therefore 1 \text{ yd. costs } \$\frac{50}{25} \text{ or } \$2;$$

$$\therefore 37 \text{ yds. cost } \$(37 \times 2) = \$74.$$

Ex. 2. **If a** workman earn £17. 6s. in $102\frac{1}{2}$ days, **how long will he be in earning** 50 guineas?

Here the required quantity is *time*, and as the given quantity of that kind is $102\frac{1}{2}$ days, **we** must place that as the third term in the proportion. The earning of 50 guineas will require a longer time than the earning of £17. 6s.: we must therefore place **the 50** guineas as **the** second term, and the £17. 6s. as the first.

Therefore the proportion **is**

£17. 6s. : $50g$. :: $102\frac{1}{2}$ days : required number of days,

346s. : 1050s. :: 205 half-days : required **number of half-days**;

$$\therefore \text{required number of half days} = \frac{1050 \times 205}{346} = 622\tfrac{38}{346}.$$

$$\therefore \text{required number } \textbf{of days} = 622\tfrac{38}{346} \div 2 = 311\tfrac{19}{346}.$$

Independent method.

A man earns £17. 6s. or 346s. in $102\frac{1}{2}$ days, or $\frac{205}{2}$ days;

$$\therefore \ldots\ldots\ldots\ldots\ldots\ldots\ldots\ldots 1s. \text{ in } \ldots\ldots\ldots\ldots \frac{205}{2 \times 346} \text{ days};$$

$$\therefore \ldots\ldots\ldots\ldots\ldots\ldots\ldots 1050s. \text{ in } \ldots\ldots\ldots\ldots \frac{205}{2 \times 346} \times 1050$$

$$= 311\tfrac{19}{346} \text{ days.}$$

Ex. 3. **If the tax on** $936 be $69.12, what will be the tax **on** $4195.20?

The $69.12 being of **the** same nature with the sum required, must be placed as the third **term in** the proportion; and as the required tax must clearly be greater than $69.12, we must place $4195.20 as the second, and $936 as the first term.

$$\$936 : \$4195.20 :: \$69.12 : \text{the required tax,}$$

$$\therefore \text{ the required tax} = \$ \frac{4195 \cdot 20 \times 69 \cdot 12}{936}$$

$$= \$309.80 \text{ nearly.}$$

Independent method.

The tax on $936 is $69.12;

$$\therefore \ldots\ldots\ldots\ldots \$1 \ldots\ldots \$\frac{69 \cdot 12}{936} \text{ or } \$\frac{\cdot 96}{13};$$

$$\therefore \ldots\ldots\ldots \$4195.20 \ldots\ldots \$\frac{\cdot 96}{13} \times 4195 \cdot 2$$

$$= \$309.80 \text{ nearly.}$$

Ex. 4. If I can travel 198 miles by railway for $11.88, how far at the same rate of charge ought I to be carried for $38.61?

$$\$11.88 : \$38.61 :: 198 \text{ m.} ; \text{ required distance.}$$

$$\therefore \text{Required distance} = \frac{3861 \times 198}{1188} \text{ miles.} = \frac{3861}{6} \text{ miles (cancelling by 198)}$$

$$643\tfrac{3}{6} \text{ miles} = 642\tfrac{1}{2} \text{ miles.}$$

NOTE. There are certain examples in which at first sight more than three terms appear to be given, but they nevertheless in certain cases come under this rule, as in the following instances :

Ex. 5. If the carriage of 5 cwt. 7 lbs. for 84 miles cost me $16.90, what will it cost me to have 21 cwt. 1 qr. 13 lbs. carried the same distance?

The 84 miles may evidently be left out of consideration, since the distance in both cases is the same.

Proceeding then according to our Rule,

5 cwt. 7 lbs. : 21 cwt. 1 qr. 14 lbs. :: $16.90 . required cost; whence it will be found that

$$\text{Required cost} = \$71.30.$$

Ex. 6. If a piece of cloth is 20 yards in length and $\tfrac{3}{4}$ yards in breadth, how broad is another piece which is 12 yards long, and which contains as much cloth as the other?

As the length of the second piece is less than that of the first, its breadth must necessarily be greater, in order that the content may be the same. Therefore in this case a less length requires a greater breadth, and so the example belongs to Rule of Three Inverse.

We have the *breadth* of the second piece to find. That of the first piece is $\frac{3}{4}$ yard: place this therefore as the third term. Now the required breadth is to be greater than this. Therefore place the 20 yards as the second term, and 12 yards as the first.

$$12 \text{ yds.} : 20 \text{ yds.} :: \tfrac{3}{4} \text{ yd.} : \text{required breadth in yds.}$$

$$\therefore \text{Required breadth} = \frac{20 \times \tfrac{3}{4}}{12} \text{ yds.} = \tfrac{5}{4} \text{ yds.} = 1\tfrac{1}{4} \text{ yds.}$$

Ex. 7. If 12 men can reap a field in 4 days, in what time can the same work be performed by 32 men?

It is clear that 32 men can perform the work in less time than 12 men, and so the time required will be less than 4 days, the third term in our proportion. We must therefore place the 12 as the second term and the 32 as the first.

$$32 : 12 :: 4 \text{ days} : \text{required time in days.}$$

$$\text{Required time} = \frac{12 \times 4}{32} \text{ days} = \tfrac{12}{8} \text{ days} = \tfrac{3}{2} \text{ days} = 1\tfrac{1}{2} \text{ days.}$$

Independent method.

12 can reap the field in 4 days;
$\therefore$ 1 man .... in 48 days;
$\therefore$ 32 men .... in $\frac{48}{32}$ days,
$= 1\tfrac{1}{2}$ days.

Ex. 8. What was the price of wheat per bushel when the penny loaf weighed 8 ounces; the statute being that it must weigh 10 oz. when wheat is at 12*s*. a bushel?

Here are two numbers, viz. 1 bushel and 1 penny, which can evidently have no effect on the answer, for if any other measure had been named in place of the bushel, and any other loaf in place of the penny loaf, the answer would be the same.

Now as wheat is dearer, or as the price is more, the weight of any given loaf is less, and conversely, as the weight of a given loaf is less, the price of wheat is greater; so that the price required must clearly be greater than 12s., which according to our Rule must be the third term of the proportion. Therefore the 10 oz. must be the second term, and the 8 oz. the first.

$$8 \text{ oz.} : 10 \text{ oz.} :: 12s.$$

$$\text{Required price} = \frac{10 \times 12}{8} s. = \frac{10 \times 3}{2} s. = 15s.$$

NOTE. Examples, such as the following, are easily worked out by the Rule of Three.

Ex. 9. A gentleman, after paying an income tax of 7d. in the £, has £248. 10s. 8d. left; what was his gross annual income?

For every 19s. 5d. which he now has, he had £1. before he paid his income-tax:

∴ 19s. 5d. : £248. 10s. 8d. :: £1. : required income,

whence, required income = £256.

Ex. 10. A hare, pursued by a greyhound, was 130 yards before him at starting; whilst the hare ran 5 yards the dog ran 7 yards: how far had the hare gone when she was caught by the greyhound?

For every 5 yards the hare runs, the dog gains 2 yards, and when he has gained 130 yards he will have caught her.

∴ 2 yds. : 130 yds. :: 5 yds. : required number of yards.

whence, required number of yards = 325.

Ex. 11. Two places, A and B, are distant from each other 324 miles by railway. A train leaves A for B at the same time that a train leaves B for A ; the trains meet at the end of 6 hours, the train from A to B having travelled 16 miles an hour more than the other. How many miles did each travel an hour.

Each train is supposed to run with uniform speed: when the trains meet, the whole distance must have been passed over by them.

∴ 6 hrs. : 1 hr. :: 324 miles : miles passed over by both trains in 1 hr.,

whence, miles passed over by both trains in 1 hr. = 54,

therefore by question, $(54 - 16) \div 2$, or $19 =$ miles travelled per hour by one train, and therefore $54 - 19$, or $35 =$ miles travelled per hour by the other.

Ex. 12. A clock, which is 4 min. $8\frac{3}{32}$ sec. too fast at half-past nine A.M. on Tuesday, loses 2 min. 45 sec. daily; what will be the time in-

RULE OF THREE. 177

dicated by the clock at a quarter-past five P.M. on the following Friday?

From $9\frac{1}{4}$ A.M. on Tuesday, till $5\frac{1}{4}$ P.M. on Friday, there are $79\frac{1}{4}$ hours.

$$\therefore 24 \text{ hrs.} : 79\frac{1}{4} \text{ hours} :: 2'.45'' : \text{time lost by clock.}$$

whence, time lost by clock $= 9'.8\tfrac{9}{32}''$;

$\therefore$ time by the clock at $5\frac{1}{4}$ P.M. on Friday.

$= 4'.8\tfrac{9}{32}'' + 5 \text{ hrs. } 15' - 9'.8\tfrac{9}{32}'' = 5 \text{ hrs. } 10 \text{ min.}$

Ex. XLIX.

(1) If 4 yards of cloth cost $2.88, what will 96 yards of the same cloth cost?

(2) If 9 yards of cloth cost $26.88, how many yards can be bought for $215.04?

(3) If 7 bushels of wheat be worth $8.82 what will be the value of 3 bushels of the same quality?

(4) The rent of 42 acres of land is $63, how many acres of the same quality of land ought to be rented for $273.

(5) If the cost of 72 tons of coals be $432, what will be the cost of 54 tons?

(6) How long will a person be saving $14.40, if he put by 30 cents per week?

(7) Find a number which shall bear the same ratio to 9 which 20 does to 15.

(8) If 2 cwt., 3 qrs., 14 lbs. of sugar cost $28.90, what quantity of the same quality of sugar can be bought for $142.80?

(9) If 3 cwt., 3 qrs. cost $33.75, what will be the price of 2 cwt, 2 qrs.?

(10) Find the value of 23 yds., 1 ft. of cloth, supposing 4 yds., 31 in. of the same quality to cost $18.

(11) What will be the income-tax at $1\frac{1}{2}$ cents in the dollar, on $257.50?

(12) What is the tax upon $1450.46, when $2061.18 is rated at $3.24?

(13) If one bushel of malt cost $1.40, how much can I buy for $129.60?

(14) Find the price of 2 tons, 3 cwt., 14 lbs. at $2.11 per quarter

12

(15) *A* pays half-yearly an income-tax of £10. 1s. 3d.; find his income, the tax being 7d. in the £.

(16) Find the amount of a servant's wages for 215 days at 55 cents a day.

(17) A bankrupt's debts amount to $983.04, and his assets to $860.16; how much in the dollar can he pay?

(18) A bankrupt pays 56⅔ cents in the dollar and his assets amount to $4560; find the amount of his debts.

(19) If a farm containing 400 ac., 2 ro., 20 po. be let at $1201.87½ for the year, what is the rent per acre?

(20) Find a fourth proportional to the numbers 3, 3·75, and 40·

(21) If 10 men can mow a field in 12 days, in how many days will 15 men mow it?

(22) If a man walk 62 miles in 3 days, in how many days will he walk 80 miles?

(23) How many yards worth 87 cents a yard must be given in exchange for 935½ yards worth $4.35 per yard?

(24) A bankrupt pays 5 fl. 7 c. 5 m. in the pound; what sum will be lost on a debt of £11793 5 fl.?

(25) Find the price of 2 tons, 16 cwt., 17 lbs. of sugar at 20 cents for 2½ lbs. (cwt.= 112 lbs.)

(26) If a person travelling 12 hours a day perform a journey in 24 days, in what length of time will he perform the same journey if he travel 16 hours a day?

(27) If 3⅗ oz. Avoir. cost $1.68, what will 30¾ lbs. cost.

(28) How many men must be employed to finish a piece of work in 15 days, which 5 men can do in 24 days?

(29) If 356 ac., 3 ro. 39½ po. be rented at £951. 19s. 10d., what is the rent of 2 acres?

(30) If 27 bus., 2 pks. cost £10. 7s. 2¼d., what is the price of 16½ bus.?

(31) How many yards of drugget an ell wide will cover 40 yards of carpet ¾ yd. wide?

(32) *A* borrowed of *B* 400 dollars for 6½ months, afterwards *A* would requite *B*'s kindness by lending him $910; how long should he lend it?

(33) A field is 121 yds. long, and 86 yds. broad; what will be its value at $80 an acre.

RULE OF THREE. 179

(34) If the price of 1 lb. of sugar be $0.0625, what is the **value** o ·75 **of a cwt.?**

(35) If $3\frac{1}{4}$ shares in a mine cost $54, what will $28\frac{3}{4}$ shares cost?

(36) If $34\frac{1}{2}$ yards of cloth cost £12. **7s. 11$\frac{3}{8}$d.**, how many **yards can** be bought for £3. 19s. 0$\frac{3}{4}$d.?

(37) Find **the** rent at $7.20 **an acre of a** rectangular field whose sides are respectively **50 chains 40 links, and** 56 chains 25 links.

(38) In what time will **25 men do a piece of** work which **12 men** can do in **3 days?**

(39) If ·3 of 4·5 **cwt. cost $11.55, what is the price per lb.?**

(40) **A piece** of Gold at £3. **17s. 10$\frac{1}{4}$d. per oz. is worth** £150; what will be **the** worth of a piece of **silver of equal weight at** 54s. 6d. per lb.?

(41) **If** a piece **of building** land 375 ft. **6 in. by 75 ft. 6 in.** cost $566.40, what will be **the price of** a piece of **similar land 278 ft. 9 in. by 151 feet?**

(42) A servant enters on **a** situation at **12 o'clock at noon on Jan. 1, 1870,** at a yearly salary of **$224,** he leaves at noon **on the 27th of May** following; what ought he to receive for his services?

(43) A was owner of a $\frac{4}{17}$ of a vessel, and sold $\frac{3}{11}$ of $\frac{2}{3}$ of his **share for** $1600; what was the value of $\frac{1\frac{2}{3}}{4\frac{1}{4}}$ of $\frac{2}{3}$ of the vessel?

(44) A exchanged with B 60 yards **of** silk worth $1.68 a yard **for** 48 yards of velvet; what was **the price of** the velvet **a yard?**

(45) A person, after paying 3 cents in the $ for income-tax on **his** income, has $7838.12 remaining; **what** had he **at** first?

(46) A watch is 10 minutes too fast at 12 o'clock (noon) on Monday, and it gains 3', 10" a day; what will be **the** time by the watch at a quarter past 10 o'clock A.M. on the following Saturday?

(47) The circumference of a circle is to its diameter as 3·1416 : 1; find (in feet and inches) the circumference of a circle whose diameter is 22$\frac{1}{2}$ feet.

(48) If the carriage **of** 3 cwt. cost $2.40 for 40 miles, how much ought to be carried for the same price for 25$\frac{3}{4}$ miles?

(49) If I spend 20 dollars in a fortnight, what must my income be that I may lay by 200 dollars in the year 1855?

(50) The house-tax upon a house rated at 175 guineas is £6. 17s. 9$\frac{3}{4}$d.; what will be the **tax** upon one rated at £120?

(51) A silver tankard, which weighs 1 lb., 10 oz., 10 dwt. cost $29.70; what is the value of the silver per ounce?

(52) A man, working 7½ hours a day does a piece of work in 9 days; how many hours a day must he work to finish it in 4½ days?

(53) If a pound of silver costs $15.84, what is the price of a salver which weighs 7 lbs., 7 oz., 10 dwt., subject to a duty of 36 cts. per ounce, and an additional charge of 44 cts. per ounce for the workmanship?

(54) How much did a person spend in 63 days, who with an annual income of $3925 is 90 dollars in debt at the end of a year?

(55) If 15 men, 12 women, and 9 boys, can complete a piece of work in 50 days, what time would 9 men, 15 women, and 18 boys take to do four times as much, the parts done by each in the same time being as the numbers, 3, 2, and 1?

(56) A person possesses $800 a year; how much may he spend per day in order to save $48.25 after paying a tax of $5 on every $100 of income?

(57) If 3 cows or 7 horses can eat the produce of a field in 29 days, in how many days will 7 cows and 3 horses eat it up?

(58) How many yards of carpet ¾ yard wide will cover a room whose width is 16 feet, and length 27½ feet?

(59) A person buys 100 eggs at the rate of 2 a penny, and 100 more at the rate of 3 a penny; what does he gain or lose by selling them at the rate of 5 for 2d.?

(60) A church-clock is set at 12 o'clock on Saturday night; at noon on Tuesday it is 3 minutes too fast: supposing its rate regular, what will be the true time when the clock strikes four on Thursday afternoon?

(61) A person after paying a poor's rate of 4 cents in the dollar has $7200 remaining; what had he at first?

(62) If a piece of work can be done in 50 days by 35 men working at it together, and if, after working together for 12 days, 16 of the men were to leave the work; find the number of days in which the remaining men could finish the work.

(63) A regiment of 1000 men are to have new coats; each coat is to contain 2½ yards of cloth 1¼ yards wide; and it is to be lined with shalloon of ¾ yard wide; how many yards of shalloon will be required?

RULE OF THREE.

(64) **If 5 ounces of silk can be** spun into a thread two furlongs and a half long, what weight of **silk would** supply a thread sufficient to reach to the **Moon,** a distance of 240,000 miles?

(65) How **many** revolutions will **a** carriage-wheel, whose diameter **is 3 feet,** make **in 4 miles?** (See Ex. **47.)**

(66) If 8 oz. of sugar be worth $0.0525, what is the value **of ·75 of a ton?**

(67) The price of **·0625 lbs.** of tea is **·4583s.**; what quantity can be bought **for £61 12s.?**

(68) **Two watches, one of which gains as much as the other loses,** viz. 2′. 5″ **daily, are set right at 9 o'clock A.M., on Monday; when will** there **be a difference of one hour in the times denoted by them?**

(69) How many yards of matting, 2·5 feet broad, **will cover a room 9 yards** long, and 20 feet **broad?**

(70) A person bought **1008 gallons of** spirits for **$3072;** 48 gallons leaked **out: at what rate must he sell** the remainder **per gallon so as not to lose by his bargain?**

(71) **If a soldier be allowed 12 lbs. of bread in 8 days,** how much **will serve a** regiment of 850 **men** for the year 1856?

(72) **If** 2000 men have provisions for 95 days, **and if after 15 days 400** men **go away; find** how **long the** remaining **provisions will serve the number left.**

(73) A gentleman has 10000 acres; **what is his yearly** rental, **if his** weekly rental for 20 square poles **be 3 cents? (1 year = 52 weeks.)**

(74) **If an** ounce of gold be **worth £4·189583, what is the value of** ·36822916 **lbs.?**

(75) **If 1000** men have provisions for **85** days, and if after **17** days **150 of** the men go away; **find how long the remaining** provisions will serve the number left.

(76) What is **the** quarter's rent of 182·3 **acres of land, at £4·65 per acre for a year?**

(77) A grocer bought **2 tons, 3 cwt., 3** qrs. of goods for **$576,** and paid $12 for expenses; **what must he sell** the goods at per cwt. in order to clear $294 on the **outlay? (cwt. = 112 lbs.)**

(78) What must be the breadth of a piece of ground whose length is 40¼ **yards, in order that** it may be twice as great as another piece of **ground** whose length is 14⅝ yards, and whose breadth is 13·3⁄₅ yards?

(79) **If** 3·75 **yards of** cloth cost **$3·825, what will 38 yds., 2 qrs., 3 nails cost?**

182   ARITHMETIC.

(80) Four horses and 6 cows together find sufficient grass on a certain field; and 7 cows eat as much as 9 horses; what must be the size of a field relatively to the former, which will support 18 horses and 9 cows?

(81) A alone can reap a field in 5 days, and B in 6 days, working 11 hours a day; find in what time A and B can reap it together working 10 hours a day.

DOUBLE RULE OF THREE.

158. There are many questions, which are of the same nature with those belonging to the Rule of Three, but which if worked out by means of that Rule as before-given, would require two or more distinct applications of it. Every such question, in fact, may be considered to contain two or more distinct questions belonging to the Rule of Three, and when each of those questions has been worked out by means of the Rule, the answer obtained for the last of them will be the answer to the original question.

159. The following example may serve to illustrate the preceding observations. "If the carriage of 15 cwt. for 17 miles cost me $20.40, what would the carriage of 21 cwt. for 16 miles cost me?"

The above questions may be resolved into the following two.

The first question may be this: "If the carriage of 15 cwt. for 17 miles cost me $20.40, what would the carriage of 21 cwt. for 17 miles cost me?" In this question the 17 miles would have no effect upon the answer, because the distance is the same in both parts of the question, and the answer would clearly remain unaltered, if any other number of miles, or if the words "a certain distance," had been used instead of the 17 miles. This number may therefore be neglected as superfluous, and we have then three terms of a proportion remaining, and the fourth is to be found. Solving the question by the Rule of Three, we find that the answer will be $28.56.

The second question may be this: "If the carriage of 21 cwt. for 17 miles cost me $28.56, what will the carriage of 21 cwt. for 16 miles cost me?" In this question, for reasons similar to those before given, the 21 cwt. will be a superfluous quantity. Applying the Rule of Three to the question, we find the answer to be $26.88.

DOUBLE RULE OF THREE. 183

From the connection of the two questions with that originally proposed, we observe that $26.88, thus obtained through two distinct applications of the Rule of Three, must be the answer to the original question.

160. The DOUBLE RULE OF THREE is a shorter method of working out such questions as would require two or more applications of the Rule of Three; and it is sometimes called the RULE OF FIVE, from the circumstance, that in the practical questions to which it is applied there are commonly five quantities given to find a sixth.

161. For the sake of convenience, we may divide each question into two parts, the *supposition*, and the *demand;* the former being the part which expresses the conditions of the question, and the latter the part which mentions the thing demanded or sought. In the question, "If the carriage of 15 cwt. for 17 miles cost me $20.40, what would the carriage of 21 cwt. for 16 miles cost me?" the words "if the carriage of 15 cwt. for 17 miles cost $20.40," form the supposition; and the words, "what would the carriage of 21 cwt. for 16 miles cost me?" form the demand. Adopting this distinction we may give the following rule for working out examples in the Double Rule of Three.

162. RULE. Take from the supposition that quantity which corresponds to the quantity sought in the demand; and write it down as a third term. Then take one of the other quantities in the supposition and the corresponding quantity in the demand, and consider them with reference to the third term *only* (regarding each other quantity in the supposition and its corresponding quantity in the demand as being equal to each other); when the two quantities are so considered, if from the nature of the case, the fourth term would be greater than the third; then, as in the Rule of Three, put the larger of the two quantities in the second term, and the smaller in the first term; but if less, put the smaller in the second term, and the larger in the first term.

Again, take another of the quantities given in the supposition, and the corresponding quantity in the demand; and retaining the same third term, proceed in the same way to make one of those quantities a first term and the other a second term.

If there be other quantities in the supposition and demand, proceed in like manner with them.

In each of these statings reduce the first and the second terms to

184   ARITHMETIC.

the same denomination. Let the common third term be also reduced to a single denomination if it be not already in that state. The terms may then be treated as abstract numbers.

Multiply all the first terms together for a final first term, and all the second terms together for a final second term, and retain the former third term. In this final stating multiply the second and third terms together and divide the product by the first. The quotient will be the answer to the question in the denomination to which the third term was reduced.

NOTE. In dealing with the final statement obtained by our Rule, note 2, p. 170, will often be found useful.

Ex. 1. If a tradesman with a capital of $2000 gain $50 in 3 months, how long will it take him with a capital of $3000 to gain $175?

The 3 months in the supposition correspond with the quantity sought in the demand. We make the 3 months therefore the third term. Then taking the capital of $2000 in the supposition, and that of $3000 in the demand, and considering them with reference to the time in the third term, we see that if the amount of capital be increased, the time in which a given gain would be produced would be diminished, so that a fourth term would be less than the third; therefore we place $3000 as a first term and $2000 as a second. Again, taking the gain of $50 from the supposition, and that of $175 from the demand, and considering them in like manner with reference to the time in the third term, we see that if the amount of gain be increased, the time in which a given capital would produce it, must be increased also, so that here the fourth term would be greater than the third; and therefore we place the $50 as a first term, and the $175 as a second term; thus we have the following statements:

$$\left.\begin{array}{l}\$3000 : \$2000 \\ \$50 : \$175\end{array}\right\} :: 3m.$$

Proceeding according to our Rule, we have the following statement:
$$3000 \times 50 : 2000 \times 175 :: 3,$$

and the required number of months $= \dfrac{2000 \times 175 \times 3}{3000 \times 50} = 7.$

The required answer is therefore 7 months.

Reason for the above process.

The tradesman, with a capital of $2000 gains $50 in 3 months.

DOUBLE RULE OF THREE.   185

Let us first find by the Rule of Three, how long he would be in gaining $175 with the *same* capital. Thus

$$\$50 : \$175 :: 3\text{ m.} : \text{required time.}$$

$$\text{Required time} = \left(\frac{175 \times 3}{50}\right) \text{months.}$$

Since then the tradesmen with a capital of $2000 would gain $175 in $\left(\frac{175 \times 3}{50}\right)$ months let us next find, by the Rule of Three, how long it would take him to gain the same sum with a capital of $3000, **and we must have** the answer to the original question. Thus

$$\$3000 : \$2000 :: \frac{175 \times 3}{50} \text{ months : required time.}$$

$$\text{Required time in months} = \left(\frac{175 \times 3}{50} \times 2000\right) \div 3000$$

$$= \frac{175 \times 3 \times 2000}{50 \times 3000}$$

$$= \frac{2000 \times 3 \times 175}{3000 \times 50};$$

whence it appears that if we arrange **the quantities given by the** question as follows :

$$\left.\begin{array}{l}\$3000 : \$2000 \\ \$50 : \$175\end{array}\right\} :: 3m,$$

and treat the numbers **as abstract; and** then multiply the two first terms together for a single **first** term, **and** the two second terms together for a single second term; and then divide the product of the second and third terms by the first, we shall obtain the answer in that denomination to which the third **term** was reduced.

Independent method of working above example.

A capital of $2000 gains $50 in 3 months,
.... $1   ...... $50 in (3×2000) months,
.... $1   ...... $1 in $\left(\frac{3 \times 2000}{50}\right)$ months,
.... $3000... ... $1 in $\left(\frac{3 \times 2000}{50 \times 3000}\right)$ months,

A capital of $3000 gains $175 in $\left(\dfrac{3 \times 2000 \times 175}{50 \times 3000}\right)$ months,

or $\left(\dfrac{2000 \times 175 \times 3}{3000 \times 50}\right)$ months;

that is, if we arrange the given quantities as follows,

$$\left.\begin{array}{c}\$3000 : \$2000 \\ \$50 : \$175\end{array}\right\} :: 3m.,$$

we obtain the required time in months by multiplying the two first terms together for a final first term, the two second terms together for a final second term; and then dividing the product of the second and third terms by the first term.

Ex. 2. If 7 horses be kept 20 days for $14, how many will be kept 7 days for $28?

$\left.\begin{array}{c}7 \text{ days} : 20 \text{ days} \\ \$14 : \$28\end{array}\right\} :: 7 \text{ horses} ;$ *More* horses can be kept for a *given sum of money* for 7 days than for 20 days, and more horses can be kept for a given number of days for $28 than for $14.

$$\text{the required number of horses} = \dfrac{20 \times 28 \times 7}{7 \times 4}$$
$$= 40.$$

Ex. 3. If 20 men can perform a piece of work in 12 days, find the number of men who could perform another piece of work 3 times as great in $\frac{1}{5}$th of the time.

The first piece of work being reckoned as 1, the second must be reckoned as 3.

$$\left.\begin{array}{c}1 : 3 \\ \tfrac{12}{5} \text{ days} : 12 \text{ days}\end{array}\right\} :: 20 \text{ men}.$$

$$\therefore \text{req}^\text{d} \text{ number of men} = \dfrac{3 \times 12 \times 20}{\tfrac{12}{5}} = 300.$$

Independent method.

In 12 days work is done by 20 men,
∴ In 1 day .... (20×12) men,
∴ In 1 ...... 3 times work ... $(20 \times 12 \times 3)$ men,
∴ In $\tfrac{12}{5}$ day .... $\dfrac{20 \times 12 \times 3}{\tfrac{12}{5}}$ men,

or 300 men.

DOUBLE RULE OF THREE. 187

Ex. 4. If 252 men can dig a trench 210 yards long, 3 wide, and 2 deep, in 5 days of 11 hours each; in how many days of 9 hours each will 22 men dig a trench of 420 yds. long, 5 wide, and 3 deep?

The first **trench contains** $(210 \times 3 \times 2)$ cubic yds.
$$= 1260 \text{ cubic yds.}$$
The second .... $(420 \times 5 \times 3)$ cubic yds.
$$= 6300 \text{ cubic yds.}$$

On the supposition, therefore, that 252 men can remove 1260 cubic yds. of earth in 55 hours, we have to find in how many hours 22 men can remove 6300 cubic yds.

Then we have the following statements.

$$\left. \begin{array}{c} 22 \text{ men} : 252 \text{ men} \\ 1260 \text{ cub. yds.} : 6300 \text{ cub. yds.} \end{array} \right\} :: 55 \text{ hrs.}$$

$$\therefore \text{req}^d \text{ time} = \frac{252 \times 6300 \times 55}{22 \times 1260} \text{ working hours}$$

$$= 350 \text{ days of } 9 \text{ hours each.}$$

Ex. 5. If 560 flag-stones, **each** $1\frac{1}{2}$ feet square will pave **a court-yard**, how many will be required **for a** yard twice the size, each flag-stone being 14 in. by 9 in.?

Superficial content **of each of** former flag-stones
$$= (1\tfrac{1}{2} \times 1\tfrac{1}{2}) \text{ sq. ft.} = (\tfrac{3}{2} \times \tfrac{3}{2}) \text{ sq. ft.} = \tfrac{9}{4} \text{ sq. ft.}$$
Superficial content **of each of** the latter flag-stones
$$= (\tfrac{14}{12} \times \tfrac{9}{12}) \text{ sq. ft.} = (\tfrac{7}{6} \times \tfrac{3}{4}) \text{ sq. ft.} = \tfrac{7}{8} \text{ sq. ft.}$$

Considering the first court-yard as 1, and therefore the second as 2, our statements **will be**

$$\left. \begin{array}{c} \tfrac{7}{8} \text{ sq. ft.} : \tfrac{9}{4} \text{ sq. ft.} \\ 1 : 2 \end{array} \right\} :: 560 \text{ flag-stones,}$$

which by our Rule, will give us the following single statement:

$$\tfrac{7}{8} : \tfrac{9}{4} \times 2 :: 560,$$

$$\therefore \text{req}^d \text{ number of flag-stones} = (\tfrac{9}{4} \times 2 \times 560) \div \tfrac{7}{8}$$

$$= (\tfrac{9}{2} \times 560 \times \tfrac{8}{7}) = \frac{9 \times 560 \times 8}{2 \times 7} = 2880.$$

Ex. 6. A town which is defended by 1200 men, with provisions enough to sustain them 42 days, supposing each man to receive 18 oz. a **day,** obtains an increase of 200 men **to** its garrison; what must now be the allowance to each man in order that the provisions may serve the whole garrison for 54 days?

$$1400 \text{ men} : 1200 \text{ men} \brace 54 \text{ days} : 42 \text{ days} \; :: 18 \text{ oz.}$$

$$\therefore \text{ number of oz. req}^d. = \frac{1200 \times 42 \times 18}{1400 \times 54} = 12.$$

Independent method.

1200 men for 42 days have each daily 18 oz.;

$\therefore$ 1 man .... (18 × 1200) oz.,

$\therefore$ 1 .... 1 day has daily (18 × 1200 × 42) oz.,

$\therefore$ 1 .... 54 days has daily $\dfrac{18 \times 1200 \times 42}{54}$ oz.,

$\therefore$ 1400 men... ...have each daily $\dfrac{18 \times 1200 \times 42}{1400 \times 54}$ oz.,

or 12 oz.

Ex. L.

1. If 7 men can reap 6 acres in 12 hours, how many men will reap 15 acres in 14 hours?

2. If 3 men earn $75 in 20 days, how many men will earn $78.75 in 9 days, at the same rate?

3. If 16 horses eat 96 bushels of corn in 42 days, in how many days will 7 horses eat 66 bushels?

4. If 800 soldiers consume 5 sacks of flour in six days, how many will consume 15 sacks in 2 days?

5. If 17 bushels be consumed by 6 horses in 13 days, what quantity will 8 horses eat in 11 days, at the same rate?

6. 16 horses can plough 1280 acres in 8 days, how many acres will 12 horses plough in 5 days?

7. If 11 cwt. can be carried 12 miles for $1.50, how far can 36 cwt. 23 lbs. be carried for $5.25?

8. If the carriage of 8 cwt. of goods for 124 miles be $30.24, what weight ought to be carried 53 miles for half the money?

9. If 5 men on a tour of 11 months, spend $1540, how much at the same rate would it cost a party of 7 men for 4 months?

10. If with a capital of $1000 a tradesman gain $100 in 5 months, in what time will he gain $49.50 with a capital of $225?

DOUBLE RULE OF THREE.

11. If it cost $84 to keep 3 horses for 7 months, what will it cost to keep 2 horses for 11 months?

12. The carriage of 4 cwt., 3 qrs., for 160 miles costs $3.85; what weight ought to be carried 100 miles for $30?

13. If one man can reap 345⅔ sq. yds. in an hour, how long will 7 such men take to reap 6 acres?

14. If 20 men in 3 weeks earned $900, in what time will 12 men earn $1500?

15. If the carriage of 1 cwt., 3 qrs., 21 lbs. for 52½ miles come to 17s. 5d., what will be charged for 2¼ tons for 46½ miles? (cwt. = 112 lbs.)

16. If 10 men can reap a field of 7½ acres in 3 days of 12 hours each, how long will it take 8 men to reap 9 acres, working 16 hours a day?

17. If 25 men can do a piece of work in 24 days, working 8 hours a day, how many hours a day would 30 men have to work in order to do the same piece of work in 16 days?

18. If the rent of a farm of 17 ac., 3 ro., 2 po., be £39 4s. 7d., what would be the rent of another farm, containing 26 ac., 2 ro., 23 po., if 6 acres of the former be worth 7 acres of the latter?

19. If 1500 copies of a book of 11 sheets require 66 reams of paper, how much paper will be required for 5000 copies of a book of 25 sheets, of the same size as the former?

20. If 5 men can reap a rectangular field whose length is 800 ft. and breadth 700 ft. in 3½ days of 14 hours each; in how many days of 12 hours each can 7 men reap a field whose length is 1800 ft. and breadth 960 ft.?

21. If a thousand men besieged in a town, with provisions for 5 weeks, allowing each man 16 oz. a day, be reinforced with 500 men more, and have their daily allowance reduced to 6⅔ oz.; how long will the provisions last them?

22. If 20 masons build a wall 50 feet long, 2 feet thick, and 14 feet high, in 12 days of 7 hrs. each, in how many days of 10 hrs. each will 60 masons build a wall 500 feet long, 4 thick, and 16 high?

23. If 10 men can perform a piece of work in 24 days, how many men will perform another piece of work 7 times as great, in one-fifth of the time?

24. If 125 men can make an embankment 100 yards long, 20 feet wide, and 4 feet high, in 4 days, working 12 hours a day, how many

190   ARITHMETIC.

men must be employed to make an embankment 1000 yards long, 16 feet wide, and 6 feet high, in 3 days, working 10 hours a day?

25. What is the weight of a block of stone 12 ft. 6 in. long, 6 ft. 6 in. broad, and 8 ft. 3 in. deep, when a block of the same stone 5 ft. long, 3 ft. 9 in. broad, and 2 ft. 6 in. deep, weighs 7500 lbs.? (112.= cwt.)

26. If 100 men drink $96 worth of wine at $1.08 per bottle, how many men will drink $345.60 worth at $1.20 per bottle, in the same time, at the same rate of drinking?

27. If 5 horses require as much corn as 8 ponies, and 15 quarters last 12 ponies for 64 days, how long may 25 horses be kept for $205.15 when corn is 55 cents a bushel?

28. If 42½ yds. of cloth which is 18 in. wide cost £59. 14s. 2d., what will 118¼ yds. of yard-wide cloth of the same quality cost?

29. 124 men dig a trench 110 yds. long, 3 ft. wide, and 4 ft. deep, in 5 days of 11 hours each; another trench is dug by half the number of men in 7 days of 9 hours each; how many feet of water is it capable of holding?

30. If the 8 cent loaf weigh 3·35 lbs. when wheat is at $1.14 a bushel, what ought to be paid for 47½ lbs. of bread when wheat is at $1.60 a bushel?

31. A pit 24 ft. deep, 14 sq. ft. horizontal section cost $14.40 to dig out; how deep will a pit be of horizontal section 7 ft. by 9 ft. which costs $21.60?

32. The value of the paper required for papering a room, supposing it ¾ yard wide, and 9 cts. a yard, is $10.35; what would it come to, if it were 2 feet wide and 8 cts. a yard?

33. 7 men working 16 days can mow a **field of corn 1320 yards long and 880 wide**; what will be the length of the side of a field 1320 yards broad which four men can mow in 42 days?

34. A beam 16 feet long, 2¼ feet broad, and 8 inches thick, weighs 1280 lbs.; what must be the length of another beam of the same material, whose breadth is 3¼ feet, thickness 7½ inches, and weight 2028 lbs.?

35. If 12 oxen and 85 sheep eat 12 tons, 12 cwt. of hay in 8 days, how much will it cost per month (of 28 days) to feed 9 oxen and 12 sheep, the price of hay being $20.16 a ton, and 3 oxen being supposed to eat as much as 7 sheep?

36. If 1 man and 2 women do a piece of work in 10 days, find in

how long a time 2 men and 1 woman will do a piece of work 4 times as great, the rates of working of a man and woman being as 3 to 2.

37. A person is able to perform a journey of 142·2 miles in 4½ days when the day is 10·164 hours long; how many days will he be in travelling 505·6 miles when the days are 8·4 hours long?

38. If the sixpenny loaf weighs 4·35 lbs. when wheat is **at 5·75s.** per bushel, **what** weight of **bread, when wheat is at 18·4s per** bushel, ought **to** be purchased for **18·13s.?**

39. If a family of 9 **people can live** comfortably in England **for** $7862.40 a year, what **will it cost a family of 8** to live in Canada in the same style for seven **months**, prices being supposed to be ⅗ of what they would **be in** England?

INTEREST.

163. INTEREST is the sum of money paid for the loan **or use of** some other **sum of money, lent for a certain time at a fixed rate; generally at so much for each $100 for one year.**

The money lent **is called the** PRINCIPAL.

The interest of $100 **for a year** is called the **RATE PER CENT.**

The principal + the interest is called the **AMOUNT.**

Interest is divided into Simple and **Compound.** When interest is reckoned **only** on the **original** principal, **it is called SIMPLE** INTEREST.

When **the interest at** the **end of the** first period, instead of being paid by the **borrower, is** retained by him and added on as principal to the former principal, interest **being calculated on** the new principal for the next period, and this interest **again, instead** of being paid, is retained and added on to the last principal for a new principal, and so on; it is called COMPOUND INTEREST.

SIMPLE INTEREST.

164. To *find the Interest of* **a given** *sum of money at a given rate per cent. for a year.*

RULE. Multiply **the** principal by the rate per cent., and divide the product by 100.

NOTE 1. The interest for any given number of years will of course be found by multiplying the interest for one year by the number of years; and the interest for any parts of a year may be found from the interest for one year, by Practice, or by the Rule of Three.

NOTE 1. If the interest has to be calculated from one given day to another, as for instance from the 30th of January to the 7th of February, the 30th of January must be left out in the calculation, and the 7th of February must be taken into account, for the borrower will not have had the use of the money for one day till the 31st of January.

NOTE 2. If the amount be required, the interest has first to be found for the given time, and the principal has then to be added to it.

Ex. Find the simple interest of $250 for one year at 8 per cent. per annum.

Proceeding according to the Rule given above,

$$\begin{array}{r} \$250 \\ 8 \\ \hline \$20.00 \end{array}$$

therefore the interest is $20.

Reason for the process.

The sum of $100 must have the same relation in respect of magnitude to $250 as the simple interest of $100 for a year has to the simple interest of $250 for a year; and thus the $100, $250, $8, and the required interest must form a proportion. (Art. 148.)

We have then

$$\$100 : \$250 :: \$8 : \text{required interest}.$$

whence, required interest $= \$\dfrac{250 \times 8}{100}$ (Art. 156.),

which agrees with the rule given above.

Independent method.

$100 for 1 year gives $8 int.

∴ $1 .... $\dfrac{8}{100}$ int.

∴ $250 .... $\left(250 \times \dfrac{8}{100}\right)$ int., or $20.

SIMPLE INTEREST.

Examples worked out.

Ex. 1. Find the simple interest and amount of £417. 7s. 9d. for 1 year, 10 months, at $4\frac{3}{8}$ per cent.

```
   £.  s.  d.                    £.   s.  d.
  417 . 7 . 9                   417 . 7 . 9
         4¾                            3
 ─────────────              8)1252 . 3 . 3
 1669 . 11 . 0                 ─────────────
  156 . 10 . 4⅞                 156 . 10 . 4⅞
 ─────────────
 £18·26 . 1 . 4⅞
        20
 ─────────
  5·21s.
        12
 ─────────
  2·56d.
```

$$\therefore \text{Int. for 1 year} = 18 . 5 . 2\frac{56\frac{7}{8}}{100}$$
$$= 18 . 5 . 2\frac{91}{160}$$

Int. for 6 mo., or ½ of 1 year $= 9 . 2 . 7\frac{91}{320}$
Int. for 4 mo., or ⅓ of 1 year $= 6 . 1 . 8\frac{137}{180}$
$\therefore$ Int. for 1 yr., 10 mo. $= 33 . 9 . 6\frac{227}{320}$
$\therefore$ amount $= £417. 7s. 9d. + £33. 9s. 6\frac{227}{320}d.$
$= £450. 17s. 3\frac{227}{320}d.$

NOTE. In examples like the above we may reckon 12 months to the year; but if calendar months are given, the interest will then be best found by the Rule of Three; as for instance in the following example:

Ex. 2. Find the simple interest and the amount of $106 from June 15, 1843, to Sept 18, 1843, at $7\frac{1}{2}$ per cent.

```
   $106         The number of days from June 15 to Sept. 18
     7½
   ─────
    742                       = 15 + 31 + 31 + 18
     53
   ─────
   $7.95                      = 95
```

$\therefore$ $7.95 is the interest for 1 year.

Hence, 365 days : 95 days :: $7.95 : interest required, whence, it will be found, that interest required $= $2.06\frac{67}{73}$.

$\therefore$ amount $= $106 + $2.06\frac{67}{73} = $108.06\frac{67}{73}.$

Ex. LI.

Find the simple Interest.

(1) On $85 for 1 year at 8 per cent.
(2) On $310 for 1 year at 7 per cent.
(3) On $1000 for 1 year at $6\frac{1}{2}$ per cent.
(4) On $475 for 3 years at $7\frac{1}{2}$ per cent.
(5) On $936.50 for 2 years at 6 per cent.
(6) On $556.75 for 6 years at 8 per cent.
(7) On $945.40 for 2 years at 7 per cent.
(8) On £198. 6s. 8d. for 1 year at $3\frac{1}{2}$ per cent.
(9) On £236. 6s. 8d. for $2\frac{1}{2}$ years at 3 per cent.
(10) On £98. 15s. 10d. for $\frac{1}{2}$ year at $2\frac{1}{2}$ per cent.

Find the amount

(11) Of $1000 for 2 years at 7 per. cent.
(12) Of $2833.25 for $4\frac{1}{2}$ years at 6 per cent.
(13) Of £1050. 6 fl. 2 c. 5 m. for 6 years at 8 per cent.
(14) Of $139.80 for $3\frac{1}{4}$ years at $7\frac{1}{2}$ per cent.
(15) Of $1895 for $4\frac{1}{4}$ years at $6\frac{7}{8}$ per cent.
(16) Of £1534. 6s. 3d. for $1\frac{3}{4}$ years at 8 per cent.

Find the Simple Interest and Amount

(17) Of $375 for 3 years, 8 months, at 7 per cent.
(18) Of $446.50 for 3 years, 3 months, at 8 per cent.
(19) Of $220 for 7 months at $7\frac{1}{2}$ per cent.
(20) Of $243.80 for 2 years, 5 months, at 8 per cent.
(21) Of 40 dollars from March 16, 1850, to Jan. 23, 1851, at 8 per cent.
(22) Of $320.75 for 2 years, 35 days, at 7 per cent.
(23) Of £34. 10s. from August 10 to October 21, at 6 per cent.

165. *In all questions of Interest, if any three of the four (principal, rate per cent., time, amount) be given, the fourth may be found, as, for instance, in the following examples.*

Ex. 1. Find the amount of $225 for 4 years at 8 per cent. per annum.

$100 for 1 year gives......$8 int.,

∴ $1 ....$$\frac{8}{100}$ int.,

SIMPLE INTEREST

$\therefore$ \$225 for 1 year gives......$\$\left(\dfrac{8}{100} \times 225\right)$ int.,

$\therefore$ \$225 for 4 years............$\$\left(\dfrac{8}{100} \times 225 \times 4\right)$ int.,

or \$72 int.;

$\therefore$ Amount = \$225 + \$72 = \$297.

Ex. 2. In what time will \$225 amount to \$297 at 8 per cent. simple interest?

\$297 − \$225 = \$72, which is the interest to be obtained on \$225 in order that it may amount to \$297?

But Int. of \$225 for 1 year = \$18; which must have the same relation in respect of magnitude to the \$72 as the 1 year has to the required time;

$\therefore$ \$18 : \$72 :: 1 year : required number of years,

whence, required number of years = 4.

Ex. 3. At what rate per cent., simple interest, will \$225 amount to \$297 in 4 years?

In other words, at what rate per cent. will \$225 give \$72 for interest in 4 years, or $\dfrac{\$72}{4}$, or \$18 in one year?

Then \$225 : \$100 :: \$18 : required rate per cent.,

whence, required rate per cent. = \$8.

Ex. 4. What sum of money will amount to \$297 in 4 years, at 8 per cent. simple interest?

\$100 in 4 yrs. at 8 per cent. amounts to \$100 + (8 × 4)\$, or \$132; and this \$132 must be to the \$297 as the \$100 is to the required sum of money;

$\therefore$ \$132 : \$297 :: \$100 : required number of dollars,

whence, required number of dollars = \$225.

Ex. LII.

1. **What sum will** amount to \$150 in 4 years, at 8 per cent. simple interest?

2. At what rate per cent. will $540 amount to $928.80 in 9 years, at simple interest?

3. In what time will $350 amount to $448, at 7 per cent. simple interest?

4. At what rate per cent. will $325.25 amount to $393·5525 in 3½ years, at simple interest?

5. In what time will $142.50 amount to $242.25 at 7 per cent. simple interest?

6. At what rate will $157 amount to $392.50 in 25 years at simple interest?

7. What sum will produce for interest $87.75 in 2¼ years at 6¼ per cent. simple interest?

8. What sum will amount to $1014.67¼ in 3½ years at 7 per cent. simple interest?

9. What sum will amount to £387. 7s. 7½d. in 3 years at 4 per cent. simple interest?

10. In what time will £1275 amount to £1549. 11s. at 3¼ per cent. simple interest?

11. At what rate per cent., simple interest, will £936. 13s. 4d. amount to £1157. 7s. 4½d., in 4⅞ years?

12. In what time will $125 double itself at 5 per cent. simple interest?

13. What sum will amount to £425. 19s. 4½d. in 10 years at 3¼ per cent. simple interest, and in how many more years will it amount to £453. 11s. 7d.?

14. What sum of principal money, lent out at 10 per cent. per annum, simple interest, will produce in 4 years the same amount of interest as $250, lent out at 6 per cent. per annum, will produce in 6 years?

NOTE. Though questions are given in Simple Interest, in which the *time* is for some years, or several payments are made; yet, in all such cases Compound Interest is the only *fair* method to both lender and borrower, and is the method employed by Building Societies, Insurance Companies, &c.

COMPOUND INTEREST.

166. *To find the Compound **Interest** of a given sum of money at a given rate per cent. for any number of years.*

RULE. At the end of each year **add the** interest of that year, found by Art. (164), to the principal at **the** beginning of it; this will be the principal for the next year; proceed **in the** same way **as** far as may be required by the question. Add **together the** interests so arising in the several years, and the result will **be the** compound **interest for** the given period.

*The reason for **the** above Rule* is clear from what has been stated in Arts. (**163** and 164).

Ex. 1. Find the Compound Interest and Amount of $2000 for 2 years at 5 per cent. per annum.

By the Rule,

$2000
5
───────
$100.00
∴ $100 = Int. for 1st yr.;
∴ $2100 = Prin¹. **for 2nd yr.**;
5
───────
$105.00
∴ $105 = Int. for 2nd yr.;
∴ **$100 + $105** or $205 = Compound Int. for 2 years,
and $2100 + $105 or $2205 = Amount.

or thus, since $\frac{5}{100} = \frac{1}{20}$

$\frac{1}{20}$ of $2000 = $ | 2000 = 1st yr.'s Prin¹.
                          | 100 = ... Int.
                          | 2100 = 2nd yr.'s Prin¹.
$\frac{1}{20}$ of $2100 = $ | 105 = ... Int.
                          | $2205 = 3rd yr.'s Prin¹., or
                          | Amount;

Comp^d. int. = $2205 − $2000 = $205.

Decimals may advantageously **be** employed in working questions in Interest.

Ex. 2. Find the **Compound Interest** of $2000, for 2 years at 6½ per cent. per annum.

$6½ int. on $100 is $\frac{6\frac{1}{2}}{100}$ or $·065 int. on $1.

∴ Int. on $2000 = (2000 × ·065)$ = $130;

∴ 2 year's Prin¹. = $2000 + $130 = $2130.

Int. on $2130 = (2130 × ·065)$ = $138.45;
∴ 3rd year's Prin¹. = $2130 + $138.45 = $2268.45.
Hence, Compound Int. = $130 + $138.45 = $268.45.

NOTE 1. It is customary, if the compound interest be required for any number of entire years and a part of a year (for instance for $5\frac{3}{4}$ years), to find the compound interest for the 6th year, and then take $\frac{3}{4}$ths of the last interest for the $\frac{3}{4}$ths of the 6th year.

NOTE 2. If the interest be payable half-yearly, or quarterly, it is clear that the compound interest of a given sum for a given time will be greater as the length of each given period is less; the simple interest will not be affected by the length of each period.

Ex. LIII.

(1) Find the compound interest of $2000 in 2 years at 6 per cent. per annum.

(2) Find the amount of $800 in 3 years at 7 per cent., allowing compound interest.

(3) Find the compound interest of $270 in 2 years at 8 per cent.

(4) Find the amount of $690 for 3 years at 7 per cent., compound interest.

(5) Find the amount of $230.75 for 3 years at 6 per cent., compound interest.

(6) Find the difference in the amount of $415.50, put out for 4 years at 7 per cent., 1st at simple, 2nd at compound interest.

(7) Find the compound interest of $130 in 3 years at 8 per cent. (interest being payable half-yearly).

(8) What will $1760.50 amount to in $2\frac{1}{2}$ years, allowing 8 per cent. compound interest?

(9) A person lays by $230 at the end of each year, and employs the money at 7 per cent. compound interest; what will he be worth at the end of 3 years?

(10) Find the difference between the simple and compound interest of $416 for 2 years at 6 per cent.

(11) What is the difference between the simple and the compound interest of $13,338 for 5 years at 5 per cent.

(12) Find the amount of $180 in 3 years at 8 per cent. compound interest (interest being payable quarterly).

PRESENT WORTH. 199

(13) What sum o. money put out to compound interest for 2 years at 7 per cent. will amount to $100?

(14) What sum at 8 per cent. compound interest will amount in 2 years to $264?

(15) A and B each lend £256 for 3 years at $4\frac{1}{2}$ per cent. per annum, one at simple interest, the other at compound interest: find the difference in the amount of interest they respectively receive.

PRESENT WORTH AND DISCOUNT.

167. A owes B $500, which is to be paid at the end of 9 months from the present time: it is clear that, if the debt be discharged at once (interest being reckoned, we will suppose, at 7 per cent. per annum), B ought to receive a less sum of money than $500; in fact, such a sum of money as will, being now put out at 7 per cent. interest, amount to $500 at the end of 9 months. The sum which B ought to receive now is called the Present Worth of the $500 due 9 months hence, and the sum to be deducted from the $500, in consequence of immediate payment, which is in fact the interest of the Present Worth, is called the Discount of the $500 discharged 9 months before it is due.

We may therefore define PRESENT WORTH to be the actual worth at the present time of a sum of money due some time hence, at a given rate of interest; and we may define the Discount o. a sum of money to be the interest of the Present Worth of that sum, calculated from the present time to the time when the sum would be properly payable.

PRESENT WORTH.

168. RULE. Find the interest of $100 for the given time at the given rate per cent., and state thus:

$100 + its interest for the given time at the given rate per cent. : given sum :: $100 : present worth required.

Ex. 1. Find the present worth of $500, due 9 months hence, at 8 per cent. per annum.

Proceeding according to the above Rule,
     Interest of $100 for 9 months at 8 per cent. is $6,
∴ $106 : $500 :: $100 : required present worth,
     whence required present worth = 471.69\frac{42}{53}$.

200 ARITHMETIC.

The reason for the above process is clear from the consideration, that $100 in 9 months at 8 per cent. interest would amount to $106, and therefore $100 is the present value of $106 due 9 months hence: and consequently we have

1st debt : 2nd debt :: 1st present worth : 2nd present worth.

Independent method,

Since Interest on $100 for 9 mo. at 8 per cent.= $6
∴ P. W. of $106 due 9 mo. hence at 8 per cent.=$100

∴ .... $1 .... $=\$\frac{100}{106}$

∴ .... $500 .... $=\$\frac{100}{106}\times 500$

$=\$471.69\frac{43}{53}$.

Ex. 2. Find the present worth of $838, due 19 months hence, at 6 per cent. simple interest.

Since the interest of $100 for 19 months, at 6 per cent.
$=\$(\frac{19}{12}\times 6)=\$1\frac{9}{2}=\$9\frac{1}{2}$,
∴ $109½ : $838 :: $100 : required present worth,
whence, required present worth=765.29\frac{143}{219}$.

DISCOUNT.

169. Rule. Find the interest of $100 for the given time at the given rate per cent., and state thus:

$100+its interest for the given time at the given rate per cent. : given sum :: interest of $100 for the given time at the given rate per cent. : discount required.

Ex. 1. Find the discount of $500, due 9 months hence, at 8 per cent. per annum.

Proceeding according to the above Rule,
The interest of $100 for 9 months at 8 per cent.= $6;
therefore, $106 : $500 :: $6 : required discount,
whence, required discount=28.30\frac{10}{53}$.

The reason for the above process is clear from the consideration, that

$6 is the interest for 9 months, at 8 per cent., of $100, the present worth of $106 due at the end of that time; and consequently we have

1st debt : 2nd debt :: discount on 1st debt : discount on 2nd debt.

Ex. 2. Find the discount on $1000, due 15 months hence, at 5 per cent. per annum.

Interest of $100 for 15 months at 6 per cent. $= 1\frac{1}{4}$ of $6 = 7\frac{1}{2}$.

$\therefore$ $107\frac{1}{2}$: $1000 :: $7\frac{1}{2}$: Discount required;

$\therefore$ Discount required $= \$\dfrac{1000 \times 7\frac{1}{2}}{107\frac{1}{2}} = \$69.76\frac{32}{43}$.

Ex. 3. Find the discount on £127. 2s. for half a year at 6 per cent.
£100$\frac{3}{5}$: £127$\frac{1}{10}$ :: £$\frac{3}{5}$: required discount;
whence, required discount $=$ £3. 2s.

NOTE 1. Discount $=$ given sum *less* Present Worth; Present Worth $=$ given sum *less* Discount.

NOTE 2. In the discharge of a tradesman's bill it is usual to deduct interest instead of discount; thus, if B contracts with A a debt of $100, A giving 12 months' credit, it is usual in business, if the interest of money be reckoned at 5 per cent. per annum, and the bill be discharged at once, for A to throw off $5, or for A to receive $95 instead of $100; but if A were to put out the $95 at 5 per cent. interest it will not amount to $100 in 12 months; therefore such a proceeding is to the advantage of B; the sum of money which in strictness ought to have been deducted, was not $5, the interest on the whole debt, but 4.76\frac{4}{21}$, the interest of the present worth of the debt, *i. e.*, the discount.

NOTE 3. Bankers and Merchants in discounting bills calculate interest, instead of discount on the sum drawn for in the bill, from the time of their discounting it to the time when it becomes due, adding THREE DAYS OF GRACE, which days are allowed usually after the time a bill is NOMINALLY due, before it is LEGALLY due; which is, of course, an additional advantage. When a bill is payable on demand, the days of grace are not allowed.

NOTE 4. If a bill, without the days of grace, should appear to be due on the 31st of any month which contains only 30 days, the last day of that month, and not the first day of the next, is considered as the day on which the bill is due. Thus a bill drawn on the 31st of October, at 4 months, would be really due, adding in the days of grace, on the 3rd of March. Also bills which fall due on a Sunday, are paid on the previous Saturday.

Ex. A bill of £1000 is drawn on Feb. 16th, 1851, at 7 months' date; it is discounted on the 8th of July at 5 per cent. What does the banker gain by the transaction?

The bill is legally due on Sept. 19; and from July 8 to Sept. 19 are 73 days.

$$\begin{array}{rl} & \text{£. s.} \\ \text{The interest of £1000 for that time} = & 10\ .\ 0 \\ \text{The true discount} \ldots\ldots\ldots\ldots\ldots\ldots\ldots = & 9\ .\ 18\tfrac{3}{101} \\ \therefore \text{the banker's gain} \ldots\ldots\ldots\ldots\ldots & 1\tfrac{99}{101}s. \end{array}$$

Ex. LIV.

Find the Present Worth of

(1) $321   due 1 yr. hence, at 7 per cent. per annm., simp¹. int
(2) $251.56   .... 6 ....
(3) $683.28   ..... 6 months ...... 8 ....
(4) $944.92   ..... 6 .... 7 ....
(5) $463.50   ..... 6 .... 6 ....
(6) £390   ..... 7 .... $6\tfrac{1}{2}$ ....
(7) $856.96   ..... 8 .... 6 ....
(8) $1252.40   ..... 1 .... 12 ....
(9) $1250   ..... 3 .... 7 ....
(10) £2110   ..... 11 .... 6 ....
(11) £275. 6s. 8d.   ..... 15 .... 4 ....
(12) £918   ..... 4 years ...... 5 ....
(13) $500   ..... 19 months ..... $5\tfrac{1}{4}$ ....
(14) £2197   ..... 3 years ...... 4 .... compound interest

Find the Discount on

(15) $64   due 4 months hence, at 6 per cent. per annum, simp¹. int
(16) $1380   ..... 9 .... 8 ....
(17) $107.25   ..... 6 .... 5 ....
(18) $125.46   ..... 3 .... 8 ....
(19) $487   ..... 5 .... 7 ....
(20) $340   ..... 5 .... 6 ....
(21) £3640   ..... 10 .... $4\tfrac{1}{2}$ ....
(22) £813. 9s.   ..... $1\tfrac{1}{2}$ .... $4\tfrac{3}{4}$ ....
(23) $250.75   ..... 17 months .... 8 ....
(24) $102   ..... 146 days .... 5 ....

(25) A bill of £649 is dated on June 23, 1853, at 6 months, and is discounted on July 8, at 3¼ per cent.; what does the banker gain thereby?

(26) Find the true discount on a bill drawn March 17, 1853, at 3 months, and discounted May 2, at 8 per cent. (days of grace allowed.)

(27) Find the simple interest on $545 in 2 years, at 7 per cent. per annum; and the discount on $621.30 due two years hence, at the same rate of interest. Explain clearly why these two sums are identical.

(28) Explain the difference between Discount and Interest. Five volumes of a work can be bought for a certain sum, payable at the end of a year: and six volumes of the same work can be bought for the same sum in ready money; what is the rate of discount?

(29) A tradesman marks his goods with two prices, one for ready money, and the other for one year's credit allowing discount at 5 per cent.; if the credit price be marked at £2. 9s., what ought to be the cash price?

STOCKS.

170. If the 6 per cent. "Dominion of Canada" stock be quoted in the money market at 105½, the meaning is, that for $105½ of money a person can purchase $100 of such stock, for which he will receive a document which will entitle him to half-yearly payments of Interest or Dividends, as they are called, from the Government of the country, at the rate of 6 per cent. per annum, on the stock held by him, until the Government choose to pay off the debt.

Similarly, if shares in any trading company, which were originally fixed at any given amount, say $100 each, be advertised in the share-market at 86, the meaning is, that for $86 of money *one* share can be obtained, and the holder of such share will receive a dividend at the end of each half-year upon the $100 share according to the state of the finances of the company.

STOCK may therefore be defined to be the capital of trading companies; or to be the money borrowed by our or any other Government, at so much per cent., to defray the expenses of the nation.

The amount of debt owing by the Government is called the NATIONAL DEBT, or the FUNDS. The Government reserves to itself the option of paying off the principal or debt at any future time, pledging itself, however, to pay the interest on it regularly at fixed periods in the mean time.

From a variety of causes the price of stock is continually varying. A fundholder can at any time sell his stock, and so convert it into money, and it will depend upon the price at which he disposes of it, as compared with the price at which he bought it, whether he will gain or lose by the transaction.

NOTE 1. Purchases or sales of stock are made through Brokers, who generally charge $\$\frac{1}{8}$, or $12\frac{1}{2}$ cts. per cent., upon the stock bought or sold : so that, when stock is bought by any party, every $100 stock costs that party $\$\frac{1}{8}$ more than the market-price of the stock : and when stock is sold, the seller gets $\$\frac{1}{8}$ less for every $100 stock sold than the market-price.

Thus, the actual cost of $100 stock in the 3 per cents. at $94\frac{1}{8}$, is $\$(94\frac{1}{8}+\frac{1}{8})$, or $\$94\frac{2}{8}$. The actual sum received for $100 stock in the 3 per cents. at $94\frac{1}{8}$, is $\$(94\frac{1}{8}-\frac{1}{8})$ or $94.

Unless the brokerage is mentioned, it need not be noticed in working examples in stocks.

NOTE 2. When $100 stock costs $100 in money, the stock is said to be at *par*.

When $100 stock costs more than $100 in money, the stock is said to be at a *premium*.

When $100 stock costs less than $100 in money, the stock is said to be at a *discount*.

All Examples in Stocks depend on the principles of Proportion, *and may therefore be worked by the Rule of Three.*

Those of most frequent occurrence will now be given.

Ex. 1. What sum of money will purchase $3600 6 per cent. stock at 94?

$100 stock (st.) costs $94 in money;

∴ $100 st. : $3600 st. :: $94 : req^d. sum;

∴ required sum $= \dfrac{3600 \times 94}{100} = \$3384$.

STOCKS. 205

Independent method.

$100 stock is bought for $94 ;

∴ $1 .... $$\frac{94}{100}$$.

∴ $3600 .... $$\$\frac{94}{100} \times 3600$$

= $3384.

NOTE. The student is strongly advised to work the questions by *independent method.*

Ex. 2. Find the cost of £2353 3 per cent. Consols at $90\frac{3}{8}$, brokerage being $\frac{1}{8}$ per cent.

£100 st. costs £$(90\frac{3}{8} + \frac{1}{8})$, or £$90\frac{1}{2}$;
∴ £100 st. : £2353 st. :: £$90\frac{1}{2}$: req^d. cost ;
∴ req^d. cost = £$\frac{2353 \times 90\frac{1}{2}}{100}$ = £2129. 9s. $3\frac{1}{2}d$. $\frac{2}{5}q$.

Independent method.

£100 stock costs £$(90\frac{3}{8} + \frac{1}{8})$, or £$90\frac{1}{2}$;

∴ £100 stock costs £90·5,

∴ £1 .... £$\frac{90 \cdot 5}{100}$,

∴ £2353 .... £$\frac{90 \cdot 5}{100} \times 2353$,

or £2129. 9s. $3\frac{1}{2}d$. $\frac{2}{5}q$.

Ex. 3. A person who has $10000 Bank-stock, sells out when it is 40 per cent. premium ; what amount of money does he receive, brokerage being $\frac{1}{8}$ per cent. ?

$100 st. sells for $$\left(140 - \frac{1}{8}\right)$, or $$139\frac{7}{8}$ money ;

∴ $100 st. : $10000 st. :: $$139\frac{7}{8}$: req^d. am^t. of money ;
∴ req^d. am^t. = $$\frac{10000 \times 139\frac{7}{8}}{100}$ = $13987.50.

Independent method.

$100 stock sells for $$(140 - \frac{1}{8})$, or $$139\frac{7}{8}$;

∴ $100 stock sells for $139·875,

∴ $1 .... $$\frac{139 \cdot 875}{100}$

$$\therefore \$10000 \ldots\ldots\ldots \$\frac{139 \cdot 875}{100} \times 10000,$$

$$\text{or } \$13987.50.$$

Ex. 4. What incomes will $8500 of 7 per cent. stock, and $8500 invested in the 7 per cent. stock at $102\frac{2}{3}$, respectively produce?

1st, since every $100 stock gives $7 int.;

$$\therefore \text{income from } \$8500 \text{ of 7 per cent. stock} = \$\frac{8500 \times 7}{100} = \$595.$$

2d, since $100 stock, which gives $7 int., costs $102\frac{2}{3}$;

$\therefore$ every $102\frac{2}{3}$ gives $7 int.;

$$\therefore \$102\frac{2}{3} : \$8500 :: \$7 : \text{req}^d. \text{ income}$$

$$\therefore \text{req}^d. \text{ income} = \$\frac{8500 \times 7}{102\frac{2}{3}} = \$579 \cdot 54\frac{6}{17}.$$

Ex. 5. One person buys £800 Consols at $90\frac{1}{3}$ and sells out at 93; another invests £800 in Consols at $90\frac{1}{3}$ and sells out at 93; what sum of money does each gain?

1st man gains £$(93 - 90\frac{1}{3})$, or £$2\frac{2}{3}$, on every £100 stock;

$$\therefore \text{his whole gain} = £(2\frac{2}{3} \times 8) = £21. \ 6s. \ 8d.$$

2d man gains £$2\frac{2}{3}$ on every £100 stock, *i.e.* on every £$90\frac{1}{3}$ of his money which he invests;

$$\therefore £90\frac{1}{3} : £800 :: £2\frac{2}{3} : \text{whole gain};$$

$$\therefore \text{whole gain} = £\frac{800 \times 2\frac{2}{3}}{90\frac{1}{3}} = £23. \ 12s. \ 4d. \text{ nearly.}$$

Ex. 6. A person invested some money in the 3 per cent. Consols when they were at 90, and some money when they were at 80; find the rate of interest he obtained in each case, and the advantage per cent. of the second purchase over the first.

$$£90 : £100 :: £3 : \text{rate per cent. in 1st case,}$$
$$£80 : £100 :: £3 : \text{rate per cent. in 2nd case,}$$

$$\therefore \text{rate per cent. in 1st case} = £\frac{100 \times 3}{90} = £3. \ 6s. \ 8d.;$$

$$\therefore \ldots\ldots\ldots\ldots 2d \ldots\ldots = £\left(\frac{100 \times 3}{80}\right) = £3. \ 15s.;$$

$$\therefore \text{advantage} = £3. \ 15s. - £3. \ 6s. \ 8d. = 8s. \ 4d.$$

Ex. 7. A person invests £1037. 10s. in the 3 per cents. at 83; the funds rise 1 per cent.; he then transfers his capital to the 4 per cents. at 96: find the alteration in his income.

£83 : £1037. 10s. :: £100 : quantity of 3 per cent. st. ;

∴ quantity of 3 per cent. st. bought $= £\dfrac{1037\frac{1}{2} \times 100}{83} = £1250$.

The funds have risen 1 per cent., therefore **to transfer £1250 stock** from the funds at 84 to the funds **at 96,**

£96 : £84 :: £1250 stock : quantity of 4 per cent. stock, (since the higher the price of the stock the less will be the amount purchased);

∴ **quantity of 4** per cent. stock $= £\dfrac{1250 \times 84}{96} = £1093$. **15s.**

$$\text{1st Income} = £\dfrac{1250 \times 3}{100} = £37.\ 10s.$$

$$\text{2nd Income} = £\dfrac{1093\frac{3}{4} \times 4}{100} = £43.\ 15s.\ ;$$

∴ alteration in income $= £43.\ 15s. - £37.\ 10s. = £6.\ 5s.$

Ex. 8. Which **is the best stock** to invest £1000 in, the 3 per **cents. at** $89\frac{1}{2}$, or the $3\frac{1}{2}$ per cents. at $98\frac{1}{2}$?

In the first case,

every £$89\frac{1}{2}$ of money gives £3 interest;

∴ every £1 of money gives $£\dfrac{3}{89\frac{1}{2}}$, or $£\dfrac{6}{179}$ interest.

In the second case,

every £$98\frac{1}{2}$ of money gives £$3\frac{1}{2}$ interest;

∴ **every £1 of** money gives $£\dfrac{3\frac{1}{2}}{98\frac{1}{2}}$, or $£\dfrac{7}{197}$, interest;

and comparing the fractions $\dfrac{6}{179}$ and $\dfrac{7}{197}$.

since 7×179 is $> 6 \times 197$,

the 2nd fraction is greater than the 1st, and therefore the 2nd investment is the best.

Ex. LV.

Find the quantity of stock purchased by investing:

(1) $2850 in the 6 per cents. at 75.

(2) $712 in the 7 per cents. at 89.

(3) \$504 in the 8 per cents. at 96.
(4) \$3741 in the 7 per cents. at 87.
(5) \$500 in the 6 per cents. at $83\frac{3}{4}$.
(6) \$800 in the 8 per cents. at $75\frac{1}{2}$.
(7) £4311. 8*s*. 9*d*. in the $3\frac{1}{2}$ per cents. at $85\frac{3}{8}$.
(8) \$2353 in the 6 per cents. at $90\frac{3}{8}$, brokerage $\frac{1}{8}$ per cent.
(9) £3277 in the 4 per cents. at $105\frac{7}{8}$, brokerage $\frac{1}{8}$ per cent.

Find the money value of

(10) \$2600 in the 7 per cents. at 93.
(11) \$1920 in the 6 per cents. at $77\frac{1}{2}$.
(12) \$3000 in the $7\frac{1}{2}$ per cents. at $92\frac{1}{2}$.
(13) \$2240 in the $6\frac{1}{2}$ per cents. at $81\frac{7}{8}$.
(14) £1000 4 per cent. stock at $97\frac{3}{8}$ per cent., brokerage $\frac{1}{8}$ per cent.
(15) £2153. 10*s*. bank stock at $188\frac{1}{8}$ per cent., brokerage $\frac{1}{8}$ per cent.

Find the yearly income arising from the investment of

(16) \$1008 in the 6 per cents. at 84.
(17) \$5580 in the 8 per cents. at 93.
(18) \$1638 in the 7 per cents. at $93\frac{3}{5}$.
(19) \$2000 in the 6 per cents. at $88\frac{1}{2}$.
(20) £3425. 15*s*. 2*d*. in the 3 per cents. at $91\frac{3}{4}$.
(21) £4788 in the $3\frac{1}{2}$ per cents. at 105.
(22) £3500 in the 3 per cent. consols at $94\frac{1}{8}$, brokerage $\frac{1}{8}$ per cent.

What sums of money must be invested in the undermentioned stocks in order to produce the following incomes?

(23) \$120 in the 6 per cents. at 85.
(24) \$288 in the 6 per cents. at 67.
(25) \$170 in the 7 per cents. at 90.
(26) £87 in the 3 per cents. at $74\frac{1}{2}$, brokerage $\frac{1}{8}$ per cent.
(27) £37. 10*s*. in the 4 per cents. at $93\frac{1}{4}$, brokerage $\frac{1}{8}$ per cent.

At what rate per cent. will a person receive interest who invests his capital?

(28) In the 6 per cents. at 91.
(29) In the 7 per cents. at 94.
(30) In the 8 per cents. at $96\frac{1}{2}$, brokerage $\frac{1}{4}$ per cent.
(31) In the 7 per cents. at $102\frac{7}{8}$, brokerage $\frac{1}{8}$ per cent.
(32) If \$7927.50 be laid out in purchasing Canadian Bank of Com-

STOCKS. 209

merce Stock at 105, yielding annual dividends of 8 per cent. per annum; what yearly income will be derived from this investment after deducting an income tax of 1½ cents in the dollar?

(33) A person invested money in Royal Canadian Bank Stock at 90, and some more at 80; find the rate of interest he obtained in each case, and the advantage per cent. of the second purchase over the first. The bank's yearly dividends being 7 per cent.

(34) If a person receives 8 per cent. on his capital by investing in Bank Stock yielding 7 per cent. per annum, what is the price of the stock, and how much stock can be purchased for $1200?

(35) How much money must a broker invest in Bank of Montreal Stock at 157, so as to procure the same income as if he had invested $5500 when the stock was at 163?

(36) A person buys $4000 Royal Canadian Bank Stock at 55, and sells out at 63; what does he gain by the transaction?

(37) A person invests $9000 in Bank Stock at 163, which pays yearly dividends of 12 per cent., and sells out when it has sunk to 157; how much does he lose by the transaction?

(38) When £100 stock may be bought in the 3 per cents. for £89¼, at what rate may the same quantity of stock be bought in the 3½ per cents, with equal advantage?

(39) A person invests his share of a legacy of $1200, which is a third in Toronto debentures at 92, paying 7 per cent. per annum interest; find his half-yearly dividends.

(40) A person transfers $5000 from the Bank of Montreal stock at 160, to Bank of Commerce stock at 107; find alteration in his income, the half-yearly dividends of the said stocks being 6 and 4 per cent., respectively.

(41) What incomes will $5000 stock paying half-yearly 4 per cent., and $5000 invested in 6 per cents. Dominion Stock at 106, respectively produce?

(42) Find income produced by $7000 Merchants' Bank stock, paying half-yearly dividends of 4 per cent., and its value when the stock is at 105.

(43) A person transfers £3000 stock from the 3 per cent. consols at 89¼, to the reduced 3½ per cents. at 98¼; find what quantity of the latter he will hold, and the alteration in his income.

(44) The stocks of the Canadian Bank of Commerce and the Quebec

14

Bank, are at 101 and 106 respectively; the former pays 4 per cent., the latter $3\frac{1}{2}$ per cent., half-yearly. Which is the better investment?

(45) The Dominion of Canada 6 per cents., which mature in 1878 being at 104; how much money must be invested in them to produce an annual income of $600, after deducting an income tax of 2 cents in the dollar?

(46) A person invests £1037. 10s. in the 3 per cents. at 83, and when the funds have risen 1 per cent. he transfers his capital to the 4 per cents. at 96; find the alteration in his income.

(47) How much in the 3 per cents. at 96 must be sold out to pay a bill of £1654, 9 months before it becomes due, real discount being allowed at $4\frac{1}{2}$ per cent. per annum?

(48) The dividends on a certain amount of 3 per cent. stock accumulated in 13 years to £3081. How much stock was there, and what will it be worth if the stock be sold at $79\frac{7}{8}$?

(49) If I lay out £1911 in the purchase of 3 per cent. consols, when they are at $79\frac{1}{2}$, at what price should I sell out my stock again in order to realise on the whole a gain of £150, after having paid $\frac{1}{8}$ per cent. for commission on each transaction?

(50) A person had £10,000 in the 3 per cent. South Sea Annuities, and the Government offered to give £110 bearing interest at the rate of $2\frac{1}{2}$ per cent. for every £100 of these annuities, or to pay the £10,000 in cash on a certain day. The latter proposal was preferred, and on the money being paid it was re-invested in consols at 93. How much would he have lost in income had he accepted the first proposal, and what will he now gain by the new investment?

(51) What sum would be saved annually if the interest on a public debt of £4,000,000 were reduced from $3\frac{1}{2}$ per cent. to 3 per cent.? If in consequence the price of this stock fell from £101 to £95$\frac{3}{4}$, how much would the whole property of the fundholders be diminished?

APPLICATIONS OF THE TERM "PER CENT."

171. There are many other cases in which the term PER CENT. occurs besides those already mentioned; we will mention certain cases and give examples in each by way of illustration.

COMMISSION is the sum of money which a merchant charges for buying or selling goods for another.

APPLICATIONS OF THE TERM PER CENT.

BROKERAGE is of the same nature as Commission, but has relation to money transactions, rather than dealings in goods or merchandise.

INSURANCE is a contract, by which one party, on being paid **a certain** sum or *Premium* by another party on property, which is subject to risk, **undertakes, in case** of loss, **to** make good to **the owner the** value of that property. **The document which expresses the contract is** called *the Policy of* **Insurance.**

LIFE ASSURANCE is a contract for **the payment of a certain sum of** money on the death of a person, in consideration **of an** annual premium to be continued during the life **of** *the Assured,* **or for a** certain number of years.

Questions on Commission, **Brokerage, and Insurance, these charges** being usually **made at so much per** cent., amount **to** the same thing as finding the interest of **a given sum** of money **at** a given rate **for one** year, and may therefore **be worked by** the Rule for Simple **Interest or** by the Rule **of** Three.

Ex. 1. A Commission merchant sold 30270 bushels **of wheat at $1.15 per bushel;** the Commission being 2 per **cent.: how much** will he receive?

  Amount obtained from the sale of the wheat **is $34810.50;**

  Therefore, $100 : $34810.50 :: $2 : Commission required;

  ∴ commission required $= \$\dfrac{34810.50 \times 2}{100} = \$696.21.$

Independent method.

  Commission on   **$100** is $2;

∴ .... $1 ... $\$\dfrac{2}{100},$

∴ ....$34810.50 ... $\$\dfrac{2}{100} \times 34810.50,$

    or $696.21.

Ex. 2. What is **the** brokerage on the purchase of $7250 6 **per** cent. Toronto debentures at ½ per cent.?

  $100 : $7250 :: $½ : brokerage required;

  ∴ brokerage required $= \$\dfrac{7250 \times \frac{1}{2}}{100} = \$36.25.$

Ex. 3. What is the premium on a policy of insurance for a house and barn valued at $2700 at $\frac{3}{4}$ per cent.?

$$\$100 : \$2700 :: \$\tfrac{3}{4} : \text{premium required};$$
$$\therefore \text{premium required} = \$27 \times \tfrac{3}{4}$$
$$= \$20.25.$$

Ex. 4. In standard gold 11 parts out of 12 parts are pure gold; how much per cent. is dross?

In every 12 parts 1 part is dross;
$$\therefore 12 : 100 :: 1 : \text{percentage of dross};$$
$$\therefore \text{percentage of dross} = \frac{100 \times 1}{12} = 8\tfrac{1}{3}.$$

Ex. 5. Archimedes discovered that the crown made for King Hiero consisted of gold and silver in the ratio of 2 : 1. How much per cent. was gold, and how much per cent. silver?

Out of every 3 parts, 2 were gold and 1 silver;
$$\therefore 3 : 100 :: 2 : \text{percentage of gold};$$
$$\therefore \text{percentage of gold} = \frac{100 \times 2}{3} = 66\tfrac{2}{3};$$
and percentage of silver $= 33\tfrac{1}{3}.$

*172. All questions which relate to gain and loss in mercantile transactions fall under the head of *PROFIT AND LOSS*.

Tradesmen measure their Profit or Loss by the actual amount gained or lost, or by the amount gained or lost on every $100 of the capital they invest.

Ex. 6. If a cask of wine containing 84 gallons cost $210, what is gained by selling it at $3.50 per gallon?

The gain = selling price *less* first cost;
the selling price $= \$(3\tfrac{1}{2} \times 84) = \$294$;
therefore the gain $= \$294 - \$210 = \$84.$

Ex. 7. A ream of paper cost me $5.20, what must I sell it at, so as to realize 20 per cent.?

The reasoning in this case is, if \$100 gain \$20, or produce \$120, what will \$5.20 produce?

∴ \$100 : \$5.20 :: \$120 : required amount in dollars,
whence, required amount = \$6.24.

Ex. 8. If I buy hay at £4. 16*s*. a ton, what must I sell it at to lose 15 per cent.?

In this case every £100 would realize £(100 − 15), or £85;

∴ £100 : £4. 16*s*. :: £85 : required amount in pounds,
whence, required amount = £4. 1*s*. $7\frac{1}{5}$*d*.

Ex. 9. A person buys shares in a railway when they are at £$19\frac{1}{2}$, £15 having been paid, and sells them at £32. 9*s*. when £25 has been paid: how much per cent. does he gain?

He buys each share at £$19\frac{1}{2}$, and he afterwards pays upon it £(25 − 15), or £10, therefore at the time he sells, he has paid on each share £29. 10*s*.; therefore by selling at £32. 9*s*. he gains on each £29. 10*s*. which he has paid (£32. 9*s*. − £29. 10*s*.) = £2. 19*s*.;

∴ £$29\frac{1}{2}$: £100 :: £$2\frac{19}{20}$: gain per cent. in pounds,
whence, gain per cent. = £10, or gain is 10 per cent.

Ex. 10. What was the prime cost of an article, which when sold for \$2.88 realized a profit of 20 per cent.?

Here what cost \$100 would be sold for \$120;

∴ \$120 : \$2.88 :: \$100 : prime cost in dollars,
whence prime cost = \$2.40.

If the above example had been, "What was the prime cost of an article which when sold for \$2.88, entails a loss of 20 per cent?"
then \$80 : \$2.88 :: \$100 : prime cost in dollars,
whence, prime cost = \$3.60.

The following method of working such as the above examples may be adopted:

Since 20 is the $\frac{1}{5}$ of 100,
therefore, $1 + \frac{1}{5}$, or $\frac{6}{5}$ = selling price,
" $\frac{5}{6}$ of selling price = prime cost,
" $\frac{5}{6}$ of \$2.88 = \$2.40,
\$2.40 is therefore cost price.

Again, since 20 is $\frac{1}{5}$ of 100;

therefore $1 - \frac{1}{5}$, or $\frac{4}{5}$ = selling price,
" $\frac{5}{4}$ selling price = prime cost,
or $\frac{5}{4}$ of $2.88 = $3.60.
$3.60 is therefore cost price.

Ex. 11. If by selling a horse for £40 I lose 20 per cent., what must I have sold him for so as to gain 10 per cent. ?

Here what would cost me £100 must be sold in one case for £80, and in the other for £110 ; and therefore we get this statement ; selling price of £100 in 1st case : selling price of horse in 1st case :: selling price of £100 in 2nd case : selling price of horse in 2nd case ;

or £80 : £40 :: £110 : selling price in pounds ;
whence, selling price = £55.

Ex. 12. A grocer buys 3 cwt. of sugar at 6d. a lb., 2 cwt. of sugar at $10\frac{1}{2}$d. a lb., and $2\frac{1}{2}$ qrs. of sugar at 1s. a lb. ; and mixes them : he sells 4 cwt. of the mixture at 9d. a lb. What must he sell the remainder at, in order to gain 25 per cent. on his outlay ?

	£	s.	d.
3 cwt., or 336 lbs., at 6d. a lb., cost	8	8	0
2 cwt., or 224 lbs., at $10\frac{1}{2}$d. a lb., cost	9	16	0
$2\frac{1}{2}$ qrs., or 70 lbs., at 1s. a lb., cost	3	10	0
∴ 630 lbs. cost	21	14	0

In order to gain 25 per cent. on £21. 14s., it must realize £27. 2s. 6d.;

	£	s.	d.
∴ he must sell 630 lbs. for	27	2	6
but he sells 448 lbs. for	16	16	0
∴ by Subtn he must sell 182 lbs. for	10	6	6

∴ he must sell 1 lb. for $\frac{£10. 6s. 6d.}{182}$, or $13\frac{8}{12}$d.

173. Tables respecting the increase or decrease of Population, &c., are constructed with reference to the increase or decrease on every 100 of such population; Education returns are constructed in the same way; and so are other *Statistical Tables*.

Ex. 13. In a school section of 150 children, 125 learn to write. What is the percentage ?

APPLICATIONS OF THE TERM PER CENT. 215

In other words, what number bears the same ratio to 100, which 125 bears to 150 ?

$$\therefore 150 : 100 :: 125 : \text{percentage};$$
$$\therefore \text{percentage} = \frac{12500}{150} = 83\tfrac{1}{3}$$

Ex. 14. Between the years 1851 and 1861 the population of the city of Toronto increased about 78 per cent., and in the latter year it was 44821. What was it in 1851 ?

For every 178 persons in 1861 there were 100 persons in 1851 ;

$$\therefore 178 : 44821 :: 100 : \text{number required};$$
$$\therefore \text{number required} = \frac{44821 \times 100}{178} = 25180 \text{ nearly.}$$

Ex. 15. In 1842 the number of the members of the University of Cambridge was 5853, and in 1852 the number was 6397 ; find the increase per cent.

Subtracting 5853 from 6397 we obtain 544, the increase on 5853 members; the question then is this; if 5853 members give an increase of 544, what increase do 100 members give ?

$$\therefore 5853 : 100 :: 544 : \text{increase per cent.};$$
$$\therefore \text{increase per cent.} = \frac{54400}{5853} = 9\frac{1723}{5853}$$

Ex. 16. The number of male and female criminals are 1235 and 988 respectively; while the decrease in the former is 4·6 per cent., the increase in the latter is 9·8 per cent.; find the increase or decrease per cent. in the whole number of criminals.

1st. $100 : 1235 :: 4·6 : $ whole decrease of male criminals ;
$$\therefore \text{whole decrease of male criminals} = \frac{1235 \times 4·6}{100} = 56·81.$$

2nd. $100 : 988 :: 9·8 : $ whole increase of female criminals ;
$$\therefore \text{whole increase of female criminals} = \frac{988 \times 9·8}{100} = 96·824;$$

$\therefore$ in $(1235 + 988)$ or 2223 persons there is an increase of $(96·824 - 56·81)$ or 40·014 persons.

$$\therefore 2223 : 100 :: 40·014 : \text{percentage required};$$
$$\therefore \text{percentage required} = \frac{4001·4}{2223} = 1·8.$$

Ex. LVI.

(1) What is the percentage on 56394 at $\frac{1}{4}$; $\frac{3}{8}$; 4; $7\frac{2}{3}$; 10; $150\frac{1}{2}$?

(2) How much per cent. is 15 of 96; 19 of 81; 23 of 256; $185\frac{1}{4}$ of 7321·75; 5·3 of 11080·5?

(3) Write in a decimal form $\frac{1}{2}$; $2\frac{3}{4}$; $4\frac{1}{3}$; $5\frac{5}{8}$; $26\frac{1}{4}$; 230·05; 500·0138 per cent.

(4) Bought 200 cords of wood at $4.25 per cord, and sold it again for $6 per cord. What was the gain upon the whole?

(5) If 5 cwt., 3 qrs., 14 lbs. be bought for £9. 8s. and sold for £11. 18s. 11d., what is the rate of gain per cwt.? (cwt.= 112lbs.).

(6) Find the total value of 43 articles at $4.40 each, 57 at $9.20 each, and 4 at $12.50 each. What is gained or lost by selling them at the rate of 3 for $29.76?

(7) A cask, which contained 2005 gallons, leaked 27 per cent., how much remained in the cask?

(8) A maltster malts 7500 bushels of barley, which in the process increases $12\frac{1}{2}$ per cent., how many bushels of malt has he?

(9) A grocer uses for a 1lb. weight one which only weighs 15·75 oz., what does he gain per cent. by his dishonesty?

(10) A person buys 400 yards of silk at $384, and sells 300 yards at $1.32 a yard, and the rest, which is damaged, at 48 cts. a yard; find how much per cent. he gains or loses.

(11) A grocer buys 2 cwt. of sugar at 12 cts. per pound, and 4 cwt. at 9 cts.; he sells 3 cwt. at 11 cts. per pound; at what rate per pound will he be able to sell the remainder so as neither to gain nor lose by the bargain?

(12) If a commodity be bought for $16.42 a cwt. and sold for 16 cts. a lb., find the rate of profit per cent.

(13) Bought goods at 13 cts. per pound, and sold them at $21.60 per cwt.: what is the gain or loss per cent.?

(14) Out of 14804 cases of Small-Pox 1588 persons died, and out of 2422 cases of Scarlet Fever 211 persons died: find the rate per cent. of mortality in each case, also the rate per cent. of mortality in the whole number of sick people.

(15) The population of Ireland was 7767401 in 1831, 8175124 in 1841, 6515794 in 1851. Find the increase per cent. in the first ten

APPLICATIONS OF THE TERM PER CENT. 217

years, the **decrease** per cent. in the second ten years, and the decrease per cent. in the 20 years from 1831 to 1851.

(16) The population of a city **is a** million; it rises $1\frac{1}{2}$ per cent. for **3 years** successively; find the population at the end of 3 years.

(17) A school contains 383 **scholars**, 3 **are** of the **age of 18 years; 5 per cent.** of the remainder **are between the ages of 15 years and 18 years; 10 per cent.** between **12 and 15; 35 per cent.** between 10 and 12, **and the** remainder **under that age**; find the number **of each** class.

(18) An article which cost 84 cents is sold **for 93 cents; find the** gain per cent.

(19) If a tradesman gain **$1.32 on an article which he sells for** $5.28, what is his gain **per cent. ?**

(20) A man **sells a horse for $135**, and loses $20 per **cent. on** what the horse cost him; **what was** the original cost?

(21) Sugar being composed **of 49·856 per cent.** of **oxygen**, 48·265 **per cent.** of carbon, and the **remainder hydrogen; find how many** pounds of each **of** these materials there **are in one** ton of sugar.

(22) In 1853 **the number of the graduates** of the University **of Toronto was** 70, **in 1861 the number was 213**, in 1869 the number was 541, find the increase per **cent. for each of those** periods and **for** the 16 years from 1853 to 1869.

(23) A merchant **buys 13600** bushels **of wheat at $1.05** a bushel. $2\frac{1}{2}$ per cent. of it is **wasted; he** sells **56 per cent.** of the remainder at $1 a bushel, 20 per cent. **at $1.05 a bushel, and** the rest at $1.25 a bushel; what does he **gain or lose** by the transaction?

(24) If the increase in **the number of** male and female criminals be 1·8 per **cent.**, while the **decrease in the** number of males alone is 4·6 per cent., **and the increase in the number** of females is 9·8. Compare the number of male and female criminals respectively.

(25) By selling an article for **5s. a person loses 5** per cent.; what was the prime cost, and what must **he sell it** at to gain $4\frac{1}{2}$ per cent.?

(26) The cost price of a book is **$1.60**; the expense of sale 5 per cent. upon the cost price; and the profit 25 per cent. upon the whole outlay; **find** the selling price of the book.

(27) If, by selling an article for £25. 10s., **8 per cent. be lost, what per cent. is gained or lost if it be sold at £38 ?**

218 ARITHMETIC.

(28) I bought 500 sheep at $6 a-head; their food cost me $1.25 a-head: I then sold them at $10 a-head. Find my whole gain, and also my gain per cent.

(29) A person having bought goods for £40 sells half of them at a gain of 5 per cent.; for how much must he sell the remainder so as to gain 20 per cent. on the whole?

(30) A vintner buys a cask of wine containing 36 gallons at $2.40 per gallon; he keeps it for four years, and then finds that he has lost 6 gallons by leakage; at what price per gallon must he sell the remainder in order that he may realize 20 per cent. upon his outlay?

(31) A person rents a piece of land for £120 a year. He lays out £625 in buying 50 bullocks. At the end of the year he sells them, having expended £12. 10s. in labour. How much per head must he gain by them in order to realize his rent and expenses, and 10 per cent. upon his original outlay?

(32) A grocer mixes two kinds of tea which cost him 38 cents and 44 cents per lb. respectively; what must be the selling price of the mixture in order that he may gain 15 per cent. on his outlay?

(33) A stationer sold quills at 11s. a thousand, by which he cleared ⅔ of the money; he raises the price to 13s. 6d. What does he clear per cent. by the latter price?

(34) A smuggler buys 6 cwt. of tobacco at 1s. 3d. per lb.; he meets with a revenue officer, who seizes ½ of it: at what rate per lb. must he sell the remainder, so as, 1st, neither to gain nor lose; 2nd, to gain 5 guineas; and 3rd, to gain cent. per cent.?

(35) A farm is let for £96 and the value of a certain number of quarters of wheat. When wheat is 38s. a quarter, the whole rent is 15 per cent. lower than when it is 56s. a quarter. Find the number of quarters of wheat which are paid as part of the rent.

(36) A person bought an American watch, bearing a duty of 25 per cent., and sold it at a loss of 5 per cent.; had he sold it for $3 more, he would have cleared 1 per cent. on his bargain. What had the first party for it?

174. Questions are often given, in which the term "Average" occurs; a few examples of such a kind will now be worked by way of illustration, and others subjoined for practice.

Ex. 1. A gentleman in each of the following years expended the

APPLICATIONS OF THE TERM **PER CENT.** 219

following sums: **in** 1858 $500, in 1859 $600, in 1860 $600, in 1861 $600 in 1862 $700, in 1863 $700, in 1864 $700. Find his **yearly average** expenditure.

The object is to find that fixed sum which he might have spent in each of the seven years, so that his total expenditure in that case might be the same as his total expenditure was in the above question.

Adding the various **sums** together we **obtain** the total expenditure which equals $4400; **this sum** divided **by 7 gives** $626.59¼ as the average yearly expenditure.

Ex. 2. In a school **of 27 boys, 1 of** the boys is **of the age of 17 years,** 2 others **of** 16, 4 others of 15½, 1 of 14¾, 2 of 14½, **5 of** 13¾, 10 of 12¼, and 2 of 10 ; find the average age **of** the boys.

The object is **to** find, what must be the age of each boy supposing all to be of the same age, that the **sum of** their ages may = the **sum of** the ages in the question.

sum of **ages in** question = $17 + 32 + 62 + 14\frac{3}{4} + 29 + 68\frac{3}{4} + 122\frac{1}{2} + 20 = 366$;

$$\therefore \text{average age} = \frac{366}{27} = 13\frac{5}{9} \text{ years.}$$

Ex. 3. In a class of 25 children, 19 have attended during the week. Days attended by children: 5 for 5 days, 6 for 4½, 3 for 4, 2 for 3½, **1** for 3, 1 for 2, 1 for ½ day. Find the average number of **days attended** by each child.

The whole number of days attended **by class**

$= (5 \times 5 + 6 \times 4\frac{1}{2} + 3 \times 4 + 2 \times 3\frac{1}{2} + 1 \times 3 + 1 \times 2 + 1 \times \frac{1}{2})$

$= 25 + 27 + 12 + 7 + 3 + 2 + \frac{1}{2} = 76\frac{1}{2}$ **days** ;

$\therefore \text{average attendance} = \frac{76\frac{1}{2}}{25} = \frac{153}{50} = \frac{306}{100} = 3.06 \text{ days.}$

Ex. 4. In a school **the** numbers **for** the week were:—Monday morning 67, Tuesday morn. 60, Wednesday morn. 65, Thursday morn. 68, Friday morn. 62, Monday afternoon 5 more than the average of Monday and Tuesday mornings, Tuesday aft. 59, Wednesday aft. 5 less than the average of Tuesday, Thursday the average of Monday morn. and Tuesday aft., Friday aft. 60. Find the average attendance for the week.

ARITHMETIC.

Number of children who attended on

$$\begin{aligned}
\text{Monday} &= 67 + 64\,; \\
\text{Tuesday} &= 60 + 59\,; \\
\text{Wednesday} &= 65 + 59\,; \\
\text{Thursday} &= 68 + 63\,; \\
\text{Friday} &= 62 + 60\,;
\end{aligned}$$

∴ the total number of children who attended on the 10 occasions =627;

$$\therefore \text{ average attendance} = \frac{627}{10} = 62{\cdot}7$$

Ex. 5. A farm of 500 acres is let at a corn-rent equally apportioned between wheat and barley; it is valued at £930 a year when the average price of wheat is 6s. a bushel, and that of barley 4s. a bushel; find the rent when wheat rises to the average price of 7s. 6d. per bushel, and barley to that of 5s. 3d. per bushel.

First we must find the number of bushels of wheat and barley at the given rent of £930.

$$\frac{£930}{2} = £465 \text{ the sum to be raised by each kind of grain;}$$

$$\therefore \frac{465 \times 20}{6} = 155 \times 10 = 1550 \text{ bushels of wheat;}$$

$$\therefore \frac{465 \times 20}{4} = 465 \times 5 = 2325 \text{ bushels of barley;}$$

∴ rent in latter case = $(1550 \times 7\frac{1}{2} + 2325 \times 5\frac{1}{4})s.$

= £1191. 11s. 3d.

Ex. LVII.

1. During the year 1865, the highest salary paid to a teacher in a county was $630; the lowest. $84: the highest salary paid in a city was $1350; the lowest, $200: the highest in a town, $1000; the lowest, $140: the highest in a village, $600; the lowest, $270. Find (1) the average of the highest salaries, (2) the average of the lowest, (3) the average salary of a teacher for the year 1865, in Ontario.

2. The number of quarters of grain imported into a country in 11 successive years were 2679438, 2958272, 3080293, 3474802, 2243151.

2327782, 2855525, 2538234, 3206482, 2801204, 3251901; find the average importation during that period.

3. If 50 quarters of wheat are sold for $8.40 per quarter and 100 quarters for $8.80 per quarter; what is the average price per bushel?

4. In a class of 23 children, 8 are boys, 15 girls. The age of the boys—4 of 8, 2 of 11, 2 of 12. Of the girls—5 the average age of the boys, 4 of 9, 2 of 10, 4 of 13. Find the average age of (1) the boys, (2) the girls, (3) the whole class.

5. There are 25 children on the register of one class in a school. 19 have been present at one time or other during the week. The sum of days on which the children have attended is 84¼. What is the average number of days per week attended by each child ever present during the week, there being no school on Saturday or Sunday? Give the answer in decimals.

6. In a school of 7 classes, the average number of days attended by each child in Class I. is 4·5; Class II., 4; Class III., 3·9; Class IV., 4·1; Class V., 3·6; Class VI., 4·2; Class VII., 3·3. Find the average number of days attended by each child in the school.

7. A Farm is valued at the yearly rental of $1812; one-third of the rent is payable in money, one-fourth in wheat, and the rest in barley, the average prices being as follows: wheat $1.51 a bushel, and barley 75½ cents a bushel. What will the rent amount to when the average prices of wheat and barley are $1.75 and 85 cents per bushel respectively?

8. A tithe-rent of £310 per annum is commuted in equal parts into a corn-rent consisting of wheat at 56s. per qr., barley at 32s. per qr., and oats at 22s. per qr.; find its value when wheat is at 64s. per qr., barley at 44s. per qr., and oats at 24s. per qr.

DIVISION INTO PROPORTIONAL PARTS.

175. **To** divide a given number into parts **which** shall be proportional to certain **other** given numbers.

Questions of this kind may be solved **by the** method employed in Art (156), or by the following.

RULE. As the sum of the given parts : any one of them :: the entire quantity to be divided : the corresponding part **of it**.

This statement must be repeated for each of the parts, or at all events for all but the last part, which of course may either be found by the Rule, or by subtracting the sum of the values of the other parts from the entire quantity to be divided.

Ex. Divide $128 among A, B, and C, so that their portions may be as 7, 11, and 14 respectively.

Proceeding according to the Rule given above,

$$32 : 7 :: \$128 : A\text{'s share};$$
$$32 : 11 :: \$127 : B\text{'s share};$$

whence A's share $= \$28$, and B's share $= \$44$.

C's share may be found from the proportion

$$32 : 14 :: \$128 : C\text{'s share};$$

whence C's share $= \$56$;

or by subtracting $28 + $44, or $72 from $128 which leaves $56, as above.

The reason for the above process is clear from the consideration, that $128 is to be divided into 32 equal parts, of which A is to have 7 parts, B 11, and C 14.

Independent method.

$128 is to be divided into 32 equal parts;

therefore $\frac{1}{32}$ of $128 = \$4$,

" $\frac{7}{32}$ of $128 = \$28$,

" $\frac{11}{32}$ of $128 = \$44$,

" $\frac{14}{32}$ of $128 = \$56$.

Ex. 2. Divide £11000 among 4 persons, A, B, C, D, in the proportions of $\frac{1}{2}$, $\frac{1}{3}$, $\frac{1}{4}$, and $\frac{1}{5}$.

Sum of shares $= \frac{77}{60}$;

$\therefore \frac{77}{60} : \frac{1}{2} :: £11000 : A$'s share in pounds;

whence A's share $= £4285.\ 14s.\ 3\frac{3}{7}d.$

FELLOWSHIP AND PARTNERSHIP.

Similarly,
B's share $= £2857.\ 2s.\ 10\frac{2}{7}d.$, C's share $= £2142.\ 17s.\ 1\frac{4}{7}d.$
D's share $= £1714.\ 5s.\ 8\frac{4}{7}d.$

Ex. 3. Divide $45000 among $A, B, C,$ and D, so that A's share : B's share :: $1:2$, B's : C's :: $3:4$, and C's : D's :: $4:5$.

In this case,
B's share $= 2\ A$'s share, $3\ C$'s share $= 4\ B$'s share,
$4\ D$'s share $= 5\ C$'s share;
$\therefore$ we have C's share $= \frac{4}{3} B$'s share $= \frac{8}{3} A$'s share,
and D's share $= \frac{5}{4} C$'s share $= \frac{10}{3} A$'s share;
$\therefore A$'s share $+ B$'s share $+ C$'s share $+ D$'s share
$= A$'s share $(1 + 2 + \frac{8}{3} + \frac{10}{3})$,
$= 9\ A$'s share;
$\therefore A$'s share $= \$5000$, B's $= \$10000$, C's $= \$13333.33\frac{1}{3}$,
D's $= \$16666.66\frac{2}{3}$.

FELLOWSHIP OR PARTNERSHIP.

176. FELLOWSHIP or PARTNERSHIP is a method by which the respective gains or losses of partners in any mercantile transactions are determined.

Fellowship is divided into SIMPLE and COMPOUND FELLOWSHIP: in the former, the sums of money put in by the several partners continue in the business for the same time; in the latter, for different periods of time.

SIMPLE FELLOWSHIP.

177. Examples in this Rule are merely particular applications of the Rule in Art. (175), and that Rule therefore applies.

Ex. 1. Two merchants, A and B, form a joint capital; A puts in $1200, and B $1800: they gain $400. How ought the gain to be divided between them?

$\$(1200 + 1800) : \$1200 :: \$400 : A$'s share in dollars,
whence, A's share $= \$160$ and $\therefore B$'s share $= \$240$.

NOTE. *The estate of a Bankrupt* may be divided among his creditors by the same method.

224   ARITHMETIC.

Ex. 2. A bankrupt owes three creditors, A, B, and C, £175, £210, and £265, respectively; his property is worth £422. 10s.: what ought they each to receive?

$$£650 : £175 :: £422\tfrac{1}{2} : A\text{'s share},$$
$$£650 : £210 :: £422\tfrac{1}{2} : B\text{'s share},$$

whence A's share = £113. 15s., B's share = £136.10s.;
C's share = £172. 5s.

COMPOUND FELLOWSHIP.

178. Rule. Reduce all the times into the same denomination, and multiply each man's stock by the time of its continuance, and then state thus:

As the sum of all the products : each particular product :: the whole quantity to be divided : the corresponding share.

Ex. 1. A and B enter into partnership; A contributes $15000 for 9 months, and B $12000 for 6 months, they gain $5750: find each man's share of the gain.

Proceeding by the Rule given above,

$\$(15000 \times 9 + 12000 \times 6) : \$15000 \times 9 :: \$5750 : A$'s share of gain,
or $207000 : $135000 :: $5750 : A's share of gain,
and $207000 : $72000 :: $5750 : B's share of gain;
whence, A's share = $3750, and B's share = $2000.

The reason for the above process is evident from the consideration, that a stock of $15000 for 9 months would be equivalent to a stock of 9 times $15000 for 1 month; and one of $12000 for 6 months, to one of 6 times $12000 for 1 month: hence, the increased stocks being considered, the question then becomes one of Simple Fellowship.

Ex. 2. There were at a feast 20 men, 30 women, and 15 servants; for every 10s. that a man paid, a woman paid 6s., and a servant 2s., the bill amounted to £41: how much did each man, woman, and servant pay?

20 men at 10s. each = 200 at 1s., 30 women at 6s. = 180 at 1s., and 15 servants at 2s. = 30 at 1s.; and 200 + 180 + 30 = 410.

Hence we have

  410 : 200 :: £41 : 20 **men's** share (in pounds);
  410 : 180 :: £41 : 30 women's share (in pounds);
  410 : 30 :: £41 : 15 **servants'** share (in pounds);
∴ 20 **men's** shares = £20, 30 **women's** shares = £18,
   and 15 servants' **shares = £3**;
∴ each man paid £1, each woman 12s., **and each servant 4s.**

EQUATION OF PAYMENTS.

179. When a person owes another several sums of money, due at different times, the Rule by which we determine the **just time when** the whole **debt may be** discharged **at one payment,** is **called the** EQUATION OF PAYMENTS.

NOTE. **It is assumed in this Rule that the sum of the interests of** the several debts for their respective times equals the interest **of the** sum of the debts for the equated **time.**

RULE. Multiply **each** debt into the time **which will** elapse before it becomes due, and **then** divide the sum of the products by the sum of the debts; the quotient **will be the equated time** required.

Ex. 1. *A* **owes** *B* $500, **whereof $200 is to be paid in 3 months,** and $300 in 5 months: **find the equated time.**

Proceeding according to the Rule given **above,**
then (200 × 3 + 300 × 5) = (200 + 300) × equated time in months,
   whence, equated time = 4½ months.

The reason for the above process, in accordance with our assumption, is clear from the consideration that the sum of the interests of $200 for 3 months, and $300 **for** 5 months, is the same as the interest of $(600 + 1500), or **$2100 for** 1 month; if therefore *A* has to pay $500 in one sum, the question is, how long ought he to hold it so that the interest on it may be the same as the interest on $2100 for 1 month. The statement therefore will be this:

  $500 : $2100 :: **1** month : required number of months :
   whence, required number of months = 4½ months;

which is evidently the equated time of payment, and agrees with the result obtained by the Rule given above.

Ex. 2. A owed B £1000, to be paid at the end of 9 months: he pays however £200 at the end of 3 months, and £300 at the end of 8 months: when was the remainder due?

In this case,

$(200 \times 3 + 300 \times 8 + 500 \times$ number of months required$) = 1000 \times 9$,

or $500 \times$ number of months required $= 6000$;

whence number of months required $= 12$.

Ex. LVIII.

1. A company of militia consisting of 72 men is to be raised from 3 towns, which contain respectively 1500, 7000, and 9500 men. How many must each town provide?

2. Divide $84.42 into two parts which shall be to each other as 5 : 16.

3. Divide 4472 into parts which shall be to each other in the ratio of 3, 5, 7, 11; and also $2400 into parts which shall be in the ratio of $\frac{1}{2}, \frac{3}{8}$, and $\frac{4}{7}$.

4. A bankrupt owes A £256. 6s. 8d., B £203. 10s., and C £141. 13s. 4d.; his estate is worth £421. 1s.; how much will A, B, and C receive respectively?

5. A mass of counterfeit metal is composed of fine gold 15 parts, silver 4 parts, and copper 3 parts: find how much of each is required in making 18 cwt. of the composition. (cwt. = 112. lbs.)

6. Two persons have gained in trade $3456; the one puts in $10560 and the other $8640; what is each person's share of the profits?

7. In a certain substance there are 11 parts tin to 100 of copper. Find the weight of tin in a piece weighing 24 cwt.?

8. A man leaves his property amounting to £13,000 to be divided amongst his children, consisting of 4 sons and 3 daughters; the three younger sons are each to have twice the share of each of the daughters, and the eldest son as much as a younger son and a daughter together; find the share of each.

9. Two persons, A and B, are partners in a mercantile concern and contribute $5760 and $9600 capital respectively; A is to have

EXCHANGE. 227

10 per cent. of the profits for managing the business, and the remaining profits to be divided in proportion to the capital contributed by each; the entire profits at the year's end is $3840; how much of it must each receive?

10. Divide $480 among A, B, C, and D, so that B may receive as much as A; C as much as A and B together; and D as much as A, B, and C together.

11. Divide £11,875 among A, B, and C, so that as often as A gets £4, B shall get £3, and as often as B gets £6, C shall get £5.

12. A commences business with a capital of $1000, two years afterwards he takes B into partnership with a capital of $15,000, and in 3 years more they divide a profit of $1500; required B's share.

13. $700 is due in 3 months, $800 in 5 months, and $500 in 10 months; find the equated time of payment.

14. Find the equated time of payment of £750, one half of which is due in 4 months, $\frac{3}{8}$ in 5 months, and the rest in 6 months.

15. A owes B a debt payable in $7\frac{1}{30}$ months, but he pays $\frac{1}{3}$ in 4 months, $\frac{1}{4}$ in 6 months, $\frac{1}{5}$ in 8 months; when ought the remainder to be paid?

16. A, B, and C rent a field for £11. 6s.; A puts in 70 cattle for 6 months; B 40 for 9 months; and C 50 for 7 months; what ought C to pay?

17. A, B, and C invest capital to the amount of $7000, $5000, and $3000 respectively; A was to have 25 per cent. of the profits, which amount to $4500; what share of the profits ought C to have?

18. A and B enter into a speculation; A puts in £50 and B puts in £45; at the end of 4 months A withdraws $\frac{1}{2}$ his capital, and at the end of 6 months B withdraws $\frac{1}{3}$ of his; C then enters with a capital of £70; at the end of 12 months their profits are £254; how ought this to be divided amongst them?

EXCHANGE.

180. EXCHANGE is the Rule by which we find how much money of one country is equivalent to a given sum of another country, according to a given *course* of Exchange.

181. By the Course of Exchange is meant the *variable* sum of the money of any place which is given in exchange for a *fixed* sum of money of another place. The Course of Exchange between any two countries will be affected by whatever causes may increase or diminish the demand for Bills of Exchange between them: thus, for instance, in London, one pound sterling, a fixed sum, is given for a variable number of French francs, more or less, according to circumstances.

182. By the Par of Exchange is meant the intrinsic value of the coin of one country as compared with a given fixed sum of money of another.

The par of exchange depends on the weight and fineness of the coins, which are known either from the Mint regulations of the different countries, or by direct assay. If the metal from which the par is calculated be not a standard of value in both countries, its market value in that country in which it is not a standard must be taken into account. Thus, in the United Kingdom gold is the *only* standard of value.

183. In order to facilitate mercantile transactions between persons residing at a distance from each other, payments are usually made by *Promissory Notes, Drafts* or *Bills of Exchange*. The holder of either of these Bills being entitled to obtain its value in gold from the party on whom it is drawn.

A Promissory Note is a written engagement to pay a sum of money after the expiration of a certain time.

FORM OF PROMISSORY NOTE.

| Stamp. |

No. 4.   Toronto, March 2, 1870.

Three months after date I promise to pay to the order of Arthur Dixon, Esquire, the sum of three hundred and fifty $\frac{00}{100}$ Dollars at the Bank of Toronto, in Toronto. Value received.

$350 $\frac{00}{100}$.   James Hope.

Due 5th June, 1870.

EXCHANGE.

Note. The *note must* be *stamped*, and on the stamp some important element of the note must be written, such as the amount of the note or the signature of the maker. **After** being signed, no alteration whatever should be made.

FORM OF DRAFT.

No. 7. Toronto, March 4th, 1870.

At sight of this Draft, pay to the order of Archibald Sinclair, Esquire, the sum of Two hundred and twelve Dollars,——value received,——and charge the same to account of

           Hope & Co.

To Thomas Brown & Co.

$212 $\frac{00}{100}$. Montreal.

FORM OF BILL OF EXCHANGE.

£800 Stg. Toronto, March 23, 1870.

Sixty days after sight of this (First) of Exchange (Second and Third of the same tenor and date unpaid), pay to the order of James Sandilands Eight hundred pounds Sterling,——value received,——and place the same with or without further advice to my account.

           James Mitchell.

To the National Bank of Scotland.

230   ARITHMETIC.

NOTE 1. Three Bills of Exchange, constituting a set, called *First*, *Second* and *Third* of *Exchange* are usually made for the same amount and sent off by different conveyances, to provide against delay, if the first should miscarry: *only one* of course is paid, the others being cancelled.

NOTE 2. The party who signs either of the above forms is called the *maker* or *drawer* of the bill; the party on whom drawn, is called the *drawee*, and after accepting it, the *acceptor*.

The usual way of accepting a bill is for the drawee to write his name across the face or back of the bill; the meaning of which is, that he undertakes to pay the bill when it becomes due. The party who buys a bill, is called the *buyer* or *remitter*; the party who has possession of it the *holder*, and the party to whom the money is to be paid, the *payee*.

If the holder or payee of a bill wishes to dispose of it either by selling or transferring it to another party, he writes his name on the back of it, *i. e.* he endorses it, and is then called the *endorser*.

If a bill is not paid, or refused to be paid, it is *protested;* protesting a bill consists in a party called a *Notary Public* notifying the maker, endorser, &c., of the non-payment, or non-acceptance of the bill.

NOTE 3. By an Act of the Parliament of the United Canadas, passed in 185-, the *dollar* was declared to be in value *one-fourth* of the *pound* currency. The pound currency to be of the weight of 101·321 grains Troy and of the standard fineness of the gold coinage of the United Kingdom.

From the above it follows, that the *legal* value of the sovereign or pound sterling $= \$4.86\frac{3}{4}$, which also is its intrinsic value.

But by an Act passed many years ago, the sovereign was declared to be only equal in value to $4.444, or £9 (sterling) $= \$40$; and this is the value which almost invariably is quoted in mercantile transactions: the premium on this depreciated value of the sovereign which will make it equal to its intrinsic value is $9\frac{1}{2}$ per cent.

Ex. 1. A merchant in Toronto has to remit to one in London £735 sterling; how many dollars will he have to give for the bill of exchange; exchange at 109 per cent., commission $\frac{1}{2}$ per cent. ?

By old statute £9 = $40

$$\therefore £1 = \$\frac{40}{9};$$

rate of exchange to the buyer is $109 + \frac{1}{2} = 109\frac{1}{2}$;

$$\therefore £1 = \$ \frac{40}{9} \times \frac{109\frac{1}{2}}{100}$$

$$= \$ \frac{73}{15}$$

$$\therefore £735 = \$ \frac{73}{15} \times 735$$

$$= \$3577.$$

Therefore the **Toronto merchant will have to pay $3577 for the Bill of Exchange.**

Ex. 2. What is the course of exchange between London and Lisbon when 594 milrees, 480 rees are received for £158. 16s. 9d. ? (1 milree = 1000 rees.)

then 594 mils., 480 rees : 1 mil, :: £158. 16s. 9d. : course of exch.;
    or 594·48 mils. : 1 mil. :: 38121d.
        whence course of exch. = 64·121...d.

that is, 64·124d., or rather more than 5s, 4d. **English money, would be paid for 1 milree of Portuguese.**

184. ARBITRATION, or COMPARISON OF EXCHANGES, is the method of fixing upon the **rate of** Exchange, called the PAR or ARBITRATION, between the first and last of a given number of places, where the course of Exchange between the first and second, the second and third, &c., of these places is known. It is called SIMPLE or COMPOUND ARBITRATION, as three or more places are concerned.

Ex. 3. Books having been purchased in Paris for the University of Toronto to the amount of 5100 francs: find whether it is better to remit directly through the Bank of British North America, exchange on Paris being at 2 per cent. discount, or by way of New York, exchange on Paris being at $3\frac{1}{2}$ per cent. premium, or by way of London, rate of exchange between Paris and London being at par, and that between London and Canada being at $107\frac{1}{2}$; commissioon either way being $\frac{1}{2}$ per cent.

        Value of franc   = 19·7 cents.
        $2\frac{1}{2}$ per cent. discount   = ·49 ......
        $\therefore$ exchange value of franc = 19·21 cents;

$$\therefore \text{ bill of exchange costs } \frac{\$5100}{100} \times 19\cdot21$$
$$= \$979.71$$

U. S. value of franc = 18·6 cents,
4 per cent. premium = ·744 ;
∴ exch. value of franc = 19·344 cents;

$$\therefore \text{ bill of exch. costs } = \$\frac{5100}{100} \times 19\cdot344$$
$$= \$986.544.$$

British value of franc = 9·75d. ;
∴ 5100 francs = £207·1875,
exchange between London and Canada, together with commission, is at 108 per cent.; therefore value of £207·1875 stg. is $994.50 Canadian currency.

Hence it is cheaper to remit by the Bank of British North America.

Ex. 4. £1 English being = 25·4 francs, 3·75 francs being = 105 kreutzers, 60 kreutzers being = 1 florin; find in English money the value of 1143 florins.

$$1143 \text{ florins} = (1143 \times 60) \text{ kreutzers,}$$
$$= \left(1143 \times 60 \times \frac{3\cdot75}{105}\right) \text{ francs,}$$
$$= £\left(1143 \times 60 \times \frac{3\cdot75}{105} \times \frac{1}{25\cdot4}\right),$$
$$= £96.\ 8s.\ 6\tfrac{2}{3}d.$$

VALUE OF FOREIGN COINS

	U.S.	Intrinsic.
Sovereign	$4.84	$4.86
Guinea	5.00	5.11
Crown, English	1.06	1.216
Shilling piece	.23	.243
Franc, France and Belgium	.186	.197
" (franc = 100 centimes)		
Florin, or Gulden, Holland	.40	.413
Florin, Austria and Augsburg	.486	.486
Florin of S. Germany States	.40	.417
Ducat, Austria (gold)		2.286

EXCHANGE.

	U. S.	Intrinsic.
Rix-dollar or Thaler, Prussia....	$0.69	$0.727
(thaler = 30 silber groschen = 360 pfennings)		
Marc Banco, Hamburg............	.35	.30
Specie-dollar, Denmark...........	1.05	1.105
Ringsbank dollar "	.52	.552
Specie-dollar, Norway...............	1.06	1.105
Milreo (1000 rees,) Portugal......	1.12	1.085
" " Brazil...........	.54	.53
Real de Vellon, Spain................	.05	.05
Real de Plata, "	.10	.133
Pillar-dollar, "	1.00	1.07
Ruble, Russia (silver)................	.75	.785
Imperial, " (gold)		7.974
Doubloon, Mexico (gold)............	15.60	15.74
Dollar, " (silver).............	1.00	1.07
Dollar, " (gold)...............	1.00	.986
Lira, Sardinia........................	.186	.197
Eagle, U. S. America (gold)........	10.00	10.00
Dollar, " "	1.00	1.00

NOTE. In the above table the column marked U. S. gives the value of the coins, as fixed by law in the United States of America; the second column gives the *intrinsic* value of the named coins Canadian currency. In ascertaining the *intrinsic* value of the coins, the valuable work by Martin and Trübner was made use of for obtaining the *weight* and *fineness* of each coin, and British standard silver taken at 5s. 2d. sterling, its average market price, per ounce.

185. New Brunswick has the same currency as Ontario and Quebec. The sovereign of the United Kingdom, in the currency of Nova Scotia, is equal to $5; the silver coins in proportion to their value of the gold coin.

In Prince Edward Island, the British Sovereign equals 30 shillings Island currency, silver coins in proportion.

The American Eagle ($10) is legal tender for £3. Island Currency.

In Newfoundland the British sovereign is legal tender for $4.80, silver coins in proportion to their value of the sovereign.

The American Eagle is legal tender for $9.85, and aliquot parts in proportion.

MONEY.

(ENGLISH.)

1869 sovereigns are coined from 40 pounds Troy standard gold, which is $\frac{11}{12}$ fine; therefore it follows that

Weight of a sovereign = 123·27447 grains,
Weight of pure gold = 113·00159 "

A pound Troy of standard silver, which is $\frac{37}{40}$ fine, is coined into 66 shillings; therefore

Weight of a shilling = 87·27273 grains
Weight of pure silver = 80·72727 "

NOTE. Mint value of an ounce standard silver is 5s. 6d., but usual market price 5s. 2d.

(FRENCH.)

The fineness of gold and silver coins in France is the same, viz. $\frac{9}{10}$. The mode of expressing the fineness of the coinage adopted by French assayers, is to state the number of parts of the pure metal which are contained in 1000 parts, and to say that the metal is so many *millièmes* fine.

One kilogramme of standard gold is coined into 3100 francs
"   "   "   "   silver...... ....:. 200 "

(UNITED STATES.)

By the Act of 1852 the weight of the Eagle was ordered to be 258 grains $\frac{9}{10}$ fine;

∴ Weight of the Eagle = 258 grains
Weight of pure gold = 232.2 "

The fineness of the silver coins is the same as that of the gold. The silver dollar coined 1857 is 412.5 grains in weight.

(CANADIAN.)

By an Act of Parliament of the United Canadas, the *pound* currency was ordered to be 101·321 grains in *weight*, of gold of the standard fineness prescribed by law for the gold coins of the United Kingdom on the first day of August, 1854.

By the law the *dollar* is defined to be one-fourth of the pound. The gold Eagle of the United States coined since 1852, is legal tender for ten dollars.

EXCHANGE. 235

Ex. LIX.

(1) A merchant in Toronto has to remit to one in Berlin (Prussia) 612 thalers; how many dollars will he have to give in order to pay the amount, commission ½ per cent., exchange at par?

(2) Convert 4750 milrees, 280 rees into English money at 64¼d. a milree, and bring the amount into Canadian currency, exchange at 108 per cent.

(3) Convert £246. 15s. 6d. into piastres and rials, exchange being at 47¼d. a piastre. (1 piastre = 8 rials.)

(4) By an Act of the U. S. Congress in 1834, it was enacted that the weight of the eagle should be 258 grains, and its fineness 899·2 millièmes. From this calculate the par of exchange between G. B. and U. S. of America.

(5) By an Act of the U. S. Congress, 1837, it was ordered that the dollar should weigh 412½ grains of silver, $\frac{9}{10}$ fine. Calculate the silver par, British standard silver being 5s. 1⅜d. per ounce.

(6) By An act of the U. S. Congress, 1853, it was enacted that in the coinage of half-dollars, quarter-dollars, &c., the half-dollar should weigh 192 grains of silver, $\frac{9}{10}$ fine. Calculate the par at 5s. 1⅜d. per oz.

(7) A merchant in London is indebted to one at St. Petersburg 15,000 rubles: the exchange between St. Petersburg and England is 50d. per ruble, between St. Petersburg and Amsterdam 91d. per ruble, and between Amsterdam and London 36s. 3d. per £ sterling: which will be the most advantageous way for the London merchant to be drawn upon?

(8) What sum in English money must be given for 500 francs, when 25·6 francs is exchanged for £1? What is the arbitrated price between London and Paris, when 3 francs = 480 rees, 400 rees = 3½s. Flemish, and 35s. Flemish = £1?

(9) A person in London owes another in St. Petersburg a debt of 460 rubles, which must be remitted through Paris. He pays the requisite sum to his broker, at a time when the exchange between London and Paris is 23 francs for £1, and between Paris and St. Petersburg 2 francs for one ruble. The remittance is delayed until the rates of exchange are 24 francs for £1, and 3 francs for 2 rubles. What does the broker gain or lose by the transaction?

(10) A gentleman has £8000 in the 3 per cents at 97½: he wishes

to sell and invest the proceeds in Canada Dominion stock at 106, yielding 7 per cent. dividends annually. Find the alteration in his income, exchange between the United Kingdom and Canada being at 8¼ per cent. premium, commission of ½ per cent. being allowed on each transaction.

SECTION VI.

SQUARE ROOT.

186. The SQUARE of a given number is the product of that number multiplied by itself. Thus 36 is the square of 6.

The square of a number is frequently denoted by placing the figure 2 above the number, a little to the right. Thus 6^2 denotes the square of 6, so that $6^2 = 36$.

187. The SQUARE ROOT of a given number is a number, which when multiplied by itself will produce the given number.

The square root of a number is sometimes denoted by placing the sign $\sqrt{}$ before the number, or by placing the fraction ½ above the number a little to the right. Thus $\sqrt{36}$ or $(36)^{\frac{1}{2}}$ denotes the square root of 36; so that $\sqrt{36}$ or $(36)^{\frac{1}{2}} = 6$.

188. The number of figures in the Square Root of any number may readily be known from the following considerations:

The square root of		
1	is	1
100	is	10
10000	is	100
1000000	is	1000
&c.	is	&c.

Hence it follows that the square root of any number between 1 and 100 must lie between 1 and 10, that is, will have one figure in its integral part; of any number between 100 and 10000, must lie between 10 and 100, that is, will have two figures in its integral part; of any number between 10000 and 1000000, must lie between 100 and 1000, that is, must have three figures in its integral part; and so on. Wherefore, if a point be placed over the units' place of the number, and thence over every second figure to the left of that place, the points

SQUARE ROOT. 237

ıber of figures in the integral part of the root. Thus
 9) consists, so far as it is integral, of *one* figure ;
o figures ; that of 176432 of *three* figures ; that of
gures ; and so on.

ıre root of   ·01   is   ·1
               ·0001   is   ·01
               ·000001   is   ·001
               ·00000001 is   ·0001
                  &c.   &c.

extracting the square root of decimals, the decimal
f all be made even in number, by affixing a cypher to
e necessary ; and then if points be placed over every
ıe right, beginning as before from the units' place of
ıe **number of** such points will show the number of
the root.

extracting the Square Root of a number.

 or dot over the **units'** place of the given number ; and
second figure to the left of that place ; and thence also
 figure to the right, when the number contains de-
a cypher when the number of decimal figures is odd ;
given number into periods. The number of points
ımbers and decimals respectively will **shew** the numbers and decimals respectively in the square root.

test number whose square is contained in the first
 this is, the first figure in the root, which place in
tient to the right of the given number. Subtract its
.rst **period,** and to the remainder bring down, on the
period.

nber thus formed, omitting the **last** figure, by twice
ot already obtained, and annex the result to the root
visor.

 the divisor, as it now stands, by the part of the root
. subtract **the** product from the number formed, as
 by the first remainder and second period.

ıre **periods to** be brought down, the operation must be

238   ARITHMETIC.

Ex. 1. Find the square root of 1369.

$$\begin{array}{r|l} & 1369\ (37 \\ & 9 \\ \hline 67 & 469 \\ & 469 \end{array}$$

After pointing, according to the Rule, we take the first period, or 13, and find the greatest number whose square is contained in it. Since the square of 3 is 9, and that of 4 is 16, it is clear that 3 is the greatest number whose square is contained in 13; therefore place 3 in the form of a quotient to the right of the given number. Square this number, and put down the square under the 13; subtract it from the 13, and to the remainder 4 affix the next period 69, thus forming the number 469. Take 2×3, or 6, for a divisor; divide the 469, omitting the last figure, that is, divide the 46 by the 6, and we obtain 7. Annex the 7 to the 3 before obtained and to the divisor 6; then multiplying the 67 by the 7 we obtain 469, which being subtracted from the 469 before formed, leaves no remainder; therefore 37 is the square root of 1369.

Reason for the above process.

Since $(37)^2 = 1369$, and therefore 37 is the square root of 1369; we have to investigate the proper Rule by which the 37 or $30 + 7$, may be obtained from the 1369.

$$\begin{aligned} \text{Now } 1369 &= 900 + 469 = 900 + 49 + 420 \\ &= (30)^2 + 7^2 + 2 \times 30 \times 7 \\ &= (30)^2 + 2 \times 30 \times 7 + 7^2 \end{aligned}$$

where we see that the 1369 is separated into parts in which the 30 and the 7, together constituting the square root, or 37, are made distinctly apparent. Treating then the number 1369 in the following form, viz.

$$(30)^2 + 2 \times 30 \times 7 + 7^2$$

we observe that the square root of the first part, or of $(30)^2$ is 30; which is one part of the required root. Subtract the square of the 30 from the whole quantity $(30)^2 + 2 \times 30 \times 7 + 7^2$, and we have $2 \times 30 \times 7 + 7^2$ remaining. Multiply the 30 before obtained by 2, and we see that the product is contained 7 times in the first part of the remainder, or in $2 \times 30 \times 7$; and adding the 7 to the 2×30, thus making $2 \times 30 + 7$ or 67, this latter quantity is contained 7 times exactly in the remaining $2 \times 30 \times 7 + 7$ or

SQUARE ROOT.

469; so that by this division we shall gain the 7, the remaining part of the root. If we had found that the $2 \times 30 + 7$ or 67, when multiplied by the 7, had produced a larger number than the 469, the 7 would have been too large, and we should have had to try a smaller number, as 6, in its place.

The process will be shewn as follows:

$$(30)^2 + 2 \times 30 \times 7 + 7^2\ (30 + 7$$
$$(30)^2$$

$$2 \times 30 + 7\ \big|\ \begin{array}{l} 2 \times 30 \times 7 + 7^2 \\ 2 \times 30 \times 7 + 7^2 \end{array}$$

This operation is clearly equivalent to the following:

$$900 + 420 + 49\ (30 + 7$$
$$900$$

$$60 + 7\ \big|\ \begin{array}{l} 420 + 49 \\ 420 + 49 \end{array}$$

This again is equivalent to the following:

$$1369\ (37$$
$$9$$

$$67\ \big|\ \begin{array}{l} 469 \\ 469 \end{array}$$

which is the mode of operation pointed out in the Rule.

NOTE 1. The reasoning will be better understood when the student has made some progress in Algebra.

NOTE 2. The divisor obtained by doubling the part of the root already obtained, is often called a *trial divisor*, because the quotient first obtained from it by the Rule in (Art. 189), will sometimes be too large. It will be readily found, in the process, whether this is the case or not, for when, according to our Rule, we have annexed the quotient to the trial divisor, and multiplied the divisor as it then stands by that quotient, the resulting number should not be greater than the number from which it ought to be subtracted. If it be, the quotient is too large, and the number next smaller should be tried in its place.

240   ARITHMETIC.

Ex. 2. Find the square root of 71690512350625.

$$
\begin{array}{r|l}
 & 71690512350625\ (8467025 \\
 & 64 \\ \hline
\{2 \times 8 = 16\} \quad\quad 164 & 769 \\
 & 656 \\ \hline
\{2 \times 84 = 168\} \quad 1686 & 11305 \\
 & 10116 \\ \hline
\{2 \times 846 = 1692\} \quad 16927 & 118912 \\
 & 118489 \\ \hline
\begin{pmatrix}(2 \times 8467 = 16934)\\(2 \times 84670 = 169340)\end{pmatrix}\ 1693402 & 4233506 \\
 & 3386804 \\ \hline
16934045 & 84670225 \\
 & 84670225
\end{array}
$$

∴ 8467025 is the required square root.

190. As the decimal notation is only an extension or continuance of the ordinary integral notation, and quite in agreement with it, the reason given for the process in whole numbers will apply also to decimals.

191. To extract the square root of a vulgar fraction, if the numerator and denominator of the fraction be perfect squares, we may find the square root of each separately, and the answer will thus be obtained as a vulgar fraction; if not, we can first reduce the fraction to a decimal, or to a whole number and decimal, and then find the root of the resulting number. The answer will thus be obtained either as a decimal, or as a whole number and decimal, according to the case. Also a mixed number may be reduced to an improper fraction, and its root extracted in the same way.

Ex. 3. Extract the square root of 53111·8116.

$$
\begin{array}{r|l}
 & 53111·8116\ (230·46 \\
 & 4 \\ \hline
43 & 131 \\
 & 129 \\ \hline
4604 & 21181 \\
 & 18416 \\ \hline
46086 & 276516 \\
 & 276516
\end{array}
$$

SQUARE ROOT.

Ex. 4. Find the square root of $\frac{5}{7}$.

This may be done by first reducing $\frac{5}{7}$ to a decimal, and then by extracting the square root of the decimal, thus $\frac{5}{7} = \cdot 714285...$

$$\cdot 7142\dot{8}5 \ (\cdot 845...$$

$$\begin{array}{r|l} & 64 \\ \hline 164 & 742 \\ & 656 \\ \hline 1685 & 8685 \\ & 8425 \\ \hline & 200 \end{array}$$

or thus, $\sqrt{\dfrac{5}{7}} = \sqrt{\left(\dfrac{5 \times 7}{7 \times 7}\right)} = \dfrac{\sqrt{35}}{7}.$

$$35 \cdot 000000 \ (5 \cdot 916$$

$$\begin{array}{r|l} & 25 \\ \hline 109 & 1000 \\ & 981 \\ \hline 1181 & 1900 \\ & 1181 \\ \hline 11826 & 71900 \\ & 70956 \\ \hline & 944 \end{array}$$

therefore $\sqrt{\dfrac{5}{7}} = \dfrac{5 \cdot 916}{7} = \cdot 845...$

Ex. LX.

Find the square roots of

(1) 289 ; 576 ; 1444 ; 4096.   (2) 6561 ; 21025 ; 173056.
(3) 98596 ; 37249 ; 11664.   (4) 998001 ; 978121 ; 824464.
   (5) 29506624 ; 14356521 ; 5345344.
   (6) 236144689 ; 282429536481 ; 282475249.
   (7) 295066240000 ; 4160580062500.
(8) 167·9616 ; 28·8369 ; 57648·01.   (9) ·3486784401 ; 39·15380329
(10) ·042849 ; ·00139876 ; ·00203401.   (11) 5774409 ; 5·774409.
   (12) 120888·68379025 ; 240398·012416.

242   ARITHMETIC.

(13) 16 ; $1{\cdot}6$; $\cdot 16$; $\cdot 016$.   (14) $235{\cdot}6$; $\cdot 1$; $\cdot 01$; 5 ; $\cdot 5$.
(15) $\cdot 0004$; $\cdot 00081$; $379{\cdot}864$.   (16) $20\tfrac{1}{4}$; $153\tfrac{7}{9}$; $\tfrac{1}{3}$; $\tfrac{2209}{9801}$.

(17) $\tfrac{3}{5}$; $\tfrac{1}{17}$; $2\tfrac{1}{2}$; $\dfrac{3\tfrac{1}{2}}{4\tfrac{1}{2}}$.   (18) $\dfrac{5{\cdot}04}{\cdot 021}$; $1\tfrac{56}{169}$; $23{\cdot}1$; 42 ;

to four places of decimals in each case where the root does not terminate.

CUBE ROOT.

192. The CUBE of a given number is the product which arises from multiplying that number by itself, and then multiplying the result again by the same number. Thus $6 \times 6 \times 6$ or 216 is the cube of 6.

The cube of a number is frequently denoted by placing the figure 3 above the number, a little to the right. Thus 6^3 denotes the cube of 6, so that $6^3 = 6 \times 6 \times 6$ or 216.

193. The CUBE ROOT of a given number is a number, which, when multiplied into itself, and the result again multiplied by it, will produce the given number. Thus 6 is the cube root of 216; for $6 \times 6 \times 6$ is $= 216$.

The cube root of a number is sometimes denoted by placing the sign $\sqrt[3]{\ }$ before the number, or placing the fraction $\tfrac{1}{3}$ above the number, a little to the right. Thus $\sqrt[3]{216}$ or $(216)^{\tfrac{1}{3}}$ denotes the cube root of 216; so that $\sqrt[3]{216}$ or $(216)^{\tfrac{1}{3}} = 6$.

194. The number of figures in the Cube Root of any number may readily be known from the following considerations:

The cube root of   1   is   1
                          1000   is   10
                          1000000   is   100
                          1000000000 is 1000

Hence it follows that the cube root of any number between 1 and 1000 must lie between 1 and 10, that is, will have one figure in its integral part; of any number between 1000 and 1000000, must lie between 10 and 100, that is, will have two figures in its integral part; of any number between 1000000 and 1000000000, must lie between 100 and 1000, that is, must have three figures in its integral part; and so on. Wherefore, if a point be placed over the units' place of the number, and thence over every third figure to the left of that place, the points will show the number of figures in the integral part of the root.

Thus the cube root of 677 consists, so far as it is integral, of *one* figure; that of 198999 of *two* figures; that of 134198999 of *three* figures; and so on.

Again, since the cube root of ·001 is ·1,
the cube root of ·000001 is ·01,
the cube root of ·000000001, is ·001,
&c. is &c.

it appears, that in extracting the cube root of decimals, the decimal places must first of all be made three, or some multiple of three in number, by affixing cyphers to the right, if this be necessary; and then if points be placed over every *third* figure to the right, beginning as before from the units' place of *whole numbers*, the number of such points will shew the number of decimal places in the cube root.

195. *Rule for extracting the Cube Root of a number.*

Place a point or dot over the units' place of the given number, and thence over every third figure to the left of that place; and thence also over every third figure to the right, when the number contains decimals, affixing one or two cyphers, when necessary, to make the number of decimal places a multiple of 3; thus dividing the given number into periods. The number of points over the whole numbers and decimals respectively will shew the number of whole numbers and decimals respectively in the cube root.

Find the greatest number whose cube is contained in the first period at the left; this is the first figure in the root, which place in the form of a quotient to the right of the given number.

Subtract its cube from the first period, and to the remainder bring down, on the right, the second period.

Divide the number thus formed, omitting the two last figures, by 3 times the square of the part of the root already obtained, and affix the result to the root.

Now calculate the value of 3 times the square of the first figure in the root (which of course has the value of so many tens) + 3 times the product of the two figures in the root + the square of the last figure in the root. Multiply the value thus found by the second figure in the root, and subtract the result from the number formed, as above mentioned, by the first remainder and the second period. If there be more periods to be brought down the operation must be repeated.

ARITHMETIC.

Ex. 1. Find the cube root of 15625.

$$15\dot{6}2\dot{5}\ (25$$
$$2^3 = 8$$

$$
\begin{array}{r|r}
3 \times 2^2 = 12 & 762 \\
3 \times (20)^2 = 3 \times 400 = 1200 & \\
3 \times 20 \times 5 = 300 & \\
5^2 = 25 & \\
\hline
1525 & \\
\text{multiply by } \quad 5 & \\
\hline
7625 & 7625 \\
\end{array}
$$

After pointing according to the Rule we take the first period, or 15, and find the greatest number whose cube is contained in it. Since the cube of 2 is 8, and that of 3 is 27, it is clear that 2 is the greatest number whose cube is contained in 15; therefore place 2 in the form of a quotient to the right of the given number.

Cube 2, and put down its cube, viz. 8, under the 15; subtract it from the 15, and to the remainder 7 affix the next period 625, thus forming the number 7625. Take 3×2^2, or 12, for a divisor; divide 76 by 12, 12 is contained 6 times in 76; but when the other terms of the divisor are brought down, 6 would be found too great, therefore take 5. Annex the 5 to the 2 before obtained; and calculate the value of $3 \times (20)^2 + 3 \times 20 \times 5 + 5^2$, which is 1525; multiplying 1525 by 5 we obtain 7625, which being subtracted from 7625 before formed leaves no remainder,

therefore 25 is the cube root required.

Reason for the above process.

Since $(25)^3 = 15625$, and therefore 25 is the cube root of 15625; we have to investigate the proper Rule by which the 25, or $20 + 5$, may be obtained from 15625.

$$
\begin{aligned}
\text{Now } 15625 &= 8000 + 7500 + 125 \\
&= 8000 + 6000 + 1500 + 125 \\
&= (20)^3 + 3 \times (20)^2 \times 5 + 3 \times 20 \times 5^2 + 5^3,
\end{aligned}
$$

where we see that the 15625 is separated into parts in which the 20 and the 5, together constituting the cube root or 25, are made distinctly

CUBE ROOT. 245

apparent. Treating then the number 15625 in the following form, viz.
$$(20)^3 + 3 \times (20)^2 \times 5 + 3 \times 20 \times 5^2 + 5^3$$
we observe that the cube root of the first part or of $(20)^3$ is 20; which is one part of the required root. Subtract the cube of the 20 from the whole quantity, and we have $3 \times (20)^2 \times 5 + 3 \times 20 \times 5^2 + 5^3$ remaining. Multiply the square of the 20 before obtained by 3, and we see that the product is contained 5 times in the first part of the remainder, or in $3 \times (20)^2 \times 5$; and adding 3 times the product of the two terms of the root + the square of the last term of the root, thus making $3 \times (20)^2 + 3 \times 20 \times 5 + 5^2$, we see that this latter quantity is contained 5 times exactly in the remainder $3 \times (20)^2 \times 5 + 3 \times 20 \times 5^2 + 5^3$, so that by this division we shall obtain the 5, the remaining part of the root.

The process will be shown as follows:

$$(20)^3 + 3 \times (20)^2 \times 5 + 3 \times (20) \times 5^2 + 5^3 \ (20+5$$
$$(20)^3$$

divisor $= 3 \times (20)^2$,

and $\dfrac{3 \times (20)^2 \times 5}{3 \times (20)^2} = 5$;

$\therefore \{3 \times (20)^2 + 3 \times 20 \times 5 + 5^2\} \times 5 =$ | $3 \times (20)^2 \times 5 + 3 \times 20 \times 5^2 + 5^3$

This operation is clearly equivalent to the following:

$$8000 + 6000 + 1500 + 125 \ (20+5$$
$$8000$$

$3 \times (20)^2 = 1200$, and $\tfrac{6000}{1200} = 5$

$(1200 + 300 + 25) \times 5 =$ | $6000 + 1500 + 125$
| $6000 + 1500 + 125$

This again is equivalent to the following:

$$15625 \ (25$$
$$8$$
———
7625

$3 \times 2^2 = 3 \times 4 = 12$, and $\tfrac{76}{12} = 5$

$3 \times (20)^2 \ \ = 1200$
$3 \times 20 \times 5 \ = 300$
$\ \ \ \ \ +5^2 \ = \ \ 25$
————
$\ \ \ \ \ \ \ \ \ \ \ \ 1525$
$\ \ \ \ \ \ \ \ \ \ \ \ \ \ \ 5$
————
$\ \ \ \ \ \ \ \ \ \ \ \ 7625$ | 7625

which is the mode of operation pointed out in the Rule.

246   ARITHMETIC.

NOTE 1. The reasoning will be better understood when the student has made some progress in Algebra.

NOTE 2. The divisor which is obtained according to the Rule given in (Art. 195) is sometimes called a *trial* divisor, because the number from the division may be too large, as was the case in the above Example, in which case we must try a smaller number. We shall readily ascertain whether the number obtained from the division is too large or not, because if it be too large, the quantity which we ought to subtract from the number formed by a remainder and a period will turn out in that case to be larger than that number, which of course it ought not to be, and so we must try a smaller number.

NOTE. 3. If at any point of the operation, the number to be divided by the trial divisor be less than it; we affix a cypher to the root, two cyphers to the trial divisor, bring down the next period, and proceed according to the Rule.

Ex. 3. Find the cube root of 223648543.

$$223648543\ (607$$
$$6^3 = 216$$

trial divisor = 3×6^2. = 108
trial divisor = $3 \times (60)^2$ = 10800
$3 \times (600)^2 = 1080000$
$3 \times 600 \times 7 = 12600$
$7^2 = 49$
          1092649
              7
          7648543

| 7648 |
| 7648543 |
| |
| |
| |
| |
| |
| 7648543 |

76 is not divisible by 108; bring down the next period and affix 0 to the root; $\frac{76485}{10800}$ goes 7 times, and 7 seems likely to be the figure required; since $7^3 = 343$, and 3 is the final figure in the remainder.

Therefore 607 is the cube root required

196. As the decimal notation is only an extension or continuance of the ordinary integral notation, and quite in agreement with it, the reason given for the process in whole numbers, will apply also to decimals.

197. To extract the cube root of a vulgar fraction, if the numerator and denominator of the fraction be perfect cubes we may find the cube root of each separately; and the answer will thus be obtained as a

CUBE ROOT.

vulgar fraction; if not, we can first reduce the fraction to a decimal, or to a whole number and decimal, and then find the root of the resulting number. The answer will thus be obtained either as a decimal, or as a whole number and decimal, according to the case. Also a mixed number may be reduced to an improper fraction, and its root extracted in the same way.

Ex. 4. Find the cube root of ·000007 to three places of decimals.

$$\cdot 000\dot{0}07\dot{0}00 \ (\cdot 01\dot{9}$$

$$
\begin{array}{r|l}
 & 1 \\
3\times 1^2 = 3 & 6000 \\
\end{array}
$$

$$
\begin{aligned}
3\times(10)^2 &= 300 \\
3\times 10\times 9 &= 270 \\
9^2 &= 81 \\
\hline
 & 651 \\
 & 9 \\
\hline
 & 5859
\end{aligned}
$$

$$
\begin{array}{r|l}
 & 5859 \\
\hline
 & 141
\end{array}
$$

Ex. 5. Find the cube root of $\tfrac{5}{9}$ to three places of decimals.

$$\tfrac{5}{9} = \cdot 555555555\ldots$$

$$\cdot \dot{5}5555\dot{5}55 \ (\cdot 822$$

$$
\begin{array}{r|l}
8^3 = 512 & \\
8\times 8^2 = 192 & 43555 \\
\end{array}
$$

$$
\begin{aligned}
3\times(80)^2 &= 19200 \\
3\times 80\times 2 &= 480 \\
2^2 &= 4 \\
\hline
 & 19684 \\
 & 2 \\
\hline
 & 39368
\end{aligned}
$$

$$
\begin{array}{r|l}
 & 39368 \\
8\times(82)^2 = 20172 & 4187555 \\
\end{array}
$$

$$
\begin{aligned}
3\times(820)^2 &= 2017200 \\
3\times 820\times 2 &= 4920 \\
2^2 &= 4 \\
\hline
 & 2022124 \\
 & 2 \\
\hline
 & 4044248
\end{aligned}
$$

$$
\begin{array}{r}
4044248 \\
\hline
143307
\end{array}
$$

248 ARITHMETIC.

198. Higher roots than the square and cube can sometimes be extracted by means of the Rules for square and cube root; thus the 4th root is found by taking the square root of the square root; the 6th root by taking the square root of the cube root, and so on.

Ex. LXI.

Find the cube roots of

(1) 1728; 3375; 29791.   (2) 54872; 110592; 300763.
(3) 681472; 804357; 941192.   (4) 2406104; 69426531; 8365427.
(5) 251239591; 28372625; 48228544.
(6) 17173512; 259694072; 926859375.
(7) 27054036008; 219365327791.
(8) ·389017; 32·461759; 95443·993; ·000912673; ·001906624; ·000024389.

(9) 3, ·3, ·03.   (10) $\dfrac{8}{27}$; $\dfrac{250}{686}$; 44·6.

(11) $405\frac{28}{125}$; $7\frac{1}{5}$; 3·00415.   (12) ·0001; $\dfrac{1257\cdot728}{16384}$,

to three places of decimals, in those cases where the root does not terminate.

(13) Find the cube root of 233·744896, and also the cube root of the last-mentioned number multiplied by ·008.

(14) The cost of a cubic mass of metal is £10481. 1s. 4d. at 10s. 5d. a cubic inch. What are the dimensions of the mass?

(15) A cubical block of stone contains 50653 solid feet; what is the area of its side?

(16) A cube contains 56 solid feet, 568 solid inches; find its edge.

SCALES OF NOTATION.

199. From what has been said (Art. 6), it appears that numbers in the common or decimal system of notation increase in a uniform manner from right to left by a ten-fold ratio. The number 10, therefore is called the *radix* or base of the scale. It is plain that in any system of Notation, we must have as many different characters as there are units in the radix of the scale.

SCALES OF NOTATION.

Thus, in the Common Scale, we have *ten* different characters. If the radix were 2, then we would only have *two*; viz., 1, 0.

If 3, then we would have *three*, 1, 2, 0; and so on. If we select a radix larger than 10, in that case, we would have to *invent* a *new* character for each additional unit. Thus if the radix were 12, for 10 and 11, we might use the characters τ, ϵ.

200. The scales are called, according as the number 2, 3, 4, 5, 6, 7, 8, 9, 10, 11, or 12, is the radix, the *BINARY, TERNARY, QUATERNARY, QUINARY, SENARY, SEPTENARY, OCTENARY, NONARY, DENARY, UNDENARY*, and *DUODENARY* Scales.

NOTE 1. The different scales are sometimes denoted thus: 231 in the quarternary scale is written $(231)_4$; $9\tau\epsilon 45$ in the duodenary is written $(9\tau\epsilon 45)_{12}$; and similarly of other numbers.

NOTE 2. The operations of Addition, Subtraction, Multiplication, Division, Square and cube Roots, are performed on numbers in any given Scale as in the common Scale of Arithmetic, care being taken that instead of using 10 and its powers, as in the common or denary scale, we use that number and its powers, which denotes the particular scale in which we are working.

201. *To express any given number in any assigned scale.*

RULE. Divide the given number by the radix; divide again the quotient by the **radix**; so continue the division, as long as possible: finally write the remainders in reversed order; the result thus obtained is the number expressed in the proposed scale.

Ex. 1. Reduce the number 43046 **ordinary scale** to the *duodecimal* scale, and prove the truth of the result.

Proceeding by the Rule given above.

```
12 | 43046
12 |  3587–2
12 |   298–11 or ε
12 |    24–10 or τ
           2–0
```

Therefore the number is $20\tau\epsilon 2$.

ARITHMETIC.

Proof

```
10 | 20τε2
10 | 25τ8–6
10 | 2ετ –4
10 | 37–0
     ———
       4–3
```

$20\tau\epsilon 2$ is in the scale whose radix is 12; $\therefore 20 = 2\times 12 + 0 = 24$, and 10 in 24 goes 2 and 4 over; then $4\tau = 4\times 12 + 10 = 58$, and 10 in 58 goes 5 times and 8 over; then $8\epsilon = 8\times 12 + 11 = 107$, $\therefore (20\tau\epsilon 2)_{12} = (43046)_{10}$, and 10 in 107 goes 7 and 7 over; then $72 = 7\times 12 + 2 = 86$, and 10 in 86 goes 8 times and 6 over, and so on.

NOTE. The operation of transforming a number from any scale to the denary, or common scale can be more readily done as follows. Take the number in the last example, $20\tau\epsilon 2$.

$$\begin{array}{r} 20\tau\epsilon 2 \\ 12 \\ \hline 24 \ = 2\times 12 + 0 \\ 12 \\ \hline 298 \ = 2\times 12^2 + 0\times 12 + 10 \\ 12 \\ \hline 3587 \ = 2\times 12^3 + 0\times 12^2 + 10\times 12 + 11 \\ 12 \\ \hline 43046 \ = 2\times 12^4 + 0\times 12^3 + 10\times 12^2 + 11\times 12 + 2 \, ; \end{array}$$

which is the same result as was obtained above by a different method.

Ex. 2. Transform the number 5056 from the septenary to the quaternary scale.

The division must be performed in the septenary scale.

```
4 | 5056
4 | 1165–0
4 |  214–3
4 |   36–1
4 |    6–3
4 |    1–2
        0–1
```

$\therefore$ number required $= 123130$;
and since 123130
$= 1.4^5 + 2.4^4 + 3.4^3 + 1.4^2 + 3.4 + 0 = 1024 + 512 + 192 + 16 + 12 = (1756)_{10}$;
$\therefore$ from Ex. 1, we see that the result is correct.

SCALE OF NOTATION. 251

Note. 5056 might first have been expressed in the denary scale thus, $5 \times 7^3 + 0 \times 7^2 + 5 \times 7 + 6 = 1715 + 35 + 6 = 1756$, and 1756 then transformed from the denary to the quaternary thus,

$$
\begin{array}{r|r}
4 & 1756 \\ \hline
4 & 439\text{--}0 \\ \hline
4 & 109\text{--}3 \\ \hline
4 & 27\text{--}1 \\ \hline
4 & 6\text{--}3 \\ \hline
4 & 1\text{--}2 \\ \hline
  & 0\text{--}1
\end{array}
$$

∴ number in quaternary scale as before = 123180.

Ex. 3. Extract the square root of 25400544 in the senary scale.

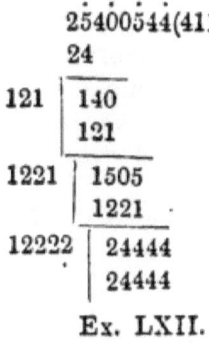

Ex. LXII.

(1) Add together 1445, 22601, 56432, 37, 577, and 6 in the octenary, and also in the nonary scale ; and the first three numbers in the septenary scale ; and the last three numbers in the undenary, and duodenary scales.

(2) Subtract 1τ864 from 7τ348 in the undenary, and also in the duodenary scale ; and 50543 from 61210 in the septenary scale.

(3) Multiply together 35 and 61 , 2064 and 312 ; 57264 and 675 in the octenary scale ; 468 and 701 ; 85τ and 734 in the undenary : 9294 and 344 ; 6τ12 and 814 ; ετ7τ8 and 2ττ9 in the duodenary ; 1456 and 6541 in the septenary ; and 30122 and 322 in the quaternary.

(4) Divide 14832216 by 6541 in the septenary scale ; 29τ96580 by 2ττ9 ; and 95088918 by ττ4 in the duodenary ; 201002 by 13 in the quaternary ; and 24510502 by 4331 in the senary.

(5) Extract the square root of 12044424 in the senary scale; 32e75721 in the duodenary; and 47610370 in the nonary.

(6) Express 1828 in the septenary scale; 7631 in the binary; 29 and 49 in the octenary; 1000 in the undenary; 1000000 in the senary; 80198 in the duodenary.

(7) Express $(62 \tau e)_{12}$ and $(1534)_6$ in the common system; $(34523)_6$ in the duodenary; $(654321)_{12}$ in the septenary.

(8) Transform 23784 and 587 from the nonary to the duodenary scale; 4321 from the quinary to the septenary; and 2304 from the quinary to the undenary scale; and prove the truth of each result.

(9) Transform $(8978)_{11}$ and $(3256)_7$ to the duodenary scale, and find their product.

202. *A proper fraction is converted from one scale into another by the following rule:*

RULE. Multiply the numerator of the fraction by the *radix* of the given scale, and divide by the denominator; repeat the same operation as often as necessary; the result is the given fraction transformed into the required scale.

Ex. 1. Express $8\tfrac{17}{49}$ in the septenary scale, and prove the truth of the result.

$$8_{10} = (11)_7;$$

By the Rule, $\dfrac{7 \times 17}{49} = 2\dfrac{3}{7}$; $\dfrac{7 \times 3}{7} = 3$; $\therefore \left(\dfrac{17}{49}\right)_{10} = (\cdot 23)_7;$

$\therefore$ number required $= 11 \cdot 23$.

Proof. $11_7 = (1 \times 7 + 1)_{10} = 8_{10},$

$$(\cdot 23)_7 = \left(\dfrac{2}{7} + \dfrac{3}{7^2}\right)_{10} = \left(\dfrac{17}{7^2}\right)_{10} = \left(\dfrac{17}{49}\right)_{10};$$

$$\therefore (11 \cdot 23)_7 = \left(8\dfrac{17}{49}\right)_{10}$$

Ex. 2. Convert $26 \cdot 5$ into the quaternary scale, and prove the result.

$(26)_{10} = (122)_4$; $\cdot 5 = \tfrac{1}{2}$; proceeding by Rule, $\dfrac{4 \times 1}{2} = 2,$

or thus, $\cdot 5 \times 4 = 2 \cdot 0$;

$\therefore (26 \cdot 5)_{10} = (122 \cdot 2)_4.$

Proof. $(122 \cdot 2)_4 = 1 \times 4^2 + 2 \times 4 + 2 + \tfrac{2}{4} = 16 + 8 + 2 + \cdot 5 = (26 \cdot 5)_{10}.$

SCALES OF NOTATION.   253

Ex. 3. Convert 323010·221122 from the quaternary to the octenary scale.

Work in quaternary scale.

```
8 | 323010              ·22112
8 | 13120 — 4               8
8 |   323 — 0          5·02300
8 |    13 — 3               8
         0 — 7         1·12000
                            8
                       3·00000
```

∴ required number is 7304·513.

NOTE. The number might have been transformed into the denary scale, and thence into the octenary.

Ex. 4. Convert $(456·16)_{12}$ to the ternary scale, and prove the truth of the result.

Work in the duodenary scale.

```
                          ·16
                            3
                         0·46
                            3
8 | 456                  1·16
8 | 15r — 0                 3
8 |  5ϵ — 1              0·46
3 |  1ϵ — 2                 3
3 |   7 — 2              1·16
8 |   2 — 1
      0 — 2
```

∴ number required is 212210·0101̈.

Work in ternary scale.

```
12 | 212210           The value of ·0̈1̈
12 |   1222 — 6
12 |     11 — 5         = $\frac{1}{22}$, (Art. 134.)
          0 — 4
                      and $\left(\frac{1}{22}\right)_3 = \left(\frac{1}{8}\right)_{10}$
```

proceeding by Rule $\frac{12 \times 1}{8} = 1\frac{1}{2}$; $\frac{12 \times 1}{2} = 6$;

∴ **number** above found in ternary scale = 456·16 in duodenary scale.

ARITHMETIC.

NOTE. Since 1 ft. = 12 in. or 12′; 1 in. = 12″, 1″ = 12‴, &c., the Duodecimal Scale is often applied to examples involving the calculation of areas of surfaces and contents of solids.

Ex. 5. Required the area of a room 17 ft. 3 in. long, and 13 ft. 10 in. broad.

$$\text{17 ft. 3 in.} = (15 \cdot 3)_{12}, \text{ and 13 ft. 10 in.} = (11 \cdot 7)_{12}$$

$$\begin{array}{r} 15\cdot 3 \\ 11\cdot 7 \\ \hline 1246 \\ 153 \\ 153 \\ \hline 17\tau\cdot 76 \end{array}$$

$$(17\tau \cdot 6)_{12} \text{ sq. ft.} = (238)_{10} \text{ sq. ft.} + \left(\frac{7}{12} + \frac{\cdot 6}{144}\right) \text{ sq. ft.}$$

$$= \left(238 + \frac{90}{144}\right) \text{ sq. ft.}$$

$$= 238 \text{ sq. ft. } 90 \text{ sq. in.}$$

Ex. 6. Find the product of 5 yds. 2 ft. 2 in. 3 pts. and 5 yds. 11 in. 7 pts.

5 yds. 2 ft. 2 in. 3 pts. = 17 ft. 2 in. 3 pts. = $(15 \cdot 23)_{12}$ ft.
5 yds. 11 in. 7 pts. = 15 ft. 11 in. 7 pts. = $(13 \cdot \epsilon 7)_{12}$ ft.

$$\begin{array}{r} 15\cdot 23 \\ 13\cdot \epsilon 7 \\ \hline \tau 039 \\ 13909 \\ 4369 \\ 1523 \\ \hline \end{array}$$

$$1\tau\tau \cdot 4\tau 09 = (274)_{10} \text{ sq. ft.} + \left(\frac{4}{12} + \frac{\tau}{12^2} + \frac{9}{12_4}\right) \text{ sq. ft.}$$

= 274 sq. ft. + 4 superficial primes + 10 superficial seconds + 9 superficial fourths.

Ex. 7. A rectangular plot of ground contains 273 sq. ft. 53 sq. in., its breadth is 14 ft. 7 in.; find its length.

$$273 \text{ sq. ft. } 53 \text{ sq. in.} = (1 \tau 9 \cdot 53)_{12} \text{ ft.}$$
$$14 \text{ ft. } 7 \text{ in} = (12 \cdot 7)_{12} \text{ ft.}$$

$$12 \cdot 7) \; 1\tau 9 \cdot 53 \; (16 \cdot 9$$
$$ \underline{127}$$
$$ 825$$
$$ \underline{736}$$
$$ \tau\epsilon 3$$
$$ \underline{\tau\epsilon 3}$$

ft.　　　ft.
$$(16 \cdot 9)_{12} = (18 + \tfrac{9}{12})_{10} = 18 \text{ ft. } 9 \text{ in.}$$

Ex. LXIII.

1. Express $7\frac{25}{26}$, $37\frac{16}{27}$, $940\frac{11}{17}$, and $\frac{125}{1728}$ in the senary and duodenary scales, and $42\frac{0}{25}$ in the quinary scale.

2. Express $123 \cdot 456$, $23 \cdot 125$, $1637 \cdot 52$, $376 \cdot 54$ in the undenary, nonary, and octenary scales.

3. Transform $345 \cdot 6273$, $7304 \cdot 513$, and $13 \cdot 454$ from the octenary to the ternary, and to the quaternary scales.

4. Divide $511173 \cdot 44$ by $\cdot 675$ in the octenary scale; and find the value of $(28 \cdot 0725)_9$.

5. The area of a rectangle is 29 ft. 4 in. and its breadth is 2 ft. 3 in. 6 parts; what is its length?

6. A room is 30 ft. 5 in. long, by 17 feet. 7 in. wide, and its height is 9 ft. 10 in.; what will the painting of it come to at 2s. 3d. per yard?

APPLICATION OF ARITHMETIC TO GEOMETRY.

203. *A Geometrical Point* is that which has no parts or no magnitude.

204. *A Geometrical Line* has length only.

205. *A Straight Line* is that which lies evenly between its extreme points

206. Other lines than straight lines are called CURVED or CROOKED LINES.

If the points A and B; C, E, and D be joined, as in the fig*. AB, CED; the lines AB and CED are CURVED or CROOKED LINES.

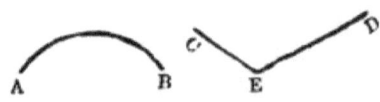

207. A line, or *linear content, is measured Arithmetically* by the number of times, or parts of a time, it contains a certain fixed line, which has been fixed upon as the unit of length or measurement. Thus, if we take one foot as the unit of measurement, and call it 1. a line of 3 yds., or 9 feet in length will be denoted by 9, a line of 2 yds, 18 in, by $7\frac{1}{2}$; a line of 1 in. by $\frac{1}{12}$ and so on.

208. A FIGURE or BODY is a portion of space enclosed by one or more boundaries.

209. THE SUPERFICIES, SURFACE, or AREA of a Body has only length and breadth, and not thickness, and may be defined to be the outward coat or face of the body. It is called a PLANE SUPERFICIES, SURFACE, or AREA, or simply a PLANE, when it is such, that whatever two points are taken in it, the straight line between them lies wholly in the superficies.

210. A superficies, surface, or area, or is *measured Arithmetically* by the number of times, or parts of a time, it contains a certain fixed area, which has been fixed upon as the unit of measurement. Thus, if 1 sq. ft. be called 1, 1 sq. yd. = 9 sq. feet will be denoted by 9, and 1 sq. in. by $\frac{1}{144}$, and so on.

211. A PLANE RECTILINEAL ANGLE is the inclination of two straight lines which meet together in a point, but are not in the same straight line.

The straight lines AB, BC meeting together at the point B, but not both of them in the same straight line BA or BC, form at the vertex B, the angle ($\angle$) ABC, or CBA.

$\angle ABC$ is said to be greater than $\angle DBC$, and less than $\angle EBC$.
The student will hence observe that the magnitude of a plane rec-

tilineal angle is not at all affected by the lengths of the straight lines which include it, but only by the greater or less inclination of the one line to the other.

212. When one straight line AB standing upon another straight line CD makes the $\angle ABC$ equal to the adjacent $\angle ABD$, then each of the $\angle$'s ABC, ABD is called a RIGHT ANGLE, and the line AB is said to be perpendicular, or at right angles to CD.

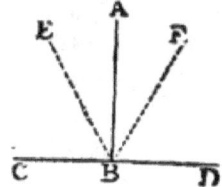

An OBTUSE ANGLE, viz. $\angle EBD$, is *greater* than a right angle.
An ACUTE ANGLE, viz. $\angle FBD$, is *less* than a right angle.

213. A CIRCLE is a plain superficies bounded by one line called the CIRCUMFERENCE and is such that all straight lines drawn from a certain point within it, called THE CENTRE, to the circumference are equal to one another.

The plane superficies, surface, or area $ABCDE$ is a circle, of which the *curved line* $ABCDE$ is the circumference or periphery, and the point F is the centre, a point such that all straight lines, FA, FB, FC, FD, &c., drawn from it to the circumference, are equal in length.

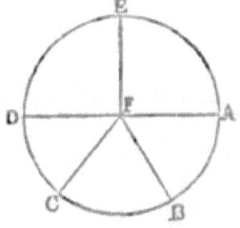

214. THE RADIUS of a circle is a straight line drawn from the centre of the circle to its circumference; in the above figure each of the lines FA, FB, FC, FD, is a *radius* of the circle $ABCDE$.

215. THE DIAMETER of a circle is a straight line drawn through the centre of the circle, and terminated both ways by the circumference; in the above figure, the straight line, AD is a *diameter* of the circle $ABCDE$.

NOTE 1. The circumference of any circle = its diameter × $\tfrac{22}{7}$.

NOTE 2. The area of any circle = the square of its radius × $\tfrac{22}{7}$.

216. RECTILINEAL FIGURES are those plane superficies, surfaces, or areas, which are contained or bounded by *straight* lines only

217. A TRIANGLE is a plain superficies, surface, or area which is bounded by *three* straight lines. The plane superficies, surface, or area contained by the straight lines *AB, AC*, and *BC*, is called the triangle *ABC*, or *CBA*, or *BAC*, whose sides are *AB, AC BC*, and whose ∠s are *ABC*, or *CBA, BCA*, or *ACB*, and *BAC* or *CAB*.

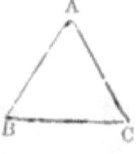

218. If one of the ∠s of a triangle, as *ACB* in the triangle *ABC*, be a *right angle*, the triangle *ABC* is said to be a *right-angled* triangle, and the side *AB* opposite to the angle *ACB* is called the *hypothenuse, BC* the *base*, and *AC* the vertical height or *altitude* of the triangle.

NOTE 1. In the right-angled triangle *ABC*, the right angle being *ACB*, square described on *AB*= square described on *AC*+square described on *BC*.

Euclid, B 1. 47.

If *one* of the angles of a triangle be *obtuse*, the triangle is said to be *obtuse angled*.

If *all* the angles of a triangle be *acute*, the triangle is said to be *acute angled*.

If *all* the angles of a triangle be *equal*, the triangle is said to be *equiangular*.

If *all* the sides of a triangle be *equal*, the triangle is said to be *equilateral*.

If *two* of the sides of a triangle be *equal*, the triangle is said to be *isosceles*.

NOTE 2. Area of any △ = ½ any side of it × perpendicular drawn upon that side, or that side produced, from the opposite angle.

If *ABC, A' B' C* be triangles, and *AD, A' D'* be drawn from *A* and *A* perpendiculars on *BC*, and on *B' C'* produced;

Then   area of △ $ABC = \frac{1}{2}$ of $BC \times AD$.
        area of △ $A'B'C' = \frac{1}{2}$ of $B'C' \times A'D'$.

Euclid, B 1. 41.

219. PARALLEL STRAIGHT LINES are such, as being produced ever so far both ways in the same plane never meet.

220. A FOUR-SIDED FIGURE, or QUADRILATERAL, is a plane superficies, surface, or area, which is bounded by *four* straight lines.

221. A PARALLELOGRAM is a plane superficies, surface, or area, which is bounded by *four* straight lines, of which each opposite two are parallel.

The plane superficies, surface, or area $ABCD$, contained by the four straight lines AB, BC, CD, and DA, of which AB is parallel to DC, and AD is parallel to BC, is called the *Parallelogram ABCD*, or *BCDA*, or *CDAB*, &c.; the side BC on which it stands is called its base, and AE or DF drawn perpendicular from A or D on BC or BC  produced is called its *altitude* or *height*. The straight lines joining A and C, B and D, are called the *diameters*, or *diagonals*, of the parallelogram.

222. A SQUARE is a parallelogram, which has all its sides equal, and each of its angles a right angle.

223. AN OBLONG or RECTANGLE is a parallelogram which has each of its angles a right angle, and not all its sides equal, but only the opposite sides equal to each other.

NOTE 1. Area of any parallelogram = its base × altitude.

NOTE 2. The perimeter of any rectilineal figure = the sum of its sides.

224. A SOLID is that which hath length, breadth and thickness.

225. THE CAPACITY, or VOLUME, or *content* of a solid is the quantity of space comprehending length, breadth, and thickness, which it contains or takes up; and it is *measured Arithmetically* by the number of times, or parts of a time, it contains a certain fixed volume, which has been fixed on as the unit of measurement.

226. A RECTANGULAR PARALLELOPIPED is a solid, contained by six right-angled quadrilateral figures, whereof every opposite two are parallel.

The figure *ABCD* is a rectangular parallelopiped.

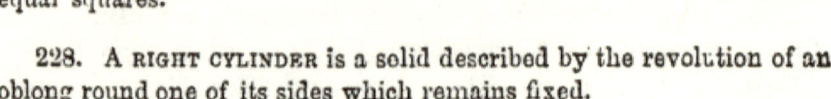

227. A CUBE is a solid figure contained by six equal squares.

228. A RIGHT CYLINDER is a solid described by the revolution of an oblong round one of its sides which remains fixed. The side of the oblong which remains fixed is called the *axis* of the cylinder. The side of the oblong opposite to the fixed side or axis traces out the *cylindrical surface*. The *circles* traced out by the other sides of the oblong are called the *bases* of the cylinder.

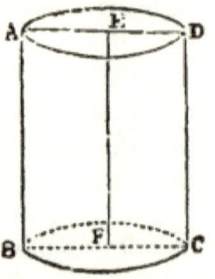

The figure *ABCD* is a right cylinder, traced out by the revolution of the oblong *ABEF* about the fixed side or axis *EF*.

229. In the Tables of Square and Cubic Measure we have seen that length multiplied by length produces area, and area multiplied by length produces capacity; the units in the products in these cases differing in kind from the units in the factors; thus, a rectangular area, whose adjacent sides are 4 and 3 feet respectively, is divisible into (4×3) or 12 equal squares, as shewn by the accompanying figure, the length of a side of each square being one linear foot. The rectangular area in this case is said to be the product of the two adjacent sides, represented respectively by numbers, the unit in the numerical product being no longer linear feet, but square feet. Similarly, if the adjacent edges of a rectangular parallelopiped be 3, 4, and 2 feet, respectively, the capacity of the solid is equivalent to 24 cubes, each containing one cubic foot; and thus the capacity of the parallelopiped is correctly expressed by the product of the three adjacent edges represented respectively by numbers, the units in the numerical product being no longer linear feet, as in the factors, but cubic feet.

APPLICATION OF ARITHMETIC TO GEOMETRY. 261

Examples in Square and Cubic Measure.

Ex. 1. Find the area of a rectangular court-yard 17 ft. 6 in. long, and 13 ft. 4 in. broad.

area = (17 ft. 6 in.) × (13 ft. 4 in.) = $17\frac{1}{2}$ ft. × $13\frac{1}{3}$ ft. = $\left(\frac{35}{2} \times \frac{40}{3}\right)$ sq. ft.

= $\frac{700}{3}$ sq. ft. = $233\frac{1}{3}$ sq. ft. = 25 sq. yds. 6 sq. ft. 48 sq. in.

Ex. 2. Find the cost of paving a floor, whose length is 33 ft. 2 in. and breadth 18 ft.: at 6s. per sq. yd.

Area of floor = (33 ft. 2 in.) × 18 ft. = $33\frac{1}{6}$ ft. × 18 ft.

$\left(\frac{199}{6} \times \frac{18}{1}\right)$ sq. ft. = $\left(\frac{199}{6} \times \frac{18}{9}\right)$ sq. yds.;

∴ cost of paving floor = $\left(6 \times \frac{199}{6} \times \frac{18}{9}\right)s.$ = 398s. = £19. 18s.

Ex. 3. A block of stone is 2 yds. 1 ft. 3 in. long, 1 ft. 7 in. broad, and 2 ft. thick: find its solid content, and its value at 2s. 3d. per cub. ft.

Cont. = (7 ft. 3 in.) × (1 ft. 7 in.) × 2 ft. = $7\frac{1}{4}$ ft. × $1\frac{7}{12}$ ft. × 2 ft.

= $\left(\frac{29}{4} \times \frac{19}{12} \times 2\right)$ cub ft. = 22 cub. ft. 1656 cub. in.

Its value = $\left(2\frac{1}{4} \times \frac{29 \times 19}{4 \times 12} \times 2\right)s.$ = $\frac{1653}{32}s.$ = £2. 11s. $7\frac{7}{8}d.$

NOTE. Since linear ft. multiplied by linear ft. give sq. ft., it follows that sq. ft. divided by linear ft. give linear ft., and so on. Thus 3 linear ft. × 4 linear ft. = 12 sq. ft., ∴ 12 sq. ft. ÷ 3 linear ft. = 4 linear ft. and 12 sq. ft. ÷ 4 linear ft. = 3 linear ft.

Again, since linear ft. multiplied by sq. ft. give cub. ft., it follows that cub. ft. divided by linear ft. give sq. ft., and cub. ft. divided by sq. ft. give linear ft. Thus 12 linear ft. × 4 sq. ft. = 48 cub. ft., ∴ 48 cub. ft. ÷ 12 linear ft. = 4 sq. ft., and 48 cub. ft. ÷ 4 sq. ft. = 12 linear ft.

Ex. 4. Find the expense of carpeting a room 15 ft. 9 in. long, and 12 ft. 5 in. broad, with carpet $\frac{3}{4}$ yd. wide, at $1 a yard

Area of floor = $(15\frac{3}{4} \times 12\frac{5}{12})$ sq. ft. = $\left(\frac{63}{4} \times \frac{149}{12}\right)$ sq. ft.

= $\left(\frac{63}{4} \times \frac{149}{12} \times \frac{1}{9}\right)$ sq. yds. = $\left(\frac{7}{4} \times \frac{149}{12}\right)$ sq. yds.

ARITHMETIC.

Now required length of carpet in linear yds. $\times \frac{3}{4}$ linear yd.

$= $ no. of sq. yds. in area of floor $= \left(\frac{7}{4} \times \frac{149}{12}\right)$ sq. yds.

$\therefore$ length of carpet in linear yds.

$$= \frac{\left(\frac{7}{4} \times \frac{149}{12}\right) \text{ sq. yds.}}{\frac{3}{4} \text{ linear yd.}} = \frac{7}{4} \times \frac{149}{12} \times \frac{4}{3}$$

$\therefore$ cost of carpet $= \left(1 \times \frac{7}{12} \times \frac{149}{3}\right) = \$28.97\frac{2}{3}$.

Ex. 5. Find the expense of painting the walls and ceiling of a room whose height, length, and breath are 17 ft. 6 in., 35 ft. 4 in., and 20 ft. respectively, at 15 cts. per sq. yd.

Area of the 2 length walls $= 2$(height $\times$ length), or $2(H \times L)$.
.... of the 2 breadth walls $= 2$(height $\times$ breadth), or $2(H \times B)$.
.... of the ceiling $\quad =$ length $\times$ breadth, or $L \times B$.
$\therefore$ area to be painted
$= 2(H \times L) + 2(H \times B) + L \times B = 2H \times \{L+B\} + L \times B$.
$= (2 \times 17\frac{1}{2})$ ft. $\times (35\frac{1}{3} + 20)$ ft. $+ (35\frac{1}{3} \times 20)$ sq. ft.

$= (2 \times 17\frac{1}{2} \times 55\frac{1}{3} + \frac{106}{3} \times 20)$ sq. ft. $= \frac{7930}{3}$ sq. ft. $= \frac{7930}{3 \times 9}$ sq. yds.

$\therefore$ expense of painting $= \left(15 \times \frac{7930}{3 \times 9}\right)$ cts. $= \$44.05\frac{5}{9}$.

230. Another method of working examples in square and cubic measure is styled CROSS MULTIPLICATION or DUODECIMALS, and it is generally employed by painters, bricklayers, &c., in measuring work. They take the dimensions of their work in feet, inches, parts, &c., decreasing from the left to the right in a twelve-fold proportion; thus, 12 inches $=1$ foot, 12 parts $=1$ inch, &c.: the inches, parts, &c., are termed primes, seconds, thirds, &c., and are distinguished by the accents ′, ″, ‴, &c., placed a little to the right above the numbers to which they belong.

SQUARE AND CUBIC MEASURE.

Rule for Cross Multiplication.

231. Rule. Write the terms of the multiplier under the corresponding terms of the multiplicand. Multiply every term in the multiplicand, beginning at the lowest, by each term of the multiplier successively, beginning with the highest; divide each product which is not of the denomination of feet by 12, add the quotient to the next product, and place the remainder under the term of the multiplicand just used, when the denomination of the multiplier is feet, one place remove to the right when it is primes, two places when it is seconds, three when it is thirds, &c. Add the products together, carrying 1 for every 12, and the sum will be the answer.

Ex. 1. Multiply 4 ft. 7 in. by 9 ft. 6 in.

By the Rule,

```
ft.
 4 .  7'
 9 .  6'
41 .  3
 2 .  3 . 6
43 .  6 . 6"
```

The product = 43 sq. ft. $+\frac{6}{12}$ths of a sq. ft. (or 6 *superficial primes* as they are called) $+\frac{6}{12}$ths of a superficial prime, *i. e.* $\frac{6}{144}$ths of a sq. ft. (or 6 *superficial seconds* as they are called.)

$\therefore$ product $= (43 + \frac{6}{12} + \frac{6}{144})$ sq. ft. $= 43$ sq. ft. $+ \left(\frac{6 \times 12 + 6}{144}\right)$ sq. ft.

$= 43$ sq. ft. $+ \frac{78}{144}$ sq. ft. $= 43$ sq. ft. 78 sq. in.

Ex. 2. Find by cross multiplication the capacity of a cube whose edge is 2 ft. 8 in.

```
ft.
 2 .  8'
 2 .  8
 5 .  4
 1 .  9 . 4"
 7 .  1 . 4"
 2 .  8
14 .  2 . 8
 4 .  8 . 10 . 8'''
18 . 11 . 6 . 8
```

The product

$= 18$ cub. ft. $+ \left(\frac{11}{12} + \frac{6}{144} + \frac{8}{1287}\right)$ cub. ft.

$= 18$ cub. ft. $+ \frac{1584 + 72 + 8}{1728}$ cub. ft.

$= 18$ cub. ft. $+ \frac{1664}{1728}$ cub. ft.

$= 18$ cub. ft. $+ 1664$ cub. in.

Ex. LXIV.

L will stand for length, B for breadth, H for height.

(1) Find the circumference of a wheel whose diameter is 4 ft. 8 in.: how many times will it turn round in $10\frac{1}{4}$ miles?

(2) How much space does a circular pond occupy, whose diameter is 15 ft.?

(3) 1. Find the diameter of a wheel which turns 4290 times in $15\frac{3}{4}$ miles?

2. A circular pond contains $2\frac{1}{2}$ acres; find its diameter.

(4) 1. A horse in turning a mill moves round at a distance from its center of 6 ft. 5 in., and makes on the average 35 circuits every $3\frac{1}{2}$ min.; how much is his pace less than 5 miles an hour?

2. A circular flower-bed, 16 ft. in diameter, has a grass border round it 4 ft. wide: find the number of sq. yds. in the border.

(5) 1. What will it cost to fence a circular bowling-green, whose radius is 52 ft. 6 in., at 84 cts. a yard?

2. A cow, tethered by a rope 7 yds long fastened to a stake in the middle of a pasture, has its rope doubled in length; how much greater space is it allowed than at first?

(6) 1. Find the hypothenuse of a right-angled triangle, whose other sides are 24 ft., and 27 ft. 6 in.

2. The hypothenuse of a triangular plot is 4 chs, 25 lks., the base is 2 chs. 55 lks.; find the other side.

(7) 1. The circumference of a circular spot is such that it encloses 1386 sq. yds.; how much is its radius less than the side of a square of the same area as the circle?

2. How long will it take a person, who walks 3 miles an hour, to walk twice round a square field containing 32 ac. 64 po.?

(8) 1. If from the extremity of a path 24 ft. wide, a ladder reaches 1 ft. 7 in. over the top of a house 45 feet high on the other side of the path; find the length of the ladder.

2. If the end of the ladder be shifted 2 feet further from the house, and then just reach to the top of a house 40 feet high on the other side of the street; find the width of the street.

(9) Two engines start from the same station, the one due North at the rate of 35 miles an hour, and the other due East at the rate of $17\frac{1}{2}$ miles an hour; how far will they be apart at the end of 4 hours?

SQUARE AND CUBIC MEASURE.

(10) Find the area of a triangle whose base is 45 feet, and altitude 17 feet.

(11) A triangular piece of ground containing $4\frac{1}{2}$ acres has a base of 135 yards; find its altitude.

(12) Find the area of each of the following rectangles, whose dimensions are: 1. $L = 36$ ft., $B = 13$ ft. 2. $L = 11$ ft. 6 in., $B = 9$ ft. 9 in. 3. $L = 20$ ft. 3 in., $B = 20$ in. 4. $L = 5$ ft. 6 in., $B = 3$ ft. 5 in. 5. $L = 8$ ft. 9 in., B 3 ft. 8 in. 6. $L = 8$ yds. 2 ft. 7 in., $B = 5$ ft. 4 in.

(13) Work by Cross Multiplication 1. 14 ft. 7′ by 5 ft. 10′. 2. 5 yds. 7′ by 1 yd. 1 ft. 11′. 3. 7 yds. 6′.10″ by 11′.7″. 4. 9 ft. 8′ by 6 ft. 10″. 5. 2 yds. 1 ft. 5′.8″ by 9 ft. 59″. 6. 11 ft. 1′.4″ by 3 ft. 4′.

(14) Find the solid content of the following: 1. $L = 13$ ft. 4 in., $B = 7$ ft. 6 in., $H = 3$ ft. 10 in. 2. $L = 20$ ft., $B = 1$ ft. 6 in., $H = 1$ ft. 2 in. 3. Of a cube whose edge is 6 ft. 7 in. Work (3) by cross multiplication.

(15) Find the length of the following rectangles: 1. Area $= 40$ sq. yds., $B = 20$ ft. 2. Area $= 46$ sq. ft. 126 sq. in., $B = 2$ ft. 6 in. 3. Area $= 6$ sq. ft., $B = 9$ in. 4. Area $= 470$ sq. yds. 5 sq. ft., $B = 25$ yds. 2 ft.

(16) Find the area of the 4 walls of the following rooms:

1. $L = 32$ ft., $B = 18$ ft., $H = 11$ ft. 2. $L = 22$ ft. 6 in., $B = 16$ ft. 8 in., $H = 12$ ft 4 in. 3. $L = 29$ ft., $B = 23\frac{1}{2}$ ft., $H = 11\frac{1}{4}$ ft. 4. $L = 32$ ft. 6 in., $B = 16$ ft. 9 in., $H = 13$ ft. 3 in. 5. $L = 27$ ft. 3 in., $B = 17$ ft. 9 in., $H = 13$ ft. 6 in., deducting in (5) for a fire-place 6 ft. square, a door-way 8 ft. by 4 ft. 3 in., and two windows each 7 ft. by 4 ft.

(17) Find the cost of painting a surface, 1. 19 ft. 6 in. by 83 ft. 4 in. at 40 cts. a square foot. 2. 25 ft. 8 in. long, and 16 ft. 9 in. wide, at 65 cts. a square foot.

(18) 1. What length of paper, $\frac{3}{4}$ yd. wide, will cover a wall 15 ft. 8 in. by 11 ft. 3 in.?

2. What length must be cut off a board, which is $7\frac{1}{2}$ in. broad, so that the area may contain 3 square feet?

(19) 1. A square field has a diagonal path across it measuring 7 chs. 35 lks.; shew that the area of the field $= 2$ ac. 3 ro. nearly.

2. Find the number of acres in a square field whose side is 4 chs. 50 lks.

3. A rectangular field is 7 chs. 35 lks. long, and 5 chs. broad; $1\frac{3}{4}$ ac. is to be cut off from it by a line parallel to its breadth; where must this line be drawn.

(20) 1. What length of carpet, 1 yd. 4 in. wide, will be required for a room whose length is 16 ft. 6 in., and width 10 ft. 8 in.?

2. A semicircular plot of ground, whose radius is 12 yards, has inside the circumference a path 2 yards wide; the rest of the space is a flower-bed; find the size of the bed.

(21) Find the cost of carpeting the following rooms:

1. $L=26$ ft. 8 in., $B=20$ ft. 3 in., with carpet $\frac{2}{3}$ yd. wide at $4s.\ 8d.$ a yd. 2. $L=20$ ft. 3 in., $B=17$ ft. 4 in., with carpet $\frac{3}{4}$ yd. wide, at $4s.\ 2d.$ a yd.

(22) How many yds. of paper, 1 ft. 4 in. wide, will be required for a square room, whose side is 18 ft. 9 in. and height 13 ft. 4 in.?

(23) What is the cost of papering a room, $L=24$ ft. 4 in., $B=26$ ft. 6 in., $H=18$ ft., with paper 28 in. broad, 5 cents per yard?

(24) Find the cost of papering a room, 19 ft. 8 in. wide, 24 ft. 4 in. long, and $13\frac{1}{2}$ ft. high, with paper $2\frac{3}{4}$ ft. wide, which costs $11s$ per piece of 12 yds.; the windows and parts not requiring paper making up a sixth of the whole surface.

(25) 1. Find the weight of water in a bath, 6 ft. long, 3 ft. wide, and 1 ft. 9 in. deep, the weight of 1 cub. ft. of water being 1000 ounces.

2. The bottom of a cistern contains 16 sq. ft. 128 sq. in.; how deep must it be to contain 1216 gallons? 1 gallon contains $277\frac{1}{4}$ cub. in. nearly.

(26) 1. A cylindrical pail is 14 in. in diameter, and 14 in. in height, how often can it be filled from a cubical cistern each of whose inside edges is 7 ft. $8\frac{2}{3}$ in.?

2. How many bushels of malt are there on the floor of a cylindrical kiln, the diameter of the floor being $6\frac{1}{4}$ yds., and the depth of the malt being 14 in.? NOTE. 1 bus.$=2218\cdot192$ cub. in.

3. The diameter of the base of the standard bushel being $18\frac{1}{2}$ in. nearly; find its height.

(27) 1. How many flag-stones each $5\cdot76$ ft. long, and $4\cdot15$ ft. wide are required for paving a cloister which encloses a rectangular court $45\cdot77$ yds. long and $41\cdot93$ yds. wide: the cloister being $12\cdot45$ ft. wide?

2. A moat of the uniform width of 15 yds., and depth of $7\frac{1}{2}$ ft., surrounding a square plot of ground containing $2\frac{1}{2}$ acres is quite full of water: how many gallons will it contain? NOTE. 1 gallon contains $277\cdot274$ cub. in.

SQUARE AND CUBIC MEASURE.

(28) 1. If 12000 copy-books be used yearly, and each book contain 20 leaves, each leaf being 7¼ in. broad and 9 in. long, find how many acres could be covered by the leaves of these copy-books spread out on the ground.

2. The area of a rectangular field whose length is four times its breadth is 8 acres 1280 yards; find its perimeter.

(29) 1. A rectangular court is 80 yards long and 50 yards broad. It has paths joining the middle points of the opposite sides 6 feet wide, and it has also paths of the same breadth running all round it on the inside. The remainder is covered with grass. If the paths cost 1s. 8d. per square foot, and the grass 3s per square yard, find the whole cost of laying out the court.

2. I have 150 books to pack, which I want to put in two boxes, whose dimensions are—the larger one, 4½ ft., 2 ft. 8 in., and 2 ft.; the smaller, 4 ft., 2½ ft., and 1½ ft. I can get 50 books into the smaller, how many will remain unpacked when I have filled both the boxes, the books being all of the same size?

(30) Find the surfaces of the cubes, whose solid contents are 1. 5 ft. 621 in. 2. 14706 ft. 216 in.

(31) 1. How many cubes whose edges are each 2¾ in. can be cut out of a cube of which edge is 22 in.?

2. What must be the height of a cylindrical column of marble, the radius of whose base is 9 inches, in order that it may contain 5½ cub. ft.?

(32) A monolith of red granite in the Isle of Mull is said to be about 108 feet in length, and to have an average transverse section of 113 sq. ft. If shaped for an obelisk it would probably loose one-third of its bulk, and then weigh about 600 tons. Determine the number of cubic yds. in such an obelisk, and the weight in pounds of a cub. ft of granite.

(33) 1. If the diameter of a cylindrical well be 5 ft. 2 in., and its depth 27 ft. 6 in.; how many cubic yds. of earth were removed in order to form it?

2. A cylindrical copper 7 ft. in height, the radius of whose base is 2 ft. 8 in. is half full of water; how many gallons does it contain?

3. How many gallons must be drawn off to make the surface sink one foot?

EXAMINATION QUESTIONS.

The following questions have been selected from the Matriculation Examination papers set for several years at the Universities of McGILL, QUEEN, TRINITY, and TORONTO.

1.

(1) What conditions must be satisfied in order that one vulgar fraction may be capable of being added to or subtracted from another? If these conditions be fulfilled, explain why it is necessary to change the forms of the fractions before performing the operations. How are these changes effected in the case of decimals?

Add together $\$5\frac{1}{4}$ and 1.35 of a £ currency, and subtract $\frac{3}{4}$ of a £ sterling. (The £ sterling to be taken as equal to £1. 4s. 4d. currency.)

(2) Three students, A, B, C, are to divide between them at the end of a term of 9 weeks a sum of $\$125\frac{1}{2}$, the share of each being proportional to the work done by him. B can do half as much again as C in the same time, A twice as much. C works steadily 8 hours a day, B works 7 hours a day for the first 2 weeks, 5 for the next 2, and 3 for the remainder, except the last week, when he works 11. During the first 7 weeks, A works only 2 hours a day for 4 days in the week; but during the last fortnight he works 14 hours a day; but he finds that in the last 4 hours of each day he gets through no more than C could. Find how much each should receive.

(3) The freight by a steamer to a certain port is $\$19.40$ per ton measurement; by a sailing vessel it is $\$6$. Insurance is at $1\frac{1}{2}$ per cent. by steamer, and $4\frac{3}{4}$ by sailing vessel. Find whether it will be more advantageous to send (and insure) by steamer or by sailing vessel a package whose dimensions are 5 ft. 6 in., 4 ft. 4 in., and 3 ft. 3 in., and value $\$780$. A ton measurement is 40 cubic ft.

(4) The pound sterling being £1. 4s. $4\frac{1}{2}d$. currency, find the exact value of the sterling shilling in cents.

(5) Water flows into a tank from two taps which running separately would in $1\frac{3}{4}$ and $2\frac{1}{2}$ hours respectively fill it up to a certain level;

at this level a waste-pipe opens; both taps running, and the tank being kept filled to this level, in what time will the waste-pipe discharge a quantity of water equal to that in the tank?

(6) A building-lot 30 feet in front by 120 in depth is sold at $100 per foot frontage; how much is that per acre?

(7) Describe the respective advantages of the use of *vulgar* and *decimal* fractions.

Divide 3654 by 2·03, explaining each step in the process.

(8) State what different units of weight occur in the English system.

Fifteen guineas weigh 4 oz. Troy, the metal consisting of 11 parts gold and 1 alloy, and the value of the alloy being $\frac{11}{235}$ths that of an equal weight of gold. Find the price per lb. Avoirdupois of the alloy.

(9) A quantity of pulp, filling a trough 3 feet deep, 10 ft. 7 in. long, and 11 inches wide, is made into paper of the same width and of such a thickness that 12 sheets would measure a quarter-inch; find the length of paper made, the pulp losing $\frac{3}{8}$ths of its bulk in the manufacture.

(10) A grocer buys 150 lbs. of coffee at 14 cts. per lb., and 39 lbs. of chicory at 6 cts. per lb.; he pays an import duty of $12\frac{1}{2}$ per cent. *ad valorem*, and mixes and sells them at 25 cts. a lb., but by the use of a false balance gains $\frac{1}{4}$ oz. on every apparent lb. sold. Find the profit per cent. made on his outlay.

(11) Give a short description of the ordinary numerical notation, and prove that every vulgar fraction can be expressed by a decimal either terminated or repeating. If the base of the system had been *two*, by how many figures would 1000 have been represented, and how would the equality, *twice two are four* have been written in figures?

(12) The mint price of pure gold is £4. 4s. $11\frac{5}{11}$d. per oz. Troy, and the gold of the coinage has one part out of 12 alloy (the value of which may be neglected); find the Avoirdupois weight of a sovereign.

II.

(1) Explain the common system of notation. Multiply 357 by 234, explaining the different steps of the process.

(2) What is the first step you take, when you wish to add two vulgar fractions together? Explain clearly why this step is necessary.

ARITHMETIC.

Simplify $\dfrac{2}{3} \times \dfrac{6}{7} \times \dfrac{14}{3} + \left(\dfrac{4\frac{1}{2}}{4\frac{1}{4}} - \dfrac{\frac{2}{3}}{\frac{5}{7}} + \dfrac{17}{60}\right) \div \dfrac{3}{5}$

(3) The population is 1842265 souls, and the revenue from customs is $3595754 by an average duty of 12½ per cent. If the duty be raised to 20 per cent. and the consumption falls off one-tenth; how much is the average taxation per head altered?

(4) State the rule for finding the quotient in division of decimal fractions.

Divide ·034695 by ·000241.

Find the number of yards, feet, and inches in ·084 of a mile.

(5) What is the interest at 7 per cent. per annum of £133. 6s. 8d. from the 1st of January, 1862, to the 15th April, 1863?

(6) Extract the square root of

    (α) 74684164.

    (β) ·03275 to 4 places.

(7) State and explain the rule for reducing a circulating decimal to a vulgar fraction.

Example, ·07̇538̇6.

Express as a decimal

$$1 + \cfrac{3}{5 + \cfrac{\frac{2}{3+\frac{1}{4}}}{2 - \cfrac{\frac{1}{5}}{\frac{1}{7} + \frac{2}{3+\frac{1}{2}}}}}$$

(8) A note for $400 at 3 months is cashed by a broker at a discount of 2 per cent. per month. At maturity it is protested for non-payment, but is renewed for a further term of 3 months, upon the maker paying the protesting charges, which amount to $1, and interest at the rate of 2½ per cent. per month in advance. Find the rate per cent. per annum which the maker on retiring the note will have paid for the money originally advanced.

(9) Out of a mass of metal consisting of 3 oz. of gold, 18 oz. of silver, and 10 oz. of nickel, a jeweler makes 40 spoons, when gold is

worth $15.60, silver 95 cents, and nickel 25 cents. He can then sell each spoon for $2.25, but afterwards the value of gold rises 15 per cent., and that of silver 12 per cent., while nickel falls to 20 cents an ounce. Find the price at which he must sell to make the same profit.

(10) Explain the reason of the rules for pointing in the multiplication and division of decimals.

Divide $0\cdot000279$ by 300000, and multiply $23\cdot4\dot{1}5\dot{9}$ by $0\cdot08\dot{3}\dot{9}$

(11) What fraction of a lb. Troy is an oz. Avoirdupois?

(12) What fraction of $19s.\ 3\frac{3}{4}d.$ is $11s.\ 11\frac{1}{4}d.$?

III.

(1) Whence does it appear that a vulgar fraction may always be reduced either to a terminated or circulating decimal? Explain how to determine which kind of a decimal any given fraction will produce. Reduce to decimals $\frac{33}{1056}$, $\frac{3}{17}$, and express as vulgar fractions in their lowest terms $3\cdot0561$, $15\cdot601\dot{3}78\dot{9}$.

(2) What is an aliquot part?

Find by "Practice" the value of

(α) 1589 bushels at $3.75 per bushel.

(β) 1 ton, 6 cwt., 2 qrs., 6 lbs., 4 oz., at $17.13 per ton.

(3) Explain what is meant by *Interest* and *Discount*.

Find the time for which the discount on a certain sum of money will be equal to the interest on the same sum for a year; the rate of interest in both cases being 5 per cent.

(4) A piece of timber is 8 ft. 6 in. wide, 5 ft. 9 in. deep, and 20 ft. 3¾ in. long; at what distance from one end must a piece be cut off so that the remainder may contain 23 cubic yards?

(5) A railroad runs through an estate for 50 miles, taking up a space 22 yds. wide, and the ground is valued at £55 an acre. The owner receives in exchange a square field worth £10 a rood, and pays as balance £3600. What is the length of the field's side?

(6) £356·3₂₁/₂₁s. is paid as the present value of a bill due seven months from that date; at the same rate per cent. interest £126 is paid as the discount of £1726 for one year and nine months. Find the amount of the first bill.

(7) How can £11. 7s. 1d. sterling be divided into crowns, half-crowns, shillings, sixpences, and pence, so that there will be an equal number of each.

(8) What is meant by saying that gold is at a premium in the United States of America?

If the premium on gold be 105, find the discount on American treasury notes.

(9) I purchase in Toronto American silver on which there is a discount of 4 per cent., and taking it to New York, where gold and silver are both at a premium of 80, I there buy American paper money with the silver; gold falling to 150, I buy gold with my paper money, and upon my return to Toronto find that I have made just enough to pay my expenses, which were $120 in Canadian currency. What was the sum originally invested?

(10) What is meant by "the Funds?" Explain why the English Funds rose on the birth of the Prince Imperial of France.

(11) A person holds stock, in the English 3 per cents. which are at 98, to the amount of £1500 sterling. This he transfers to Canadian Government 6 per cents. which are at 105: find the alteration in his income in dollars if the £1 sterling is worth $4.87.

(12) When gold is at 250 in Wall street, New York, what further rise will make a reduction of one cent in the dollar

IV.

(1) The value of the old Spanish dollar (which was the unit of exchange between America and England) was 4s. 6d. sterling, but gold became the standard of the currency of the United States of America, by the acts of 1834-7, which made the gold eagle weigh 258 grains, being nine-tenths fine. The English coinage is of metal 22 carats fine, 40 lbs. being coined into 1869 sovereigns. With these data explain why the bank par of exchange between New York and London is said to be $109\frac{1}{2}$.

(2) Add three-fifths of 4s. 7d. to seven-twentieths of 1s. $5\frac{1}{2}d$., and subtract from the result thirteen-forty-eighths of 5s.

(3) Shew that every vulgar proper fraction can be reduced to a terminating or circulating decimal; and examine the form of a frac-

tion which gives rise to a decimal consisting of p digits which do not recur and of q digits which are repeating.

Reduce $\frac{1}{13}$ to a decimal; multiply the decimal by 1·4, and divide the product by $\frac{7}{85}$.

(4) The old standard bushel was defined by statute to contain 2150 cubic inches, but on examination was found to contain only 2124. By the Act of 1824 the bushel was declared to contain 2218 cubic inches. Examine the real loss on the rental (£1075) of a farm (which was calculated on a certain percentage of the selling price of the corn grown), supposing the price per bushel to remain the same.

(5) Having 3 separate parcels of powders weighing respectively 84 lbs., 3 oz., 15 dwts., Troy; 45 lbs., 10 oz., 4 drs., 12 grs., Apothecaries; and 32 lbs., 7 oz., 3·712 drs., Avoirdupois; how can I subdivide them into parcels weighing the same integral number of grains?

(6) The link of Gunter's chain being $7\frac{22}{23}$ inches, prove that ten square chains make an acre.

The Scotch ell being 37·069 inches, and 24 ells making the Scotch chain, what difference (in square feet) is there between 55 English and 42 Scotch acres?

(7) A grocer buys a stock of tea and sells $\frac{5}{8}$ths of its nominal amount at 82 cts. per lb., thus clearing $190; he now calculates that if he sells the remainder at 85 cts. per lb., he will on the whole make 30 per cent. on his outlay; but he has forgotten to take into account a loss in weight of 2 per cent. by waste in handling. How much less cash will he receive than he expected?

(8) Explain the distinction between *Simple* and *Compound* interest.

What rate per cent. per annum interest is discount on a note for one year at 7 per cent.?

What rate per cent. per annum interest (compound) is discount on a note for half a year at $3\frac{1}{2}$ per cent.?

(9) Prove the following rule for computing interest at 6 per cent. for a period of months and days, the substance of which was given in the Leader of March 11th, 1865.

Multiply the number of months by 5, and add one-sixth the number of days; multiply this sum by the principal expressed in dollars; the result will be the interest expressed in mills.

1

(10) A tradesman who gives six months' credit abates 5 per cent. for cash; find the rate of interest in order that this may be the true discount.

(11) Shew that changing the position of the decimal point is equivalent to multiplying or dividing by a power of 10 whose index is the number of places by which the position of the point is changed.

Multiply 23·58 by ·0005, and divide the result by ·36; in each case shewing the correctness of the rule by which the position of the point is determined.

(12) A person has an income of £100 from Bank of England stock, which pays $6\frac{1}{2}$ per cent. dividend, and sells at $198\frac{1}{2}$. Find his income in dollars if he sells out and invests in the Canadian 5 per cents. at $112\frac{1}{2}$, the £ sterling being equal to $4.86.

V.

(1) Reduce the numbers 3954 and 6872 from the denary scale to the nonary; obtain their product when thus transformed, and reduce the result to the septenary.

(2) A person purchases a quantity of goods in Liverpool for £37. 10s. sterling, and sells in Montreal for £65 Canadian currency; he pays in Montreal an *ad valorem* duty of $4\frac{1}{2}$ per cent. Neglecting other incidental charges, what is the gain per cent., supposing a pound sterling to be $4.87?

(3) A vessel contains 120 gallons of wine; 20 gallons are drawn therefrom and the vessel filled with water; 15 gallons of this mixture are then drawn and the vessel again filled with water. If this operation be performed six times, 20 and 15 gallons being drawn alternately, how much wine will the mixture contain?

(4) Reduce the decimals ·21316, ·31249, and ·8934 to their equivalent vulgar fractions.

(5) A, B, and C engage in trade; A contributes £150, B £200, and C £250. At the end of two years A draws £100, and one year after B £150. When the partnership is wound up, at the end of four years, it is found that there is to the credit of the firm £1000. Are the data sufficient to enable us to make an equitable distribution? Give reasons for your answer. If sufficient, determine the amount to which each is entitled.

(6) If the decimal point in any number be moved one place to the left, and then again, and so on, and the numbers thus formed be added together, the sum is the result of dividing the original number by 9.

(7) Which of the following statements is more nearly correct?

$$\frac{10}{9 \cdot 009} = 1 \cdot 11, \text{ or } \frac{10}{1 \cdot 11} = 9 \cdot 009.$$

(8) Prove that the square of $99 \cdot 9899995$ differs from 9998 by little more than a unit in the eighth decimal-place.

(9) The distance between the earth and moon being expressed by $59 \cdot 9643$ with reference to the earth's radius as unit, and this radius being $3962 \cdot 8$ miles, each of these numbers being exact to the nearest decimal; what can be known of the moon's distance from the earth in miles?

(10) The imperial gallon is defined as containing $277 \cdot 2$ cubic inches, and as holding 10 lbs. weight of water. What would be the error in saying that a cubic foot of water weighs 1000 oz.?

(11) In the French system, a cubic centimetre of water is said to weigh one gramme. The metre is $39 \cdot 371$ inches, and the gramme $15 \cdot 434$ grains. How does this compare with the English statement, and what is the reason of the difference?

(12) From the fact that ten square chains make an acre, deduce the length of a link in inches.

VI.

(1) Sterling gold is 22 carats fine, and from 40 lbs. of it are coined 1869 sovereigns. Jewelers' gold is 18 carats fine. An ornament made of the latter, and weighing 22 oz. was sold at an advance of two-thirds on its value by weight, and the jeweler's profit was equivalent to £1$\frac{400}{869}$ per oz., on the pure gold contained in it. What was the charge for workmanship disregarding the value of the alloys?

(2) Two persons, A and B, borrow \$300 on joint mortgage from a building society, A taking \$200 and B \$100, the amount being repayable principal and interest by equal monthly instalments to which A and B contribute proportionally. After a few payments have been made, they desire each to borrow \$200 additional, and propose to

merge the old debt and the new into a single mortgage for $700, whereupon the account stands thus:

  Amount required to pay off old mortgage............. $245
   " of new loan... 400
  Surplus... 55
    Total.. $700

Discuss the interest each has in this surplus of $55, and the proportion in which they should contribute to the instalments payable in future.

(3) From "10 square chains make one acre," deduce the length of a link in inches.

(4) Find the G. C. M. of 1859 and 3042.

(5) Find the prime factors of 6300.

(6) Find the value of $1\frac{1}{2} + 2\frac{2}{3} - 3\frac{3}{4}$.

(7) Find the value of $\frac{1}{2}$ yard + $\frac{3}{4}$ foot − $\frac{2}{3}$ in.

(8) Reduce $\dfrac{\frac{2}{3} + \frac{1}{4}}{\frac{2}{3} - \frac{1}{2}}$ to its simplest form.

(9) Find the square root of 213·536 and $\frac{1}{2}$.

(10) Find the cube root of 47045·881.

(11) Divide ·558 by ·024.

VII.

(1) Define a numerical fraction. Distinguish between Vulgar and Decimal Fractions. Express 5 fur., 3 per., $3\frac{1}{2}$ yds. as a fraction of a mile in each system.

(2) Divide ·01 by ·00001, and multiply the quotient by ·3.

(3) Light travels at the rate of 192000 miles per second. How long will it take to come from the Sun to the Earth, the distance being 95000000 of miles?

(4) Extract the square root of 373·45.

(5) Show that if any number expressed in the decimal scale be divisible by 9, the sum of its digits will be divisible by 9.

(6) Find $\frac{4}{5}$ of £2. 3s. 9d., and express the result as a fraction of £1. 2s. 6d.

(7) Reduce 0·35278 to a vulgar fraction. Prove the rule.

(8) If the gas consumed by one burner cost 17s. 9d. for 40 days, what will be the charge for another burner for 56 days; 220 cubic feet

of gas being consumed by the latter, while 115 are consumed by the former?

(9) Extract the square root of 0·000008.

(10) Reduce 328 to the binary scale.

(11) Find the amount of £5 in $2\frac{1}{2}$ years at 3 per cent. compound interest; the interest payable yearly.

(12) The national debt of the United Kingdom amounted, in the year 1860, to £80147741; the interest paid on it was £26833470; calculate the average rate per cent. paid as interest.

The total revenue for the year ended June, 1861, was £71863095: how much per cent. was the total interest of the total revenue?

VIII.

(1) A book consists of $21\frac{3}{4}$ sheets of 16 pages, each page containing 38 lines; how many sheets will it run to if printed in sheets of 24 pages, each page containing 32 lines; the length of the line in the latter case being $\frac{2 3}{3 3}$ that of the former?

(2) A bankrupt pays his creditors £1915. 10s. 6d.; calculate the whole amount of his debts, the composition being 9s. 5d. in the £1.

(3) Divide 358·3 by 1·27, and from the quotient subtract $\frac{2}{3}$ of $\frac{4}{5}$ of 12.

(4) Reduce 3 furlongs, 5 yards, 2 feet, 1 inch, to the decimal of a mile.

(5) Add $\frac{3}{4} + \frac{4}{5} + 1\frac{7}{8}$, and from the result substract $\frac{7}{8}$ of 2.

(6) Reduce the circulating decimal ·564 to the equivalent vulgar fraction.

(7) If the yearly rent of 325 acres 2 roods of land be $450, what would be the rent at the same rate of a square mile?

(8) Find the interest on £485. 7s. 6d. sterling for 3 years and 8 months at 6 per cent., and reduce the result to dollars and cents; £1 sterling being worth $4.86.

(9) Extract the square root of ·075, to 4 figures.

(10) Find the value of $\frac{2}{3}$ of $\frac{4\frac{1}{2}}{2}$ of 25 cwt. 3 qrs. 1 lb., and reduce the results to the decimal of a 100 cwt.

(11) Add together the fractions $\frac{1}{4} + 2\frac{3}{4} + 5\frac{9}{7} + \frac{3}{8}$, multiply the sum by $\frac{3}{5}$, and divide the product by 4 times the third of 7.

(12) Find the interest on $657.40 for three months and 10 days, at 5 per cent. Convert the result into sterling money, a pound being worth $4·86.

IX

(1) Add together $2\frac{1}{2} + \frac{3}{4} + 7\frac{5}{9}$; subtract from the sum the half of $\frac{2}{3}$, and divide the remainder by 6.

(2) The total value of the Imports of Canada for the year 1861 was $43054836, and the total duty on them was $4768192.89. What was the average rate per cent. levied?

(3) Find the interest on $19876.54 for 3 years and 5 months at $4\frac{1}{2}$ per cent.
Convert the result into Halifax currency.

(4) Extract the square root of 2 to 4 decimal places.

(5) Express 805 yds. 2 ft. 5 in. as a decimal of a mile, and verify the result by reducing the decimal to a vulgar fraction, and finding the value of that fraction of a mile.

(6) Calculate the ratio of the English mile to the French kilomètre; the kilomètre being equal to 1000 mètres, the mètre = 39·371 inches.

(7) Find the value of $\frac{3}{4}$ of 5s. 6d., bring it to the decimal of £1 currency, and convert the result into dollars and cents.

(8) If $100 in Canadian bank-notes be worth $103.50 in United States silver, what is the value of 367 United States silver dollars in Canadian currency?

(9) Find the interest on $650 at 6 per cent., for 3 years and 8 months.

(10) Add together the sum, the difference, the product, and quotient (the greater being divided by the less) of $\frac{5}{8}$ and $\frac{4}{7}$.
Give the reasons for the rule in each process mentioned.

(11) The weight of a cubic inch of water is 252·458 grains, a gallon of water weighs 10 lbs. Avoirdupois; find the number of cubic inches in a gallon.

(12) Reduce the fractions in question (10) to decimals; solve the question then, and shew that the two results coincide.

X.

(1) Give the rule for division of decimals and the reason for it.

(2) If gold be at a premium of 49 per cent. when purchased with United States notes, what is the gold value of $357 in notes?

(3) To what sum will $500 amount in 6 years, 5 months, and 20 days at 6 per per cent. per annum, simple interest?

(4) Extract the square root of 32·56.

(5) Add together $\frac{2}{3}$ and $\frac{3}{4}$: multiply the sum by $1\frac{3}{8}$, and divide the result by $4\frac{3}{8}$.

Reduce the above vulgar fractions to decimals, perform the same operations, and show that the results obtained by the two methods coincide.

(6) If 6 men will dig a trench 15 yards long and 4 broad in three days of 12 hours each, in how many days of 8 hours each will 8 men dig a trench 20 yards long and 8 broad?

(7) Divide the sum of 10 and $\frac{1}{15}$ by the difference, and also the difference by the sum, and find the difference of the two quotients.

(8) Find the value of ·439£. + 1·256s. + 3·718d.

(9) If 21 men mow 72 acres of grass in 5 days, how many must be employed to mow 460 acres, 3 roods, 8 perches in 6 days?

(10) What sum must be put out on interest at $4\frac{1}{2}$ per cent. to amount to £4027. 19s. 4d. in $5\frac{1}{2}$ years?

(11) Reduce £557. 19s. $5\frac{1}{4}d$. sterling to dollars and cents (the value of £1 sterling being $4.867), and then convert the dollars and cents to Halifax currency.

(12) Reduce the circulating decimal ·8325 to a vulgar fraction.

XI.

(1) What is the present worth of $3560 payable in 8 months, discount being at the rate of 6 per cent. per annum.

(2) A bar of gold is 4·17 inches long, 0·64 wide, 0·31 inches deep; a bar of silver is 13·22 inches long, 1·14 inches wide, 0·65 inches deep; find the ratio of the first bar to that of the second, if the weights of any equal bulks of gold and silver be in the ratio of 19·35 to 10·51.

(3) Add $\frac{1}{4} + 3\frac{1}{2} + 6\frac{7}{8}$; reduce the result to a decimal form, and divide it by the half of $\frac{2}{3}$ of 4.

(4) Find a number such that the square of it shall be one-and-a-half times 35.

(5) Find the interest on $3450.35, for 135 days, at $6\frac{1}{2}$ per per annum.

(6) Find how much per cent. is 53 of 65?

(7) Find the greatest common measure of 1281 and 7259.

(8) From the sum of $\frac{1}{2} + 3\frac{1}{4} + 2\frac{7}{8}$; take the difference of $\frac{3}{5}$ and $\frac{1}{3}$, and divide the remainder by the half of $\frac{2}{3}$.

(9) Convert $\frac{23}{39}$ into a decimal, and divide the square of the result by ·0012.

(10) The volume of a sphere whose radius is r is $\frac{4}{3}\pi r^3$ (where $\pi = 3\cdot 14159$); find hence in lbs. Avoirdupois the weight of a hollow globe $\frac{1}{4}$ of an inch thick, the diameter of whose internal surface is 3 inches, if the weight of one cubic inch of the material be 500 grains.

(11) Calculate the ratio of the English mile to the French kilometre; the kilomètre being 1000 metres (the mètre = 39·371)

(12) What is the difference between the income arising from £2500 invested in 5 per cent. stock, when the price of the stock is 114 and the same sum invested in 3 per cent. stock at 92?

XII.

(1) The greatest amount of sea-salt which 10 lbs. of pure water can dissolve is 37 lbs. How much salt will be required to saturate to an equal degree of saltness, 2 gallons and 3 quarts?

(2) The area of a circle (radius = r) is πr^2 and the volume of a cylinder with circular base is equal to area of the base multiplied by the height. Hence, find the height of a cylindrical jar which exactly contains a gallon (10 lbs.) of water, if the diameter of the base of the jar be 8 inches, and the weight of one cubic inch of water be 252·5 grains.

(3) British standard silver contains 37 parts in 40 of fine silver and 1 lb. Troy of standard silver is coined into 66 shillings. Calculate the value of the money which can be coined from 100 lbs. Avoirdupois of fine silver.

(4) The moon revolves in her orbit round the earth in 27 days, 7 hrs., 43 min., 11 sec. Through how many degrees of her orbit does she move in 7 days?

EXAMINATION QUESTIONS. 281

(5) If one steamer sail 3000 miles in 11 days, how far will another sail in 5 days, if she can sail 8 miles for the former's 7?

(6) The population of the city of London in 1801 was 864845, and in 1841 1690084. Find the rate per cent. of the increase in ten years.

(7) At present the value of the British sovereign is $4.86⅔; it is proposed to lessen the value of the dollar by Act of Parliament, so that the value of the sovereign shall be $5.04⅕. Calculate what sum in the proposed currency would be equivalent to $2600 of the present currency.

(8) The mean distance of Mercury from the Sun is 0·38 times the Earth's distance from the Sun. Assuming the earth to move in a circle round the sun in 365·25 days at the rate of 16·8 geographical miles per second, and that 60 geographical miles equal 69¼ statute miles; find the distance of Mercury from the sun in statute miles.

(9) Out of a cubical vessel, whose side is 2 feet and which is full of water, 5 gallons are drawn. Find by how many inches the depth of the water in the vessel is lessened, assuming that a gallon of water weighs 10 lbs. and a cubic foot 1000 ounces.

(10) Add together $\frac{3}{4} + \frac{4}{5} + 2\frac{7}{8}$; divide the result by half the difference between $\frac{5}{8}$ and $\frac{3}{16}$, and reduce the quotient to a decimal.

(11) Find the interest on £476. 16s. 8d. at 5½ per cent. per annum for 7 months.

(12) What is the interest on $3678.56 for 5 months at 6½ per cent. per annum.

XIII.

(1) Troy, Avoirdupois and Apothecaries' weight. Comparative weight of Troy and Avoirdupois pounds; advantages of requiring the use of only one kind of lb.

(2) Feet and yards in a mile. Feet and yards in a pole. Square feet and square poles in an acre. Cubic inches in a cubic foot and cubic feet in a cubic yard.

(3) Add $\frac{2}{3}$, $\frac{3}{15}$, $\frac{11}{15}$. Divide $\frac{5}{7}$ of $\frac{2}{9}$ of $13\frac{1}{2}$ by $\frac{2}{9}$ of $\frac{7\frac{1}{2}}{8\frac{1}{5}}$.

(4) Divide ·025 by ·12; 594·27 by ·047.

(5) Square root of ·00089; cube root of 140·, value of $\frac{7}{8}$ of acre,

(6) 75 yds. at 8¾d.

(7) Interest of £60 at 10 per cent. for one year. Of £27. 10s. at 6 per cent. for one year.

(8) If 5 men can build a wall in 6 days, how many can build it in one?

(9) A gentleman pays in all 50s. to his work-people, to each man 1s., each woman 8d., and each boy 4d., the number of each being equal; what was the number of each?

(10) If a family of 8 persons expend £200 in 9 mos., how much will serve a family of 18 persons 12 mos.?

(11) What is the price of 60 lbs. at 2s. 6d. a lb.? at 3s. 4d. a lb.?

(12) Give the cost of 1875 lbs. at $3 a ton.

XIV.

(1) Change ·327 into a vulgar fraction.

(2) Find the least fraction which added to the sum of 1.2. ·12, ·012, and 210, will make the result a whole number.

(3) Give the square root of 1·3 to four places of decimals. Give the cube root of $\frac{5}{7}$ to two places of decimals.

(4) Divide 3 days 8 hours by 2 hours and 40 minutes.

(5) If 15 pumps working 8 hours a day can raise 1260 tons of water in 7 days, how many pumps working 12 hours a day will raise 7650 tons in 14 days?

(6) If 12 men can dig a ditch in 4 days, in what time can 32 men perform the same work?

(7) How many yards of carpeting 27 inches wide will cover a room 14 × 16?

(8) Find the present value of $1 due in 3 months at 8 per cent.

(9) A person buys goods for £5. 17s. 6d. and sells them for £9. 18s. 6d. How much per cent. does he gain?

(10) If 3 oz. of gold be mixed with 9 oz. of silver, what is the value of 1 oz. of the alloy, gold being $18 and silver $1.25 per ounce?

(11) 5 lbs. of tea at $1, 9 lbs. at 90 cts., and 14½ lbs. at 80 cts., what is a lb. of it worth? and how many lbs. of each at the above rates must be taken to make a compound worth 85 cts. a pound?

EXAMINATION QUESTIONS.

XV.

(1) Find the value of (I) $\frac{5}{16}$ of 5 hrs. 25 min. 40 sec.
(II) $(\frac{1}{2} + \frac{3}{4})£. + (\frac{1}{2} + \frac{3}{5})s. + (\frac{1}{4} + \frac{4}{5})d.$ (III) £3. 18s. 6d. × 756¾.

(2) The forewheel of a carriage is 6 ft. 6 in. round, and the hindwheel is 11 ft. 4 in.; how far must the carriage travel before each wheel shall have made a number of complete turns? How often will this happen in 10 miles?

(3) Define a decimal fraction, and give the rules for pointing in the multiplication and division of decimals.
Divide ·001 by 1 × ·01 × 100; ·20736 by 1·2 × ·012 × 120, and 98·8452864 by 76·8 × ·0987.

(4) Write down the table of time. How do we determine whether any particular year is a leap-year? Are 1864, 1900, 1950, 2000 leap-years?

(5) What is the interest at 7 per cent. per annum of £133. 6s. 8d. from January 1st, 1862, to April 15th, 1864?

(6) Extract the square root of 74684164 and ·03275 to 4 decimal places.

(7) Divide £23. 15s. 7½d. by 37, and 571 yds. 2 qrs. 1 nail by 47.

(8) A wall that is to be built to the height of 27 feet was raised 9 feet high by 12 men in 6 days; how many men must be employed to finish it in 4 days?

(9) Reduce to their lowest terms $\frac{1344}{1536}$ and $\frac{252}{364}$.

(10) Find the value of (I) $\frac{5}{8}$ of an acre. (II) $\frac{1}{5}$ of 4s. 10d
(III) ·009943 of a mile. (IV) ·625 of a shilling.

(11) Reduce (I) $\frac{2}{15}$ of a pound to the fraction of a penny.
(II) $\frac{2}{7}$ cwt. to the fraction of a lb. (III) ·26d. to decimal of a pound.
(IV) ·056 of a pole to decimal of an acre.

(12) Find (a) the interest on £547. 15s. for 3½ yrs. at 5¾ per cent. per annum.
(β) The present value of £720 due in 4 yrs., at 5 per cent. interest.

(13) A square fishpond contains an acre; find the length of a side.

XVI.

(1) Add together $\frac{3}{8}$, $2\frac{1}{4}$, $13\frac{3}{10}$, and reduce $\frac{3}{4}$ of $2s.\ 4\frac{1}{2}d.$ to the fractions of $2s.\ 6d.$

(2) Reduce $3s.\ 4\frac{1}{2}d.$ to the decimal of £1, and $3s.\ 6a.$ to the decimal of £2. 10s.

(3) Prove the rule for pointing in the extraction of the square root of the numbers. Find the square root of 534·5344, and prove that
$$\frac{3\sqrt{8}-2\sqrt{7}}{\sqrt{8}-\sqrt{7}} = 17\cdot482\ldots$$

(4) The prime cost of a cask of wine of 38 gallons is £25, and 8 gallons are lost by leakage: at what price per gallon must the remainder be sold so as to gain 10 per cent. in the whole prime cost?

(5) If in Toronto there is a discount of $\frac{1}{2}$ per cent. on English gold, when exchange in London is quoted at 112, shew that a merchant who wishes to send money to London will save nearly 2 per cent., if instead of buying exchange he sends the gold, having given that the par of exchange is $109\frac{1}{2}$, that when exchange is at par the pound sterling is worth $4.87, and that the charge for freight, insurance, &c., on gold from Toronto to London is $\frac{3}{4}$ per cent.

(6) Prove that
$$\frac{\sqrt{5}+\sqrt{3}}{2\sqrt{2}} - \frac{\sqrt{3}}{\sqrt{5}-\sqrt{3}+\sqrt{2}} = \tfrac{1}{2} \text{ and } \left(\sqrt{2+\sqrt{2}+\sqrt{2}}\right)^3 - 3\sqrt{2+\sqrt{2+\sqrt{2}}}$$
$$= \sqrt{2+\sqrt{2-\sqrt{2}}}.$$

(7) Describe Gunter's chain, and explain fully how it is used to find the acreage of a field.

(8) Define Present Value and Discount. If the discount on £567 be £34. 14s. $3\frac{3}{7}d.$, simple interest being reckoned at $4\frac{1}{2}$ per cent. per annum, when is the same due?

(9) What is meant by "course of exchange" and by "par of exchange?" Explain briefly the cause of fluctuation in the price of exchange.

Exchange between Toronto and London being quoted at $112\tfrac{1}{4}$, what must I give for a Bill of Exchange for £18. 19s. stg.?

(10) Explain the method of transforming circulating decimals into their equivalent vulgar fractions, taking as examples $\cdot 6\dot{7}$, $\cdot 1\dot{4}$.

Simplify the following expressions, briefly explaining any artifices adopted:

(I) $\cdot 928571\dot{4} + \cdot 8214285\dot{7} + \cdot 4\dot{8} + 2 \cdot \dot{8}\dot{7}$.

(II) $11\cdot 036 - 3\cdot 98\dot{7}\dot{5}$.

(III) $\cdot 1\dot{6}\dot{3} \times \cdot 0\dot{6}$.

(IV) $1\cdot 01587\dot{3} \div 1\cdot 63\dot{6}9047\dot{6}1$.

(11) The true length of the year is 365·24224 days; find in what time the error in the common reckoning will amount to a day.

(12) Define a vulgar fraction, and prove the rule for multiplying fractions together, taking as example $\tfrac{2}{3} \times \tfrac{7}{8}$. Show which is greater $\sqrt[4]{\tfrac{2}{3}}$ or $\sqrt{\tfrac{2}{3}}$ without finding their actual values.

XVII.

(1) Among how many men must £105. 8s. 4d. be divided in order that each man may have £10. 10s. 10d.?

(2) Express the fractions $\tfrac{37}{8}$, $\tfrac{75}{104}$, $\tfrac{118}{14}$ as fractions having a common denominator, and express the difference of the first two as a fraction of the difference of the second two.

(3) How many ounces are there in a hundred-weight, and how many square yards in an acre?

(4) Divide 220·8864 by 72·66 and 2·208864 by ·07266.

(5) Find the rental of 5 acres, 2 roods, 27 poles at £1. 12s. 8d. per acre.

(6) What is the amount of $200 in 5 years at 5 per cent. compound interest?

(7) 3 men, 4 women, 5 boys or 6 girls can do a piece of work in 60 days; how long will it take 1 man, 2 women, 3 boys, and 4 girls working together?

(8) Find the difference in the expense of carpeting a room 17 ft. 9 in. long and 12 ft. 6 in. broad with Brussels carpet $\tfrac{3}{4}$ of a yard wide

at 4s. 6d. per yard, and with Kidderminster $\frac{7}{8}$ of a yard wide at 3s. 6d. per yard.

(9) What sum will amount to £425. 19s. $4\frac{4}{5}d$. in 10 years at $3\frac{1}{2}$ per cent. simple interest?

(10) What is the yearly interest arising from the investment of £385. 7s. $3\frac{1}{2}d$. in the purchase of 3 per cent. stock at $94\frac{3}{8}$?

(11) Write down the tables of Troy measure, and of square measure.

(12) Divide 109839 by 35 by short division, explaining the method of finding the remainder.

XVIII.

(1) Simplify (I) $\frac{23760}{26136}$. (II) $(\frac{2}{7} \text{ of } 5\frac{2}{11}) + \frac{1}{3} \text{ of } (2\frac{1}{2} + 6\frac{1}{4})$.

(III) $\dfrac{1}{2 + \dfrac{1}{3 + \dfrac{1}{4 + \dfrac{1}{5}}}}$
   (IV) $(\frac{7}{8} \text{ of } \frac{1}{2} \text{ of } \frac{3}{10}) \div (\frac{6}{11} \text{ of } \frac{7}{9} \text{ of } \frac{9}{2})$.
   (V) $\dfrac{1\frac{3}{7} \times \frac{1}{2} - 1\frac{1}{4} \times \frac{1}{3}}{\frac{16}{21} \times \frac{1}{2} - 1\frac{3}{4} \times \frac{1}{3}}$.

(2) Find the value of
(I) $£4\frac{1}{6} + 11\frac{1}{3}s. + 7\frac{9}{12}d.$ (II) $1416\,A\;2\,R\;16\,P \div \frac{1}{8}$ of $(4ac.\;3ro.\;27po.)$.

(3) Reduce $\frac{5}{8}$ of 1s. 9d. to the fraction of 3s. 4d.

(4) Find the value of (I) $2\dot{\cdot}\dot{7} - \dot{\cdot}91\dot{3}$. (II) $91\cdot78 \times \cdot381$.
   (III) $\cdot00044406 \div \cdot 0112$.   (IV) $2\dot{\cdot}2\dot{7} \div 1\cdot13\dot{6}$.
   (V) $\frac{1}{1.2} + \frac{1}{1.2.3} + \frac{1}{1.2.3.4} + \ldots$ to 5 places of decimals.

(5) Find by practice the value of 5 yds. 2 ft. 9 in. at 5s. $3\frac{1}{2}d$. per foot.

(6) If a lb. of standard gold which is 22 carats fine be worth £46. 14s. 6d., find the value of the Mohur of Bengal of weight 7 dwt. 23 grs., and fineness 993 parts pure gold to 7 of alloy.

(7) What sum invested at 4 per cent. simple interest will amount to £111 in 5 years?

(8) If £1000 of 3 per cent. stock at 72 be transferred to the 4 per cents. at 90, find the change of income.

XIX. 1874. (ADMISSION HIGH SCHOOLS.)

(1) By what must £157 12. $10\frac{1}{2}$ be divided to give a quotient of $33\frac{1}{2}$?

(2) How much wheat is necessary to sow a field containing $7\frac{3}{4}$ acres if $\frac{3}{4}$ of an ounce is sown on every square yard?

(3) How many minutes between 12 o'clock noon May 24th, and half-past nine in the forenoon of September 3rd? and express the answer as a fraction of the year.

(4) Add $1\frac{4}{9}$ of $\frac{9}{13}$, $\frac{3}{8}$ of $1\frac{12}{18}$, $\dfrac{\frac{16}{6}}{\frac{22}{9}}$

(5) A house and lot together cost \$3,600; the value of the lot is $\frac{1}{8}$ that of the house. Find the value of each.

(6) Subtract $2\frac{7}{800}$ sq. yards from $\frac{7}{9}$ of $\frac{3}{14}$ of 3 acres.

(7) Prove that multiplying the numerator of a fraction by any number produces the same effect as dividing the denominator by the same number.

(8) Simplify $\cdot75$ of $\frac{1}{3} \div 7\cdot6$ of $\frac{5}{76} - (1\cdot875 - 1\frac{5}{8}) \times 2 + \dfrac{4\cdot875}{4\frac{7}{8}}$

(9) If $\frac{2}{3}$ of $\frac{3}{7}$ of an acre produce 42 bushels of potatoes, how many bushels will an acre produce?

(10) A man working $9\frac{3}{4}$ hours per day finishes a piece of work in 6 days: in what time would he have finished it if he had worked $8\frac{1}{3}$ hours per day?

XX. 1876.

(1) Bought $19\frac{1}{3}$ yds. Irish linen at 5/4, $16\frac{3}{4}$ yds calico at 1/8, and $16\frac{1}{2}$ yds. silk at 8/4; find the amount of the bill in dollars and cents.

(2) Add together $\frac{3}{4}$ of $\frac{5}{8}$ of £2 5*s*., $\frac{2}{3}$ of 3 guineas, and $\cdot\dot{2}\dot{7}$ of £1 18*s*. 6*d*., and reduce the result to the decimal of £25.

(3) If a pipe discharge 2 hhd. 23 gal. 2 qt. 1 pt. of water in one hour, in how many hours will it discharge 11 hhd. 25 gal. $1\frac{3}{5}$ pt., the water flowing with the same velocity?

(4) Add together, $\dfrac{16}{\frac{7}{15} \text{ of } 2\frac{8}{11} \times \frac{11}{35}}$, $\dfrac{1\frac{13}{27}}{1\frac{2}{3} \text{ of } 3\frac{9}{10}} \times \frac{1}{\frac{5}{11}}$ and divide the result by $\dfrac{3\frac{2}{3} \text{ of } 5\frac{1}{4} \text{ of } 7\frac{1}{2}}{63} - \dfrac{1}{3\frac{1}{2}} \div \dfrac{\frac{4}{17} \times \frac{3}{14}}{\frac{1}{4}}$.

(5) A man's annual income is $2400; find how much he may spend per day so that after paying a tax of 2 cents 7½ mills on every dollar of income he may save $582 a year (365 days).

(6) A room is 36 feet long and 24 feet wide; find the difference in the expense of carpeting it with carpet a yard wide at $1.40 a yard, and with carpet 27 inches wide at $1.15 a yard.

(7) If 162 gallons of water will fill a cistern 4 ft. 4 inches long, 2 ft. 8 inches broad, and 2 ft. 3 inches deep, how many cubic inches are contained in a pint?

(8) Three men can mow a field in 6 days; they mow together for 2 days and then one of them ceases work, and the other two finish the field in 7 days; find how long the man who ceased work at the end of the second day would have taken to mow the whole field by himself.

(9) A man sold two city lots for $600 each; on the one he gained ¼ of the price it cost him, and on the other he lost ¼ of the price it cost him: find his entire loss on the sale of the two lots.

(10) A drover bought a number of cattle for $4375, and sold a certain number of them for $43 a head for the total sum of $3655, gaining $680, for how much per head must he sell the remainder so as to gain $400 more?

NOTE—Ten marks for each question.

XXI. (III. CLASS. 1873.)

(1) From a pound Troy are coined $46\frac{72}{40}$ sovereigns; find (in £ s. d.) the price per. oz. of gold.

(2) Divide $29.50 between two persons, so that one shall receive half as much again as the other.

(3) Simplify $\frac{1}{2}$ of $1\frac{3}{16} - \dfrac{1\frac{2}{3}}{13\frac{1}{3}}$ of $\frac{19}{20} + \frac{3}{11}$ of $\dfrac{6\frac{6}{13}}{3\frac{2}{3}}$.

EXAMINATION QUESTIONS. 289

(4) The sum paid for 494 gallons of oil, *including* a duty on each gallon which amounts to $\frac{1}{5}$ of the cost price of a gallon, is $1719.12; find the duty on a gallon.

(5) A merchant tailor bought 27 pieces of cloth, each containing $19\frac{1}{2}$ yards, at $4.31\frac{1}{4}$ a yard, and paid freight $9.62\frac{1}{2}$; he sold so as to gain $381.87\frac{1}{2}$. At what price per yard was the cloth sold?

(6) A and B can do a work in 3 days, B and C in 6 days, A and C in 4 days. If $16 be paid for the work, what is each man worth per day?

(7) Find the value of 30 cwt. 1 qr. 15 lbs. of sugar at $10.20 per cwt. (qr. = 25 lbs).

(8) A person, after paying an income tax of 2 mills in the dollar, has $1531.93 left. Find his gross income.

(9) Find the cost of covering a room 27 feet wide and 30 feet long, with matting 2 ft. 6 in. wide and costing 1.62\frac{1}{2}$ a yard.

(10) A miller has a bin 8 ft. long, $4\frac{1}{2}$ ft. wide, and $2\frac{1}{4}$ ft. deep, holding 75 bushels; how deep must he make another bin which is to be 18 ft. long and $3\frac{5}{8}$ ft. wide, so that its capacity may be 450 bushels?

(11) A man, engaged in business with a capital of $10920, is making $12\frac{1}{2}$ per cent. per annum on his capital; but on account of ill health, he quits the business, and loans his money at $7\frac{3}{4}\%$, how much does he lose by the change in 2 years, $5\frac{1}{4}$ months?

XXII. (III. CLASS. 1876.)

(1) Find what quantity must be added to
$$\left(\frac{1\frac{1}{2} \text{ of } 3\frac{1}{4}}{3\frac{1}{4} \text{ of } 2\frac{2}{3}} \text{ of } \frac{1\frac{3}{7} \text{ of } 1\frac{1}{8}}{1\frac{2}{7} \text{ of } 32\frac{2}{3}} + \frac{2\frac{1}{5} \text{ of } 6\frac{3}{4}}{3\frac{1}{3} \text{ of } 4\frac{1}{2}} \right)$$
$$3\frac{1}{2}$$
to make it equal to $\left(\frac{1}{28\frac{1}{7}} \text{ of } 3\frac{3}{4} \text{ of } 3\frac{1}{7} \text{ of } 1\frac{5}{7} + \frac{3}{8} \right)$.

(2) Reduce to its simplest form $\dfrac{(\cdot075)^3 + (\cdot025)^3}{(\cdot075)^2 - (\cdot075)(\cdot025) + (\cdot025)^2}$; and divide $9 \cdot \dot{1}704\dot{5}$ by $3.\dot{3}\dot{6}$, giving the result to the end of the first period.

(3) Express $\tfrac{3}{10}$ of $12s.\ 6d. + \tfrac{4}{15}$ of 3 guineas $+ \tfrac{5}{12}$ of £4 $- \tfrac{3}{70}$ of $2\tfrac{1}{2}d.$, as a fraction of £5.

(4) A merchant marks his goods so that he may allow a discount of 5% and still make a profit of 15%. Find the marked price of broad cloth that cost him \$3.80 a yard.

(5) At an election in a constituency in which the number of voters was 1800, the votes polled by the candidates were in the ratio of 7 to 5, and the successful candidate was elected by a majority of 240. Find the number that did not vote.

(6) A rectangular plot of ground is 60 feet long and 50 feet wide; one pathway is made surrounding the plot on the outside, and two others intersecting at right angles in the middle of the plot; if these pathways are 5 feet wide and cost $62\tfrac{1}{2}$ cents a square yard, find their entire cost.

(7) A and B engaged in business, the former contributing \$7,500, and the latter \$4,500. The gross receipts for the first year were \$2,800, of which 5% was paid for insurance, and $14\tfrac{2}{7}$ per cent. for other expenses; of the balance, B received a certain sum for managing the business, and the remainder was divided in proportion to the capital each invested: A's share was \$1,250; find B's allowance as manager.

(8) At what rate per cent. will \$1,520 amount to \$1,733.75 in $2\tfrac{1}{4}$ years. Find also in what time \$33.40 will double itself at $6\tfrac{2}{3}$ per cent. per annum.

(9) A drover bought 400 sheep at a certain price per head. He sold $\tfrac{3}{4}$ of them at a gain of 20 per cent, $\tfrac{7}{10}$ of them at a gain of 15, and the remainder at a loss of 10 per cent., gaining on the whole \$217. How much did he pay for the 400 sheep?

(10) If three horses are worth 7 cows, and 5 cows cost as much as 30 sheep, and 16 sheep cost \$165; find the value of 12 horses.

XXIII. (II. CLASS. 1873.)

(1) When greenbacks are at a discount of $16\tfrac{2}{3}$, what is the price of gold?

(2) State and prove the rules for converting the different kinds of decimals into vulgar fractions.

(3) Water expends 10 per cent. in freezing; find the weight of water in a solid piece of ice whose dimensions are 12 ft., 8 ft., 5½ ft. (cubic foot of water weighs 1000 ounces).

(4) Show that the sum of $\sqrt{0\cdot79\dot{0}12345 67\dot{9}}$ and $\sqrt{0\cdot\dot{0}12345 67\dot{9}}$ is unity.

(5) Show (no formulas) how to find the (true) discount on a sum of money for a given time and rate. How much may be gained by hiring money at 5 per cent. to pay a debt of $6400, due 8 months hence, allowing the present worth of this debt to be reckoned by deducting 5% per annum discount?

(6) A person having $5000 Bank Stock, sells out when it is at 40% premium; what amount of money does he receive, brokerage being $\frac{1}{8}$%?

(7) If a piece of silk cost 80 cents a yard, at what price shall it be marked that the merchant may sell it at 10 per cent. less than the marked price, and still have 20% profit?

(8) A merchant in Toronto has $4800 due him in Halifax; how much more will he realize by having a draft for this sum on Halifax and selling it at ½% discount, than by having a draft on Toronto remitted to him, purchased in Halifax for this sum at ¾% premium?

(9) A and B are partners. A's capital is to B's as 5 to 8; at the end of four months A withdraws ½ of his capital, and B ⅔ of his; at the end of the year their whole gain is $4000; how much belongs to each?

(10) A commission merchant in Montreal sells for a Toronto merchant 800 bbls. flour at $6.37½, on a commission of 3%, and buys certain goods required by his Principal, on a commission of 2% on the price paid for the goods, taking his commission out of the money in hand. Find the whole amount of commission.

(11) A person sold two horses at $160 each, losing 20% on one and gaining 20% on the other. Did he gain or loose on the whole transaction, and how much?

ARITHMETIC.

(12) The side, BC, of equilateral triangle ABC, is 30 feet; lines are drawn from the angles, B, C, bisecting the opposite sides, and intersecting in D. Find the area of the triangle BDC.

NOTE.—In order to pass for a 2nd Class Certificate, the candidate must obtain at least fifty per cent. of the total value of this and the Grammar paper, and at least fifty per cent. of the aggregate of the values of all the papers.

XXIV. (II. CLASS. 1876.)

(1) Find the difference between

$$\left(\frac{\cdot 26 + \cdot 2 \text{ of } 3 \cdot 7}{\cdot 48 - \cdot 014 \text{ of } 20} - \frac{4 \cdot \dot{3} + 5 \cdot \dot{6}}{7 \cdot \dot{4} - \cdot \dot{2} \text{ of } 11} \right) \text{ of } £1 \; 10s. \; 6d.$$

and
$$\left(\frac{\frac{1}{2} \text{ of } \frac{5}{8} \text{ of } 7\frac{1}{5}}{\frac{1}{3} + 4\frac{1}{2} \text{ of } \frac{4}{27}} + \frac{\frac{4}{5} - 2\frac{1}{3} + 1\frac{8}{15}}{\frac{7}{85} + 150\frac{5}{13} - 74\frac{2}{5}} \right) \text{ of } £1 \; 5s. \; 6d.$$

(2) Show that Bank discount exceeds true discount by the simple interest on the true discount. Find the amount which a banker gains by discounting a bill of $2451.50, drawn 12th of July at 4 months, and discounted September 3rd, at 5 per cent. per annum, usual days of grace: give answer to exact fraction of a cent.

(3) A retail merchant bought a quantity of Canadian tweed, and marked it at an advance of 25 per cent. on cost, and in selling it used a yard measure which was $\frac{3}{4}$ of an inch too short, his entire gain being $124.80; find the cost price of the cloth, and the amount the merchant gained by using the false measure,

(4) A person invests a certain sum (U S. currency) in U. S. 5's 10-40 (i.e. certain bonds paying 5 per cent.), and $70\frac{10}{8}$ per cent. more than that sum in U. S. 6's 5-20, the former being at a discount of 5 per cent., and the latter at a premium of 8 per cent., and the interest on both payable in gold. His income from the two investments was $1400 in gold. Find the amount (currency) invested in each kind of bonds.

(5) Three workmen, A, B, C, did a certain piece of work and were paid daily wages according to their several degrees of skill. A's efficiency was to B's as 4 to 3, and B's to C's as 6 to 5; A worked 5 days, B 6 days, and C 8 days, and the whole amount paid for the work was 36\frac{1}{4}$. Find each man's rate of wages per day.

(6) **A** merchant in Montreal owes another in Lisbon 1623½ milrees, and he resolves to remit through London, Amsterdam, and Paris; exchange between Montreal and London is at 9½ per cent., between London and Amsterdam £1 sterling for £1 3/5 Flemish, between Amsterdam and Paris £1 Flemish per 13 francs, and between Paris and Lisbon 3 francs per 450 rees; if the expenses of this circuitous course be 2½ per cent., what will it cost the Montreal merchant to settle his Lisbon account? (1000 rees = 1 milree).

(7) I bought a hind quarter and a fore quarter of beef weighing together 252 lbs.; I paid 7¼ cents a pound for the hind quarter and 5½ cents a pound for the fore quarter, and found that I had paid 17½ cents on the whole more than if I had bought both quarters at 6¾ cents per pound: find the weight of each quarter.

(8) A person bought a piece of land for $1000, to be paid for in five years with interest at 10 per cent.; he was allowed a choice of two modes of payment, **(1)** he could leave the principal unpaid till the end of the five years, paying **the interest due** annually; (2) he could pay $200 of the principal **each year together** with the accrued interest: money being worth 10 per cent. compound interest; determine whether **one** of these modes **was** more profitable than **the other, and how much his land** ultimately cost him.

(9) A merchant bought 400 lbs. of Tea and 1600 lbs of Sugar, the cost of the latter per pound being 16⅔ per cent. that of the former; he **sold** the tea at a profit of 33⅓ per cent., and the sugar at **a** loss of 20 **per cent.** gaining however, on the whole $60; find his buying prices and his selling prices.

(10) (*a*) Two Towers 40 feet **and 50 feet** high respectively, **are standing in** the same horizontal plane **120 feet apart;** how far from each tower is that point in the **line** joining their bases, which is equally distant from their summits.

(*b*) Two adjacent sides of parallelogram are 25 **feet** and 35 feet respectively, and one **of** the diagonals is 10 $\sqrt{12}$; find the other diagonal.

XXV. (NORMAL SCHOOL. 1875.)

(1) **A** person sells $6000 Bank Stock, which pays half-yearly dividends **at** 4% at 112, and invests in American Railway stock

at $98\frac{1}{4}$. What yearly dividend should the latter stock pay in order that his income may be unchanged, gold being quoted at $112\frac{1}{4}$?

(2) Incomes below £150 a year being subject to $5d$. in the pound income tax, and incomes above £150 to $7d$. in the pound; find what income above £150 a man must have that he may be just $7\frac{3}{4}d$. a year poorer than a man who has £149. 10s. a year.

(3) The hour, minute, and second hands of a clock move on the same centre and are together at 12 o'clock; at what time will the hour hand be midway between the other two?

(4) A owes $15,000 bearing interest at 5 per cent. per annum; he pays at the end of each year for interest and part payment of principal $2500; find the amount of his debt at the end of the third year.

(5) A man began business with a certain capital; he gained 20% the first year, which he added to his capital, and $37\frac{1}{2}$ per cent. the second year, which he added to his capital; in the third year he lost 40 per cent.; had he received $600 more for the goods sold the last year, he would have cleared in the three years two per cent. of his original capital. Find the capital with which he commenced business.

(6) A man borrows £5000 and agrees to pay principal and interest in four equal semi-annual payments; find the amount of each payment, interest being 5 per cent. per half year.

(7) A merchant sold goods for which he received a 45 days' note, which he immediately discounted at the bank of Commerce at 6 per cent. The discount was $3·870$\frac{21}{23}$; find the face of the note.

(8) Exchange between Paris and Amsterdam being at the rate of 2 francs 20 centimes to the guilder, that between London and Paris at the rate of 25 francs 80 centimes to the £, and that from New York on London at $9\frac{1}{2}\%$ premium, what will be the cost of a remittance for 1000 guilders from New York to Amsterdam by bills of exchange through London and Paris.

(9) A coal company's net earnings are $5368, and it pays $4000 in dividends on 2500 shares, each $20 par value; what per cent. of dividend does it pay, and how much surplus does it retain?

(10) A broker licenses $42100 to invest in U. S. $5/20$ bonds, after reserving $\frac{1}{4}$ per cent. on the par value of the amount purchased. What was his commission, the bonds being at premium of 5 per cent?

(11) What is the difference in the cost of fencing (1) a square 10

acre field and a circular field of the same area, the fence costing $5 a rod? (2) Find the side of the largest square stick of timber that can be sawed from a log 30 inches in diameter.

XXVI. (INTERMEDIATE EXAMINATION. 1876.)

(1) Simplify $\frac{1}{3}(3\frac{1}{3}+1\frac{1}{4})£ + \frac{1\frac{1}{3}-\frac{1}{3} \text{ of } 1\frac{5}{6}}{\frac{1}{10} \text{ of } 3\frac{1}{3}+\frac{1}{72}} \times \cdot 95 \text{ of } 5s. + \frac{8 \cdot 4}{\cdot 012}d$

(2) A and B can do a piece of work in $3\frac{1}{4}$ days, A and C $5\frac{1}{2}$ days, and B and C in $5\frac{1}{2}$ days. If $15 be paid for the work, what wages does each man earn per day?

(3) A person buys a lot of land at $120 an acre, and by selling a portion in allotments he makes 90 per cent. on all he sells, so that after reserving 20 acres, he finds that he has realized on the remainder $840 more than the entire lot cost him. How many acres did he buy?

(4) A Toronto merchant owes £900 in Liverpool, G. B. He determines to remit to Paris at 5 francs 50 centimes per $1; thence to Hamburg at 185 francs per 90 marcs; thence to Amsterdam at $18\frac{1}{4}$ stivers per marc; thence to Liverpool at 220 stivers per £1 sterling: how much must he remit to discharge his debt in Liverpool, and how much does he gain over direct exchange at $9\frac{1}{4}\%$ premium?

(5) A man invests $19,450 in Bank of Montreal Stock at 194, and $19,850 in Bank of Toronto Stock at 198, paying his broker in each case, $\frac{1}{4}$ per cent. on the amount of stock purchased. If the former pays a half yearly dividend of $6\frac{1}{2}$ per cent., and the latter a half-yearly dividend of $6\frac{1}{4}$ per cent., find his total income for the half-year.

(6) Coffee, costing 35 cents per pound, is mixed with Chicory worth 10 cents a pound, in the proportion of 5 pounds of coffee to 2 pounds of chicory, and the mixture is sold for 34 cents a pound : find the gain per cent.

(7) A person invests the present worth (true discount) of $30,192 (due six months hence at 4 per cent. per annum) in Bank stock paying 6 per cent. yearly interest and selling at $92\frac{1}{2}$; his taxes amount to $6\frac{2}{3}$ per cent of his gross income from the above investment : find his net annual income.

(8) A and B invest capital in the proportion of 4 to 5 in business;

at the end of 6 months A withdraws $\frac{2}{3}$ of his capital, and B $\frac{3}{5}$ of his. At the end of the year there is found to be a gain of $4,050; how is this to be divided?

(9) 1. In multiplication, why are the successive partial products not placed directly over one another?

2. Can the multiplier be a concrete number? Explain clearly the meaning of the factors in 5 ft. × 3 ft. = 15 sq. feet.

3. Is a fraction a number? Explain fully why $\frac{3}{4}$ has the same value as $\frac{12}{8}$.

(10) Find within an inch the length of a side of a square field which contains two acres.

XXVII. (I. CLASS. 1873.)

(1) Among the candidates who presented themselves at an examination for first-class certificates, A obtained 65 per cent. of the aggregate of marks, and failed to pass; B obtained 80 per cent. of the aggregate, and thus obtained 120 marks more than the required minimum for pass. If A had made 240 marks more he would have just reached the minimum for pass. Find the aggregate of marks and the per centage required to pass.

(2) A person in London owes another in St. Petersburg a debt of 460 roubles, which must be remitted through Paris; he pays the requisite sum to his broker at a time when the exchange between London and Paris is 23 francs per £1, and between Paris and St. Petersburg 2 francs per rouble: the remittance is delayed until the rates of exchange are 24 francs per £1 between London and Paris, and $1\frac{1}{2}$ francs per rouble between Paris and St. Petersburg: what did the broker gain or lose by the transaction?

(3) Prove the rule for finding a residue of figures in the extraction of the square root. Extract the square root of 7 to 10 decimal places, and deduce $\dfrac{\sqrt{7}+1}{\sqrt{7}-1}$, $\sqrt{8+2\sqrt{7}}$.

(4) A person starts with a capital which produces him 4 per cent. per annum compound interest; he spends yearly a sum equal to twice the original interest on his capital. Find in how many years he will be ruined, having given log. 2 = ·3010300. log. 13 = 1·1139434.

(5) A man holds three notes, the first for $1000, due April 1st; the second $1600, due July 1st; the third, $1200, due September 1st; he has them exchanged for two others, one of which is for $2000, payable May 1st: find when the other note matures.

(6) If the cost of digging a trench varies as the product of the depth to which it is sunk and the quantity of earth thrown out, find the cost of digging a trench 270 feet long, 6 feet broad, and 12 feet deep, having given that a trench 4 feet broad and 9 feet deep costs 45 cents for each yard in length.

(7) An insurance company issued a policy of insurance covering 80 per cent. of the estimated value of a ship and cargo, at $4\frac{1}{2}$ per cent., and immediately re-insured 50 per cent. of the risk in another company at $3\frac{1}{2}$ per cent. During the voyage the ship was wrecked, and the second company lost $900 more than the original insurers; what did the owners lose?

(8) The expense of constructing a railway is $10,000,000, of which 40 per cent. is borrowed on mortgage at 6 per cent., and the remainder is held in shares; what must be the average weekly receipts so as to pay the shareholders 5 per cent., the working expenses being 65 per cent. of the gross receipts?

(9) Three persons, A, B and C, form a partnership, contributing to the common capital, $3500, $2200, and $2500 respectively; at settlement, A's gain is $1120, B's $880, and C's $1200: given that B's stock was in the business two months longer than A's, find the time the money of each continued in trade?

(10) If gold can be beaten out so thin that a grain will form a leaf of 56 square inches, how many square inches of such gold-leaf will be required to make a cubic inch, the weight of a cubic foot of gold being supposed to be 1200 lbs. Avoirdupois?

(11) The sides of a triangle, $A\ B\ C$, are 25, 30 and 35 feet respectively; on these sides external squares are described, $A\ C\ D\ E$, $A\ B\ K\ H$, $B\ C\ G\ F$: find the aggregate area of the squares described on the lines $G\ H$, $K\ D$, $E\ F$.

(12) The sides of a rectangle have to each other the ratio of $1 : \sqrt{3}$; and a perpendicular is let fall from one of the angles upon the diagonal: find in what ratio the diagonal is divided.

XXVIII. (I. CLASS. 1876.)

(1) Prove the rules for pointing in Multiplication and Division of Decimals.

Reduce to its simplest form $\dfrac{(\cdot 075)^5 + (\cdot 05)^5}{(\cdot 075)^4 - (\cdot 075)^2(\cdot 025)^2 + (\cdot 05)^4}$

(2) The owner of some city property allows his agent 5% for collecting his rents; the amount which he annually pays for insurance and repairs (and on which he pays no income tax) is $8\frac{1}{3}$% of his net income; his income tax at 2 cents $7\frac{1}{2}$ mills on the dollar, is $198·25: Find the gross rents from his city property.

(3) Reckoning commercial discount at 8% how many years would a bill have to run so that the holder would be willing to pay something to get it off his hands? Show that the error in computing commercial discount, instead of true discount, varies nearly as the square of the time, when the time is small, and where the discount is small compared with the debt.

The interest on a sum of money for 2 years is 71\frac{133}{180}$, and the discount for the same time is 63\frac{17}{20}$; Find the rate % and the sum of money.

(4) A Building Society wishes to realize 10 per cent. on its loans; the instalments paid to it can be reinvested at 3 per cent. per half year; extending the formulæ $A = PR^n$ to include the case of n being fractional, shew that the quarterly instalment on a loan of $1000, payable in 6 years, is
$$1{,}000\,(1\cdot 1)^6 \times \dfrac{\sqrt{1\cdot 03} - 1}{(1\cdot 03)^{12} - 1}$$

(5) A retail dealer bought a quantity of broadcloth and marked it for sale at an advance of 20% on cost; in measuring it off to his customers he used a false yard-measure, by which he gained on the entire sale an additional sum of $39, making on the whole a profit of $379.20: Find the cost price of the cloth and the length of his yardstick.

(6) By the construction of the Canada Pacific Railway, 80% is added to the debt of the Dominion; for the next fourteen years after the completion of the road $5,000,000 of the principal, in addition to the interest, is annually paid off, and at the end of that time the rate of interest on the national debt is reduced 10%; if, in spite

of these reductions, it be found that the interest on the public debt is still 20% more than before the increased debt, find the cost of the Pacific Railway.

(7) Examine the merits of the following definition: "Four quantities are said to be proportional when a part of the first is contained in the second as often as a like part of the third is contained in the fourth." Give examples of its failure.

Where do you consider that the notion of ratio is first introduced in works on arithmetic?

Given that the distance through which a body draws another in one second varies as the force of attraction; that the force of attraction is directly proportional to the mass of the first body, and inversely to the square of the distance from the centre; that the mass is proportional to the product of the density and volume; and that when the earth's volume and density are each unity, those of Jupiter are 1387·431 and ·22 respectively: Find how far a body will fall from rest in one second at the surface of Jupiter, if at the surface of the earth it fall through 16·08 feet in the same time.

(8) A person has an estate which yields a net income of £1620, after paying expenses to the extent of 10 per cent. He sells it and invests the proceeds in the 4½ per cents at 96, the income now being subject to charges of 5 per cent., and his net income is £16. 17s. 5d. less than before: Find for how many years' purchase on the gross income he sold his property.

(9) English standard gold is $\frac{1}{12}$ alloy, and 44½ guineas weigh one pound troy; the weight of a shilling is $87\frac{3}{11}$ grains troy, and pure silver is $14\frac{133}{4340}$ heavier than an equal value of pure gold. If silver were to fall one per cent. in value, find what change would have to be made in the alloy in a shilling in order that 20 shillings might still be equal to £1, the alloy being supposed of the same specific gravity as silver, and the weight of the shilling unchanged.

(10) *a.* The three sides of a triangle are 20, 30, and 25 respectively. Find the position of the point which is equally distant from the three angles.

b. Two sides of a triangle are 8 and 12½ respectively, and the line bisecting the angle they contain is 6: Find the third side.

ANSWERS TO THE EXAMPLES.

EX. I. (p. 12.)

Notation.

1. 63 ; 81 ; 99 ; 40 ; 13.   2. 200 ; 303 ; 764 ; 888.
3. 4000 ; 1471 ; 6930 ; 9009.   4. 27504 ; 33000 ; 9016.
5. 100000 ; 676050 ; 202593.
6. 7003000 ; 11108106 ; 54054088 ; 613020303.
7. 2000000000 ; 9000300021 ; 94090094904.

Numeration.

1. Forty-three; sixty; eighty-eight; ninety-seven; fifty-nine; twelve; twenty-one; nineteen.

2. Two hundred and fifty-six; four hundred and one; five hundred; nine hundred and ninety-nine; three hundred and sixty-five; five hundred and seventy-eight; eight hundred and thirty-seven.

3. Two thousand; one thousand seven hundred and twenty-four; three thousand and three; seven thousand, five hundred and eighty-four; one thousand and seventy-five; one thousand, five hundred and forty-one.

4. Thirty-seven thousand and three; forty-seven thousand and forty-nine; sixty-three thousand and ninety; eighty thousand and eight; three hundred and forty-one thousand, three hundred and twenty-three.

5. Six millions, eight hundred and fifty thousand, four hundred and six; eight millions, eighty thousand, eight hundred and eight; seven millions, eight hundred and forty nine thousand, six hundred and thirty; four hundred and eighteen thousand, two hundred and fifty-four.

6. Ten millions and one; twenty millions, two hundred and twenty thousand and twenty-two; ninety-two millions, five hundred and sixty-eight thousand, nine hundred and eighty-seven; thirty millions one hundred and eighty thousand and seventy.

7. Two billions, five hundred and sixty millions, five hundred and thirty thousand, two hundred; eight hundred millions, three hundred and nine thousand five hundred and sixty; nine billions, seven hundred and thirty-eight millions, four hundred and thirteen thousand, two hundred and eight.

8. Seven billions and seventy millions, four hundred and twenty-three; nine hundred and eighty-seven millions, six hundred and fifty-four thousand, three hundred and twenty-one; five billions, seven hundred and seven millions, sixty-eight thousand and eighty.

ANSWERS (pp. 15–29.) 301

9. One hundred trillions, one hundred and ninety-eight billions, seven hundred millions, ten thousand and ninety; forty-eight quadrillions, seven hundred and twenty-six trillions, eight hundred and seventy billions, six hundred and thirty-four millions, one hundred and three thousand, two hundred and sixty-four.

Ex. II. (p. 15.)

1. 423578. 2. 3611911. 3. 1148390. 4. 2923038.
5. 1881390. 6. 8794787869. 7. 713878539140.
8. 8565743090. 9. 16237839200306. 10. 9691400353.
11. 51463796. 12. 14547; 48829; 82391. 13. 779264; 2925615.
14. 149036957938; 16696683926; 142228910945. 15. 98929.
16. 4304268. 17. 1000002733636293. 18. $22540000.
19. $340086. 20. $80081668. 21. $126246091.

Ex. III. (p. 19.)

1. 899899. 2. 300368384. 3. 73646889.
4. 6130908; 7036970; 111232112.
5. 115849491; 2922930923; 568990634342.
6. 8087; 4936. 7. 3999996; 99700000. 8. 14515927.
9. $53524.82. 10. $1814609. 11. $567.

Ex. IV. (p. 21.)

1. XXX; XLVIII; LIX; CCXX; DC; M.DCCC.XLIII.

2. Twenty-three, 23; sixty-nine, 69; two hundred and eighteen, 218; five thousand and one, 5001; one hundred and fifty thousand, six hundred and three, 150603; two millions, one hundred, 2000100.

Ex. V. (p. 28.)

1. 4015708. 2. 949723. 3. 24612151. 4. 6235560.
5. 67248560. 6. 33075. 7. 4843162. 8. 3270069.
9. 128137428. 10. 694090141. 11. 514795033260.
12. 4222494, 6802762, 12432634;
   61964682, 87860370, 897683780;
   586289802, 2868835536, 2581382769;
   182581498641, 58943103679, 70935237485, 67108855380.

ANSWERS (pp. 29-38.)

13. 16322724; 213777000; 2361710300; 21810149152; 16340824080; 121932631112635269; 40,155,302,248,305,278,754,132.
14. 44886996200592; 2605651657240; 128572831324016; 15232906283422580; 1,630,188,053,103,649,203,285.
15. 1955470720; 684763647963885.
16. 3876; 54095923986; 440956790820.   17. 21084100; 1408008.

Ex. VI. (p. 36.)

1. 543817. 2. 18574687. 3. 930622, rem. 36.
4. 71340387. 5. 814545, rem. 17. 6. 11805559.
7. 234915. 8. 704745. 9. 8862.
10. 40930, rem 270. 11. 591863. 12. 22151337, rem. 47191.
13. 5719070. 14. 7575. 15. 65299477.
16. 243096259. 17. 3396, rem. 5094687. 18. 14830201.
19. 9000900090009, rem. 1; and 900009000090, rem. 10.
20. 3854, rem. 26167. 21. 746115, rem. 83337.
22. 6084. 23. 874359. 24. 764095.
25. 11717201, rem. 645. 26. 5771, rem. 542962567.
27. 39486, rem. 2211. 28. 35. 29. 2826863, rem. 55.
30. 68911741. 31. 9862. 32. $11\frac{13}{22}$.
33. (1) 34761.90\frac{10}{22}$, 137254.90\frac{10}{22}$. (2) $102493 nearly.
34. (1) 68559.12\frac{573}{1377}$, 70390.42\frac{316}{352}$. (2) $1831.30 nearly.

Ex. VII. (p. 38.)

1. 28944. 2. Nine millions, ninety thousand, nine hundred and nine; ninety thousand, nine hundred and nine; 9181818; 9000000.
3. 36 years. 4. 548501. 5. 3431522.

II.

2. 5 years. 3. 700400000000000000. 4. 533242. 5. 19052.

III.

2. 300 days, and 75 lines remaining. 3. 13008. 4. 9376.
5. A, B, and C score respectively 18, 57, and 33 runs.

ANSWERS (pp. 40-57.)

IV.

1. 24570. 2. 29; 71.
3. 100100101; one thousand and ten millions, one hundred and one thousand and ten; 1810. 5. 4549205.

V.

1. 69788; 48, with remainder 91. 2. 815. 3. 20000 English only 30000 French only; and 70000 both English and French. 4. 863412.
5. 524.

VI.

1. M.D.LXIII, IX. 3. 567342.
4. Two hundred and seventy thousand, one hundred and thirty; twenty six thousand seven hundred and eighty-four; 10234; 6.
5. 31, 13, 16 years are the ages of the children.

Ex. VIII. (p. 57.)

1. 87828 cts.; 102787 cts. 2. 13680d.; 3744d.
3. 201 halfpence; 975q. 4. 80425q.; 188663 halfpence.
5. 88560q.; 8550 fourpenny-pieces. 6. £5909. 18s. 9½d.
7. 200 half-crowns, 1000 sixpences, 1500 fourpences.
8. 343555 grs.; 6493 lbs., 19 dwts., 21 grs.
9. 195507 lbs., 2 oz., 17 dwts., 12 grs.; 51456 drs.; 154368 sc.
10. 249056 oz.; 14 tons, 15 cwt., 1 qr., 18 lbs., 14 oz., 9 drs.
11. 162 tons, 17 cwt., 3 qrs., 25 lbs., 9 oz.; 19439375 drs.
12. 98920 grs.; 72 oz., 4 dwts., 22 grs. 13. 6864 yds.; 36305280 in.
14. 7 lea., 4 fur., 10 po., 5 yds., 2 ft., 4 in.; 4712544 in.
15. 2661126 barleycorns; 88016½ yds. 16. 79 chains, 2 yds.; 9 ft., 9 in.
17. 1348 nails; 1124 nails. 18. 2004 nails; 880 nails.
19. 5680 po.; 273460 sq. yds. 20. 6188724 sq. in.; 138847½ sq. ft.
21. 1575000 sq. links; 312 ac., 2 ro.
22. 783 cub. ft.; 3 cub. yds., 10 cub. ft., 1031 cub. in.
23. 794153 cub. in.; 1245888 cub. in.
24. 4504 pts.; 11432 gals., 2 qts., 3 gills.
25. 24344 qts.; 89863 bus., 3 pks., 1 gal., 2 qts.
26. 9000 bush.; 1291 chald., 34 bush., 3 pks.
27. 27336 sheets; 108 reams, 9 quires, 17 sheets.
28. 22266000 sec.; 2674859 sec.

304   ANSWERS (pp. 60-64.)

29. 668190½ pts.; 334095¼ qts.; 83523 13/16 gals.; 2320 61/576 bar.
30. 873223200 sec.   31. 5025 hrs.; 18090000 sec.
32. 77976320 sq. ac.; 2169180800 sq. ac.; 2115840000 sq. ac.

Ex. IX. (p. 60.)

1. £153. 16s. 3¾d.   2. £271. 10s.   3. £3329. 8s. 1½d.
4. 143 tons, 15 cwt., 4 qrs., 21 lbs.   5. 14 lbs., 9 oz., 2 drs., 19 grs.
6. 223 ac., 3 ro., 15 po.
7. £66851. 0s. 4½d.; £79251. 16s. 0¼d.; £769861. 15s. 2½d.
8. 49 lbs., 10 oz., 13 grs.; 193 lbs., 9 oz., 19 dwts.; 1757 lbs., 1 oz., 18 dwts., 14 grs.
9. 232 lbs., 4 oz., 4 drs., 1 sc.; 246 lbs., 4 oz., 2 drs., 17 grs.
10. 2 tons, 13 cwt., 3 qrs., 24 lbs., 8 oz.; 2214 tons, 9 cwt., 2 qrs., 26 lbs.; 153 tons, 9 cwt., 2 qrs., 2 lbs., 2 oz.
11. 199 m., 2 fur.; 166 m., 7 fur., 13 po., 1 yd., 2 ft., 1 in.; 125 lea., 2 m., 4 fur., 198 yds.   12. 185 yds., 1 qr., 3 na.; 182 Eng. ells.
13. 181 ac., 16 po.; 87 ac., 2 ro., 26 po., 14½ sq. yd., 2 sq. ft., 93 sq. in.
14. 85 c. yds., 9 c. ft., 575 c. in.
15. 304 gal., 1 qt.; 47 pipes, 55 gals., 1 qt.; 403 hhds., 36 gals., 7 pts.
16. $701.47; $178937.93.
17. 28 mo., 1 wk., 19 hrs., 40 m.; 216 yrs., 39 wks., 6 d., 17 hrs., 51 m., 51 sec.
18. 35 yrs., 7 mo., 2 wks., 3 d., 13 hrs.

Ex. X. (p. 64.)

1. £331. 19s. 11½d.   2. £313. 6s. 1¾d.
3. 4 cwt., 3 qrs., 5 lbs., 7 oz.   4. 12 fur., 36 po., 1 yd.
5. 5 ac., 2 ro., 31 po.   6. 55 qrs., 5 bush., 1 pk., 1 gal.
7. £77. 17s. 10¾d.   8. 859 lbs., 4 oz., 6 dwts., 5 grs.
9. 6 tons, 16 cwt., 3 qrs., 5 lbs.   10. 15 lbs., 8 oz., 4 drs., 1 sc.
11. 30 yds., 1 qr., 2 na.   12. 1 yd., 1 ft. 10 in.
13. 1 m., 5 fur., 36 po., 5 yds.,   14. 1 ro., 28 po., 28 sq. yds., 8 sq. ft.
15. 1 cub. yd., 20 cub. ft., 1305 cub. in.   16. 15 tuns, 2 hhd., 53 gals., 1 qt., 1 pt.
17. 5 bar., 3 fir., 3 qts.   18. 2 mo., 1 wk., 3 d.   19. 7°, 51′, 25″.
20. £52. 0s. 4½d.   21. $5978.50.

ANSWERS (pp. 67-69.)

Ex. XI. (p. 67.)

1. $1738.80 ; $3042.90.   2. £79. 16s. 3d.; £95. 15s. 9d.
3. £58402. 15s. 5¼d.; £69218. 2s.
4. £379113. 9s. 2¼d.; £399745. 9s. 8¼d.; £1222447. 9s. 7½d.; £6011656. 5s. 8¼d.
5. 693 lbs., 2 oz., 11 dwts., 16 grs.; 3119 lbs., 5 oz., 12 dwts., 12 grs.
6. 33 tons, 2 cwt., 2 qrs., 22 lbs., 15 oz.; 228 tons, 15 cwt., 3 qrs., 10 lbs. 12 oz.
7. 547 lbs., 5 oz., 4 drs.; 3102 lbs., 3 oz., 1 dr., 1 sc.
8. 606 yds., 1 qr., 2 na.; 3570 yds., 3 qrs., 2 na.
9. 2 fur., 10 po., 1 yd., 1 ft., 10 in.; 1 m., 1 fur., 14 po., 1 ft., 2 in.
10. 186 ac., 3 ro., 27 po., 26¾ yds., 4 ft.
11. 4571 ac., 1 ro., 24 po.; 40380 ac., 2 ro., 32 po.
12. 577 gal., 2 qts.; 14841 gal., 3 qts.   13. 199 lds., 1 qr., 7 bus., 2 pks., 3681 lds.
14. 1 yr., 13 wks., 4 d., 8 h., 34 m.; 38 yrs., 46 wks., 2 d., 13 h., 6 m.
15. 572 tuns, 1 pipe, 35 gals., 3 pts.; 7706 tuns, 1 hhd., 10 gals., 2 qts.
16. 1691 bar., 17 gal., 2 qts., 1 pt.; 33135 bar., 30 gal., 2 qts.
17. £1053. 4s. 10¾d.   18. $25452.

Ex. XII. (p. 69.)

1. £35. 3s. 3¾d.   2. £39. 9s. 6½d.   3. 15 lbs., 10 oz., 1 dwt., 14 grs.
4. 228 lbs., 1 oz., 2 drs., 1 sc.   5. £1. 14s. 2d.
6. 2 po., 3 yds., 6 in. and 28 in. rem.   7. £1. 3s. 10¼d.
8. £3. 13s. 4d.   9. 2 c. yds., 4 c. ft., $1391\frac{584}{768}$ c. in.
10. 10s. 0½d.   11. 2 tuns, 1 pipe, 27 gals., 2 qts.
12. £77. 11s. 4¼d.   13. $283.   14. 12s. $9\frac{270}{2737}d$.
15. £13. 16s. 8¼d.   16. 17 cwt., 10 lbs., 11 oz., $8\frac{74}{75}$ drs.
17. 3 cwt., 1 qr., 12 lbs., 8 oz., $7\frac{13}{65}$ drs.   18. 7 m. 26 d.
19. 14 d., 13 h., 27 m.   20. 3 lbs., 11 oz., $14\frac{40}{67}$ drs.
21. 7 lbs., 14 oz., $4\frac{340}{903}$ drs.
22. 2 ac., 3 ro., 27 po., 19 sq. yds., 7 sq. ft., $1\frac{5}{13}$ sq. in.
23. 14 po., 12 sq. yds., 6 sq. ft., $119\frac{1}{13}$ sq. in.
24. $1\frac{543}{903}$ nails.   25. 7 bus., $1\frac{14}{107}$ pk.   26. $715.

ANSWERS (pp. 71-77.)

Ex. XIII. (p. 71.)

1. 38 times.　　2. 23 times.　　3. 105 times.　　4. 156 times.
5. $26\frac{113}{155}$ times.　6. 36 times.　　7. 66 times.　　8. 288 times.
9. $186\frac{34783}{55559}$ times.　10. 73 times.

Ex. XIV. (p. 74.)

1. 156 fl., 1560 c., 15600 m.; 63·2 c., 632 m
2. 309·5 fl., 3095 c., 30950 m.; 961·29 fl., 9612·9 c., 96129 m.
3. 180·65 fl., 1806·5 c., 18065 m.; 9·25 fl., 92·5 c., 925 m.
4. 100·01 fl., 1000·1 c., 10001 m.; 460·25 fl., 4602·5 c., 46025 m.

Ex. XV. (p. 76.)

1. £264. 1 fl. 3 c. 5 m.　　2. £552. 7 fl. 7 c. 7 m.
3. £3. 1 fl. 1 c.　　4. 1 m.　　5. £1. 5 fl. 5 m.
6. £384. 1 c. 5 m.; £4838. 2. fl. 8 c. 9 m.　　7. £16. 6 fl. 5 c.; £932. 4 fl.
8. £300760. 2 c. 5 m.; £2786492. 8 fl. 8 c.
9. £38. 9 fl. 1 c. 5 m.　　10. £678. 5 fl. 6 c. 4 m.　　11. £46. 3 fl. 6 c.

Ex. XVI. (p. 77.)

I.

1. 344.　　3. $125400.　　4. $687.57.　　5. £1712.　　6. 54 gal.

II.

1. £37. 12s. 8½d.　　2. $28732.96.　　3. 90090; 17920000.
4. £2. 4s. 3d.; £1. 9s. 6d.　　5. 56 dozens.　　6. £1. 8s.

III.

1. 68 lbs., 13 dwts., 8 grs.　　2. 116 yds.　　3. £1. 5s. 5d.
5. $4258·56.　　6. 6 dollars, 12 dollars, 36 dollars, 144 dollars.

IV.

2. Gain in 2d case × 12 = gain in 1st case × 11.　　3. 19s.
5. 19 gal.　　6. 1000 perches.　　7. 16225866.

V.

2. 4752000000.　　3. $4\frac{10}{13}$ dozens.　　4. 216000.　　5. 58 cents.
6. £384. 5 fl. 5 c. 8 m.

ANSWERS (pp. 79-88.)

VI.

1. £46. 14s. 6d. 2. 4 yds., 1 ft. 3. $20.
4. 15s. 6d. 5. $7.92; $3.96; $1.32. 6. 425 quarters.

VII.

1. 5260320 min. 2. 15 tons, 15 cwt., 1 qr., $14\frac{1}{3}$ lbs.; 48 m., 5 fur., 1 po.
3. $26\frac{2}{3}$ yds. 4. 1205 days, 15 hrs., $11'\,6\frac{138''}{192}$. 5. $17.80.
6. $92040; $220560.

VIII.

2. 7 lbs., 6 oz., $1\frac{11}{21}$ drs. 3. $844200. 4. $3724.58.
5. £1311. 1 fl. 8 c. 1 m.; £505. 9 m.; £5727. 8 fl. 8 m. 6. £108.

IX.

1. May 1, 1769. 2. 19800. 3. $20\frac{460}{571}$ sec.
4. 144540. 5. 5 hrs., 7 m., 12 sec 6. £5791. 10s.

X.

1. 36. 2. 11s. $10\frac{1}{2}d$. $\frac{870}{9134}q$. 3. $235\frac{20}{118}$ mo.
4. $547.50. 5. $1.60. 6. 6 men, 12 women, 18 boys.

Ex. XVII. (p. 84.)

1. 8.   2. 15.   3. 9.   4. 11.   5. 4.   6. 40.
7. 17.   8. 2.   9. 20.   10. 15. 11. 25. 12. 8.
13. 8.   14. 12. 15. 493. 16. 13. 17. 13. 18. 2.
19. 7.   20. 6.   21. 36. 22. 84. 23. 504. 24. 83.
25. 11. 26. 123. 27. 23. 28. 36. 29. 2223. 30. 14.2587.
31. 87. 32. 37. 33. 2.   34. 2.   35. 13. 36. 3.
         37. 23. 38. 7.   39. 4.   40. 2.

Ex. XVIII. (p. 88.)

1. 48.   2. 900.   3. 105.   4. 140.
5. 11803. 6. 18648. 7. 50337. 8. 408672.
9. 344988. 10. 2663667. 11. 10867905. 12. 11754483.
13. 72.   14. 48. 15. 30.   16. 120. 17. 1080.
18. 204. 19. 1102. 20. 192. 21. 252. 22. 3465.
23. 600. 24. 720. 25. 2520. 26. 1260. 27. 1134.
28. 7200. 29. 2520. 30. 1008. 31. 22680. 32. 2017790775.

ANSWERS (pp. 91-95.)

Ex. XIX. (p. 91.)

1. $\frac{15}{12}, \frac{45}{12}, \frac{60}{12}, \frac{180}{12}$. 2. $\frac{287}{63}, \frac{615}{63}, \frac{861}{63}, \frac{1845}{63}$.

Ex. XX. (p. 91.)

1. $\frac{5}{16}, \frac{5}{21}, \frac{5}{32}, \frac{5}{40}, \frac{5}{80}$. 2. $\frac{15}{6457}, \frac{15}{11740}, \frac{15}{14675}, \frac{15}{26415}$.

Ex. XXI. (p. 92.)

1. $\frac{21}{3}, \frac{49}{7}, \frac{154}{22}, \frac{27}{3}, \frac{63}{7}, \frac{198}{22}, \frac{33}{3}, \frac{77}{7}, \frac{242}{22}$.

2. $\frac{52}{2}, \frac{130}{5}, \frac{338}{13}, \frac{598}{23}, \frac{910}{35}, \frac{218}{2}, \frac{545}{5}, \frac{1417}{13}, \frac{2507}{23}, \frac{3815}{35}$,

$\frac{234}{2}, \frac{585}{5}, \frac{1521}{13}, \frac{2691}{23}, \frac{4095}{35}, \frac{250}{2}, \frac{625}{5}, \frac{1625}{13}, \frac{2875}{23}, \frac{4375}{35}$.

Ex. XXII. (p. 93.)

1. $3\frac{3}{4}$. 2. $12\frac{3}{5}$. 3. $6\frac{1}{3}$. 4. $12\frac{3}{5}$. 5. 9.
6. $36\frac{11}{12}$. 7. $45\frac{2}{13}$. 8. $3\frac{18}{53}$. 9. $31\frac{7}{27}$. 10. $92\frac{5}{63}$.
11. 6. 12. $37\frac{67}{157}$. 13. $60\frac{10}{111}$. 14. $1514\frac{108}{153}$. 15. $59\frac{163}{239}$.
16. $96\frac{40}{995}$. 17. $22\frac{17}{1168}$. 18. $303\frac{642}{1741}$. 19. $2631\frac{861}{36423}$. 20. $121\frac{637}{3724}$.

Ex. XXIII. (p. 94.)

1. $\frac{7}{3}$. 2. $\frac{38}{7}$. 3. $\frac{41}{9}$. 4. $\frac{37}{5}$. 5. $\frac{311}{12}$.
6. $\frac{480}{11}$. 7. $\frac{326}{13}$. 8. $\frac{223}{15}$. 9. $\frac{14022}{7}$. 10. $\frac{11152}{13}$.
11. $\frac{2482}{43}$. 12. $\frac{1100}{81}$. 13. $\frac{767}{224}$. 14. $\frac{5453}{202}$. 15. $\frac{72442}{441}$.
16. $\frac{90325}{851}$. 17. $\frac{21631}{137}$. 18. $\frac{51208}{2859}$. 19. $\frac{45689}{107}$. 20. $\frac{72413}{720}$.

Ex. XXIV. (p. 95.)

1. $\frac{8}{15}$. 2. $\frac{27}{35}$. 3. $\frac{3}{10}$. 4. $\frac{55}{72}$. 5. $\frac{245}{48}$.
6. $\frac{1}{3}$. 7. $\frac{375}{44}$. 8. $\frac{175}{8}$. 9. $\frac{14}{15}$. 10. $\frac{6399}{22}$.

ANSWERS. (pp. 96–98.) 309

Ex. XXV. (p. 96.)

1. $\dfrac{1}{2}$. 2. $\dfrac{2}{3}$. 3. $\dfrac{2}{3}$. 4. $\dfrac{2}{5}$. 5. $\dfrac{8}{9}$.

6. $\dfrac{2}{3}$. 7. $\dfrac{7}{15}$. 8. $\dfrac{6}{11}$. 9. $\dfrac{41}{57}$. 10. $\dfrac{13}{17}$.

11. $\dfrac{16}{25}$. 12. $\dfrac{5}{11}$. 13. $\dfrac{275}{903}$. 14. $\dfrac{35}{52}$. 15. $\dfrac{9}{17}$.

16. $\dfrac{117}{296}$. 17. $\dfrac{3}{4}$. 18. $\dfrac{123}{127}$. 19. $\dfrac{523}{6452}$. 20. $\dfrac{103}{136}$.

21. $\dfrac{5}{72}$. 22. $\dfrac{27}{1624}$. 23. $\dfrac{17}{31}$. 24. $\dfrac{5}{7}$. 25. $\dfrac{253}{5641}$.

26. $\dfrac{547}{741}$. 27. $\dfrac{2031}{3058}$. 28. $\dfrac{901}{1900}$. 29. $\dfrac{2}{7}$. 30. $\dfrac{13}{20}$.

31. $\dfrac{15}{29}$, 32. $\dfrac{7}{13}$.

Ex. XXVI. (p. 98.)

1. $\dfrac{15}{30}, \dfrac{20}{30}, \dfrac{24}{30}$. 2. $\dfrac{16}{40}, \dfrac{35}{40}$. 3. $\dfrac{8}{12}, \dfrac{9}{12}, \dfrac{10}{12}$. 4. $\dfrac{6}{27}, \dfrac{5}{27}$.

5. $\dfrac{12}{28}, \dfrac{10}{28}, \dfrac{11}{28}$. 6. $\dfrac{18}{36}, \dfrac{27}{36}, \dfrac{20}{36}$. 7. $\dfrac{63}{72}, \dfrac{66}{72}, \dfrac{68}{72}$.

8. $\dfrac{20}{48}, \dfrac{21}{48}, \dfrac{26}{48}$. 9. $\dfrac{25}{30}, \dfrac{27}{30}, \dfrac{28}{30}$. 10. $\dfrac{36}{90}, \dfrac{60}{90}, \dfrac{50}{90}, \dfrac{63}{90}$.

11. $\dfrac{16}{24}, \dfrac{18}{24}, \dfrac{20}{24}, \dfrac{21}{24}$. 12. $\dfrac{429}{3003}, \dfrac{1092}{3003}, \dfrac{1617}{3003}, \dfrac{2002}{3003}$.

13. $\dfrac{240}{400}, \dfrac{175}{400}, \dfrac{28}{400}$. 14. $\dfrac{147}{252}, \dfrac{216}{252}, \dfrac{80}{252}, \dfrac{39}{252}$,

15. $\dfrac{308}{396}, \dfrac{180}{396}, \dfrac{286}{396}, \dfrac{54}{396}, \dfrac{11}{396}$. 16. $\dfrac{224}{672}, \dfrac{588}{672}, \dfrac{560}{672}, \dfrac{432}{672}, \dfrac{72}{672}, \dfrac{357}{672}$.

17. $\dfrac{486}{729}, \dfrac{324}{729}, \dfrac{189}{729}, \dfrac{72}{729}, \dfrac{48}{729}, \dfrac{31}{729}$. 18. $\dfrac{9000}{10000}, \dfrac{900}{10000}, \dfrac{90}{10000}, \dfrac{9}{10000}$.

19. $\dfrac{3255}{6300}, \dfrac{1190}{6300}, \dfrac{3276}{6300}, \dfrac{60}{6300}, \dfrac{3500}{6300}$. 20. $\dfrac{434}{756}, \dfrac{297}{756}, \dfrac{636}{756}, \dfrac{189}{756}$.

Ex. XXVII. (p. 98.)

In order of value the fractions will stand thus:

1. $\dfrac{8}{9}, \dfrac{7}{10}, \dfrac{3}{5}$. 2. $\dfrac{7}{8}, \dfrac{5}{6}, \dfrac{3}{4}, \dfrac{1}{2}$. 3. $\dfrac{4}{3}$ of $\dfrac{6}{7}, \dfrac{7}{12}, \dfrac{1}{5}$ of $\dfrac{3}{8}$.

ANSWERS (pp. 100-102.)

4. $\frac{31}{60}, \frac{10}{21}, \frac{5}{12}, \frac{3}{16}$.   5. $\frac{21}{26}, \frac{8}{11}, \frac{7}{13}, \frac{3}{7}, \frac{9}{22}$.   6. $\frac{27}{32}, \frac{7}{10}, \frac{27}{40}, \frac{9}{16}, \frac{3}{8}$.

7. $\frac{3}{7}$ of $\frac{5}{8}$ of $4\frac{2}{11}$ of $\frac{3}{5}$ of 5, $\frac{14}{28}, \frac{1}{6}$ of $\frac{1}{2}$ of $4\frac{3}{4}$.

8. $\frac{15}{4}, 3\frac{1}{3}, \frac{2}{7}$ of $9\frac{2}{5}$.   9. $1\frac{4}{3}, \frac{6}{7}, \frac{5}{8}, \frac{29}{56}, \frac{13}{28}$.

10. $\frac{8}{9}, \frac{9}{22}, \frac{7}{18}, \frac{3}{11}, \frac{5}{36}$.   11. $1\frac{1}{36}, \frac{700}{748}, \frac{401}{448}, \frac{113}{152}, \frac{51}{76}$.

12. $\frac{15}{4}, 3\frac{1}{3}, \frac{2}{7}$ of $9\frac{2}{5}, \frac{2}{7}$ of $\frac{5}{9}$ of $\frac{4}{5}$.

13. $\frac{3}{4}$ and $\frac{1}{6}$.   14. $\frac{47}{48}$ and $\frac{7}{16}$.

Ex. XXVIII. (p. 100.)

The sums will be:

1. $1\frac{8}{21}$.   2. $\frac{2}{3}$.   3. $1\frac{17}{24}$.   4. $\frac{1}{3}$.   5. $\frac{17}{64}$.   6. $1\frac{11}{42}$.

7. $\frac{29}{36}$.   8. $\frac{25}{32}$.   9. $2\frac{23}{30}$.   10. $\frac{149}{156}$.   11. $11\frac{7}{40}$.   12. $14\frac{2}{3}$.

13. $2\frac{1}{12}$.   14. $1\frac{23}{105}$.   15. 1.   16. $\frac{5}{6}$.   17. $1\frac{43}{120}$.

18. $2\frac{25}{252}$.   19. $1\frac{19}{21}$.   20. $1\frac{17}{105}$.   21. $1\frac{11}{105}$.   22. $15\frac{23}{36}$.

23. $10\frac{7}{36}$.   24. $5\frac{153}{100}$.   25. $3\frac{4}{15}$.   26. $1\frac{11}{105}$.   27. $2\frac{1}{4}$.

28. $1\frac{257}{300}$.   29. $7\frac{4}{21}$.   30. $585\frac{5}{6}$.   31. $4445\frac{7}{6}$.   32. $25484\frac{41}{60}$.

33. $1\frac{2221}{10000}$.   34. $45\frac{543}{720}$.   35. $23\frac{031}{5811}$.   36. $11\frac{4}{17}$.   37. $23\frac{177}{143}$.

38. $13\frac{11}{45}$.   39. $18\frac{43}{210}$.   40. $8\frac{79}{120}$.   41. 5076.   42. $3\frac{33}{125}$.

Ex. XXIX. (p. 102.)

1. $\frac{1}{4}$.   2. $\frac{3}{8}$.   3. $\frac{17}{99}$.   4. $\frac{1}{20}$.   5. $\frac{1}{36}$.

6. $\frac{1}{12}$.   7. $1\frac{4}{15}$.   8. $4\frac{7}{125}$.   9. $1\frac{7}{8}$.   10. $31\frac{37}{150}$.

11. $3\frac{1}{48}$.   12. $11\frac{7}{12}$.   13. $137\frac{13}{720}$.   14. $641\frac{3703}{13875}$.   15. $191\frac{19}{293}$.

16. 121.   17. $31\frac{11}{24}$.   18. $\frac{1}{56}$.   19. $\frac{2}{15}$.   20. $\frac{1}{6}$.

21. By $\frac{2}{27}$.   22. $3\frac{23}{72}, 2\frac{35}{72}$.   23. $10\frac{115}{168}$.   24. $2\frac{101}{672}$.

25. The sum of the fractions is 5 times as great as their difference.

Ex. XXX. (p. 104.)

1. $\frac{10}{21}$. 2. $\frac{104}{135}$. 3. $\frac{2}{3}$. 4. $\frac{2}{9}$. 5. $\frac{4}{5}$.

6. $2\frac{1}{3}$. 7. $10\frac{2}{27}$. 8. $\frac{2}{3}$. 9. 40. 10. $5\frac{3}{8}$.

11. 1 12. $\frac{85}{1152}$. 13. $\frac{10}{29}$. 14. 2. 15. $\frac{1}{6}$.

16. $\frac{3}{425}$. 17. $3\frac{1}{205}$. 18. $242\frac{17}{24}$. 19. $\frac{1}{85}$. 20. $\frac{5}{13}$.

Ex. XXXI. (p. 106.)

1. 4. 2. $2\frac{1}{2}$ 3. $1\frac{3}{13}$. 4. $1\frac{1}{12}$.

5. $\frac{153}{608}$. 6. $\frac{8}{21}$. 7. $\frac{16}{25}$. 8. $\frac{21}{60}$.

9. $\frac{45}{352}$. 10. $\frac{1}{12}$. 11. $3\frac{8}{15}$. 12. 153.

13. $347\frac{5}{8}$. 14. $1\frac{7}{15}$. 15. $\frac{4805}{496}$ and $\frac{320}{496}$. 16. $\frac{2}{5}$:

17. $\frac{2}{9}$. 18. $5\frac{1}{4}$. 19. $5\frac{2}{5}$ 20. $\frac{2}{3}$ 21. 36. 22. $7\frac{98}{125}$.

Ex. XXXII. (p. 108.)

1. 75 cts.; \$2.; \$2.50; 75 lbs.; \$7.50.
2. £1. 2*s*. 6*d*.; £1. 6*s*. 8*d*.; $11\frac{1}{4}$*d*.; 8*s*.
3. \$4.08.; £1. 11*s*. 3*d*.; 3 tons, 8 cwt., 2 qrs., $7\frac{1}{7}$ lbs.; \$30.
4. 62 cents; £1. 16*s*. $4\frac{3}{4}$*d*. $\frac{1}{5}$*q*.; \$3.66$\frac{2}{3}$; 1*s*. $11\frac{1}{2}$*d*. $\frac{4}{5}$*q*.
5. 7*s*. $8\frac{1}{2}$*d*.; £3. 17*s*. 3*d*.; 1*s*. 4*d*.
6. 1 qr., $6\frac{1}{4}$ lbs.; 12 oz.; 6 fur., 88 yds.; 2 ro., 20 po.
7. 3 fur., 25 po., 2 yds., 1 ft., 6 in.; 7 hrs., 12 m.; 2 ft. 3 qrs., 21 lbs.
8. 7 lbs., 9 oz., $9\frac{3}{5}$ drs.; 1 lb., 9 oz.; 2 gals., 1 qt., $1\frac{1}{5}$ pt.; 4 ac., 1 ro., 2 po., 3 yds.; 1 ft. $94\frac{14}{15}$ in.
9. 3 hhds., 22 gals., 2 qts.; 2 tuns, 1 hdd., 31 gals., 2 qts.; 6 bus., 3 pks., $1\frac{5}{25}$ gals.
10. 5 hrs., 36 m.; £2; \$3.12$\frac{1}{2}$.
11. £2. 16*s*. 3*d*.; $6\frac{47}{484}$ cents.
12. \$4.; $33\frac{1}{3}$ cents.
13. 112 lbs., 4 oz., 18 dwt., 12 grs.; 2 mls., 1 fur., 22 po., $\frac{31}{44}$ ft.
14. \$2.17$\frac{1}{2}$.
15. 3*s*. $1\frac{1}{4}$*d*. $\frac{10}{21}$*q*.
16. 45 cts.
17. £4. 0*s*. $4\frac{8}{9}$*d*.
18. 1 ton, 11 cwt., 3 qrs., 20 lbs.
19. 12 cwt., 2 qrs. 13 lbs., 2 oz., $10\frac{2}{3}$ drs.
20. 1 lb., 1 oz., 12 dwts., $5\frac{1}{3}$ grs.
21. 4 fur., 39 po., 2 yds.
22. 5 cub. ft., $110\frac{3}{37}$ cub. in.
23. 4 *b*., 3 *p*., 4 m., 4 fur., 1 po., $1\frac{1}{2}$ yds., $5\frac{3}{8}$ in.
25. 2725 days, 18 hrs., 34 m.
26. 4 ac., 1 ro., 23 po., $3\frac{3}{4}$ yds.

Ex. XXXIII. (p. 109.)

1. $\frac{9}{16}; \frac{63}{100}.$ 2. $\frac{5}{12}; \frac{27}{160}.$ 3. $\frac{147}{16}; \frac{559}{651}.$ 4. $\frac{47}{1000}; \frac{495}{2}.$

5. $\frac{79}{200}; \frac{2560}{1}.$ 6. $\frac{59}{64}; \frac{3}{1100}.$ 7. $\frac{761}{10560}; \frac{2617}{11664}.$

8. $\frac{8}{15}; \frac{161}{480}.$ 9. $\frac{9}{38}; \frac{1681}{3840}.$ 10. $\frac{1159}{10368}; \frac{11}{576}.$

11. $\frac{9269}{480}; \frac{3}{8}.$ 12. $\frac{293}{2880}; \frac{79}{480}.$ 13. $\frac{3025}{6596}; \frac{36}{1}.$ 14. $\frac{3}{16}; \frac{1}{60}.$

15. $\frac{1}{3}; \frac{16}{171}.$ 16. $\frac{1}{400}; \frac{1}{3000}.$ 17. $\frac{3}{800}; \frac{1}{2112}.$ 18. $\frac{192}{175}; \frac{6}{11}.$

19. $\frac{1}{2240}; \frac{31}{184}.$ 20. $\frac{27}{80}; \frac{420}{79}.$ 21. $\frac{1225}{2304}; \frac{5}{6}.$ 22. $\frac{1}{1868}; \frac{4}{25}.$

23. $\frac{1}{160}.$ 24. $\frac{108}{125}.$ 25. $\frac{28}{81}.$ 26. $\frac{7}{5400000}.$ 27. $\frac{213}{440}.$

28. $\frac{2822400}{61}.$ 29. $\frac{39}{64}.$ 30. $\frac{11}{24}.$ 31. $\frac{233}{20}.$

Ex. XXXIV. (p. 113.)

I.

2. $4\frac{95}{136}$ and $3\frac{7}{136}$; $8\frac{1}{2}.$

3. (1) $37\frac{5}{8}.$ (2) $8\frac{16}{21}.$ (3) $\frac{17}{56}.$ (4) $4\frac{25}{48}.$ (5) $3\frac{16}{113}.$ 5. $5\frac{13}{21}.$

II.

2. $21\frac{3}{4}$ and $3\frac{19}{24}.$ 3. $\frac{764}{113}$ and $\frac{121}{456}.$

4. (1) $\frac{37}{975}.$ (2) $14\frac{73}{186}.$ (3) $\frac{9}{247}.$ (4) $\frac{1}{11}.$ 5. 16.

III.

2. (1) 5000. (2) $\frac{5}{18}.$ (3) 2. (4) $1\frac{2}{546}.$

3. $\frac{1}{3}$ of 4 is greater by $\frac{1}{12}.$ 4. $\frac{1474}{1521}.$ 5. $1\frac{841}{1575}.$

ANSWERS (pp. 114–120.)

IV.

1. $\dfrac{18}{35}$. 2. (1) $\dfrac{21}{22}$. (2) $\dfrac{7}{120}$. (3) $\dfrac{437}{584}$. (4) $\dfrac{25}{81}$. (5) $12\dfrac{3}{11}$.

3. $2\dfrac{23}{30}$ and $\dfrac{1}{945}$. 4. $\dfrac{1}{4}$.

5. The quotient is 144 times as large as the product.

V.

1. $\dfrac{7}{9}$; $\dfrac{3}{11}$. 2. (1) $3\dfrac{3}{5}$. (2) $\dfrac{55}{63}$. (3) $\dfrac{67}{80}$. (4) $\dfrac{867}{1280}$.

(5) $\dfrac{255}{364}$. (6) $\dfrac{39}{40}$. (7) $\dfrac{41}{84}$. 3. $\dfrac{2}{3}$. 4. $\dfrac{1}{360}$. 5. $\dfrac{3}{16}$.

VI.

2. (1) 1. (2) 1. (3) $\dfrac{45}{841}$. (4) 3. 3. $\dfrac{12}{47}$. 5. $1\dfrac{1}{3}$. 6. $2\dfrac{5}{23}$; $\dfrac{11}{20}$.

VII.

2. (1) $\dfrac{1424}{3725}$. (2) $13\dfrac{17}{165}$. (3) 2. (4) $\dfrac{335}{468}$. 3. $18\dfrac{3}{4}$ and $3\dfrac{43}{50}$. 4. $\dfrac{43}{1650}$.

5. The whole score was 240 runs, and the score of each 30, 24, 24, 12, 12, 12, 30, 30, 30, 30, 6.

Ex. XXXV. (p. 120.)

1. $\dfrac{3}{40}$; $\dfrac{106}{125}$; $\dfrac{151}{50}$; $\dfrac{1717}{500}$; $\dfrac{1717}{5}$; $\dfrac{1717}{50000}$; $\dfrac{10001}{200000}$;

$\dfrac{230409}{1000}$; $\dfrac{230409}{100000}$; $\dfrac{10686}{5}$; $\dfrac{11425001}{1250}$; $\dfrac{38401}{1600}$.

$\dfrac{657097353}{80000}$; $\dfrac{20819}{2500000}$; $\dfrac{10000009}{10000000}$; $\dfrac{1}{1000000000}$.

2. ·1; ·3; ·7; ·53; ·07; ·003; ·9178; 91·78; ·09178; ·0091; ·00009; 520·3; ·9; 3·0142; 6·72819; ·000672819; 6728·19.

3. 7; 70; 700; 70000. ·6; 60; 60000. 4·31; 43100.
162·01; 16201; 16201000; 9001600; 900·16.

4. ·051; ·00051; ·0000051. ·00008; ·000000008. ·005016; ·00005016; ·3780186; ·0003780186.

5. ·5; ·7; ·19; ·28; ·005; 9·7; ·000001; 14·4; 280·0004; 7·007; 100·00001; 1·0010001; ·000000005.

6. Four-tenths; twenty-five hundredths; seventy-five hundredths; seven hundred and forty-five thousandths; one-tenth; one thousandth; one hundred thousandth; twenty-three and seventy-five hundredths; two and three hundred and seventy-five thousandths; two thousand three hundred and seventy-five ten thousandths; two thousand three hundred and seventy-five hundred millionths; one and one millionth; one million and one ten millionths; one hundred millionth.

Ex. XXXVI. (p. 121.)

1. 47·09595.   2. 290·381404.   3. 6153·70427.
4. 2335·5073.   5. 418·94514.   6. 406·529522.
7. 953·77386.   8. 370·430375.   9. 62·5358119.
10. 9181·6074975.   11. 5082·3192995.   12. 1011022959·091000001.

Ex. XXXVII. (p. 122.)

1. 1·0918; 5·8345; 141·03; ·0001; ·304317.
2. 211·6875.   3. ·0421813.   4. 602·3415997.
5. 4·4954.   6. ·48553.   7. 9·1794.
8. ·09; 655·30283; 21·068124; 9788·852.   9. 6·3; ·699993; 99·706.

XXXVIII. (p. 123.)

1. 159·6; 1596; 15·96; ·0001596.   2. 173·889; ·173889; 1·7989.
3. ·0063612; 3·72812; ·12376.   4. 3·07930896.   5. 210·6144185.
6. ·00329875.   7. ·03611.   8. ·0000274104.   9. ·0006594.
10. ·00007614.   11. ·055757592.   12. ·27492.   13. ·001; ·20736.
14. 32·86164.   15. 1549795·52.

Ex. XXXIX. (p. 126.)

1. 2·1; 91·78.   2. ·025; 24·3.   3. ·00003; ·374.
4. 10, 100, 10000.   5. 250; 16·25.   6. 51472; ·0000051472
7. ·057; 813·4.   8. ·0072; 59640.   9. 10500; 137·56.
10. 5020; 543.   11. 326000; 32·6; ·0097.   12. 1·3; 13; ·13; 130.
13. ·002; ·000002; 2.   14. 2·01; 20100; ·001875.
15. 948·7096; 9487096.   16. 26153·4 : 21·4.
17. 2040000; ·00082175.   18. 7934·7; 79347; 79347000.
19. ·00002; ·000002; 20.   20. ·57; 57000.

ANSWERS (pp. 129–134.)   315

21. 3·7356; ·0117; 76·9230.   22. 1·4895; 50830·1313.
23. 320911·4782; 1·9005; 1·3157.   24. 14036019·0930; ·0011.
25. 12·8413; ·0026.   26. ·0000186; ·00186.
27. ·00256256; ·256256; 256·256.   28. 4360; 103·36; ·04545.

Ex. XL. (p. 129.)

1. ·25; ·75; ·625; ·36; ·3125; ·95.
2. ·515625; ·432; 2·85; 1·36; ·00625.
3. 6·171875; ·2375; ·05078125; ·005859375; 15·0075264.
4. ·007080078125.   5. ·84375.   6. ·0001.   7. ·661.   8. ·575.
9. ·79375.   10. ·5.   11. 11·7578125.   12. 86·497.   13. 562·926.

Ex. XLI. (p. 132.)

1. ·$\dot{5}$; ·1$\dot{8}$; ·0$\dot{2}\dot{7}$; ·$\dot{4}2857\dot{1}$.
2. ·5$\dot{6}$; ·7$\dot{4}\dot{3}$; ·$\dot{1}9753086\dot{4}$; 15·1$\dot{5}\dot{6}$.
3. ·9$\dot{1}78977\dot{2}$; 7·$\dot{2}8571\dot{4}$; ·000$\dot{1}\dot{7}$.
4. 24·00$\dot{9}$;   17·01$\dot{8}5714\dot{2}$; 2·167834...
5. ·$\dot{0}5263157894736842\dot{1}$; ·$\dot{0}434782608695652173913\dot{9}$;
   ·$\dot{0}344827586206896551724137931\dot{1}$; ·$\dot{0}3225806451612\dot{9}$.

6. $\frac{7}{9}$; $\frac{7}{90}$; $\frac{5}{22}$.   7. $\frac{289}{495}$; $\frac{5}{37}$; $\frac{79}{300}$.   8. $\frac{1}{540}$; $\frac{1007}{333}$; $\frac{17}{1375}$.

9. $\frac{1}{7}$; $\frac{191}{480}$; $\frac{51467}{134680}$.   10. $\frac{4}{13}$; $\frac{10619}{16835}$; $\frac{39}{14}$.

11. $\frac{114137}{333000}$; $\frac{1043}{33300}$; $\frac{385}{48}$.   12. $\frac{1284121}{15000}$; $\frac{51}{14}$; $\frac{4023367}{31680}$.

Ex. XLII. (p. 134.)

1. 31·371538.   2. 700·612301.   3. 6·116666; 1·681818; 308·052752.
4. 2·2$\dot{3}8461\dot{5}$; 13·7$\dot{2}61904\dot{7}$.   5. 13·$\dot{2}$; ·2$\dot{7}$.   6. 25·21$\dot{3}$; 300.
7. 363·5$\dot{7}4\dot{0}$; 245·$\dot{3}$.   8. 1·35169...; 17·4$\dot{5}$.
9. 48·7$\dot{6}$; 6·7$\dot{6}$.   10. 303·75; 2·$\dot{3}$.   11. 7; 48·734; ·0134.

Ex. XLIII. (p. 136.)

1. 45 cts.; 67½ cts.; $4·3854. 2. 5*s.* 7½*d.*; 15*s.* 11·088*d.*; 14*s.* 4½*d.*

3. 43 cts.; $5·783; $1.20.

4. 2 m., 1100 yds.; 2 d., 12 hrs., 55′. 21″; 7 oz., 4 dwt.

5. 3 qrs., 10 lbs., 1·216 oz.; 7 lbs., 5·2 oz.; 14 po., 2 yds., 7·2 in.

6. 4 tons, 3 cwt., 1 qr., 5 lbs., 8 oz.; 3 cwt., 2 qrs., 12 lbs., 8 oz.; 8 sq. po.

7. 3 lbs., 10 oz., 5·568 grs.; 2 qrs., 3 bush., 3 pks.; 14 cwt., 20 lbs., 10·816 oz.

8. 3 ac., 3 ro., 14 po.; 63 gals. 9. 37 po.; 9 d., 15 hrs.

10. $506·325; 19 qrs. 11. 7 ac., 3 ro., 20 po.; 2 m., 1150 yds., 2·052 ft.

12. 13 sq. yds., 1 sq. ft., 111·6 sq. in.; 4 m., 6 po., 1 yd., 2 ft., 11·97696 in.

13. 38⅓ cts.; $1.88⅓; 5 oz., 12 dwts., 16 grs.

14. 15*s.* 6*d.*; 1*s.* 5½*d.*; 13*s.* 4*d.*

15. 6 sq. yds., 108 sq. in.; 3 fur., 10 po., 3 yds., 2 ft.; 20 d., 6 hrs.

16. 8$\frac{19}{180}$ ac.; 20 hrs., 30 m. 17. 7*s.*; $2.52.

18. 1 cwt., 24 lbs., 13 oz., 13$\frac{13}{15}$ drs. 19. £1. 2*s.* 9¾*d.*

20. 152 wks., 5 d., 10 hrs., 54$\frac{6}{11}$ sec.

21. 1 ro., 39 po., 28¼ sq. yds., $\frac{63}{550}$ sq. in.

EX. XLIV. (p. 139.)

1. ·625; ·9375. 2. ·23125; ·796875. 3. ·503125; ·0572916.

4. $\cdot 2208\dot{3}$; $48\cdot 08\dot{3}$. 5. ·0355; ·3001875. . . 6. $\cdot 273295\dot{4}\dot{5}$; $\cdot 07291\dot{6}$.

7. $\cdot 2785\dot{4}9382716\dot{0}$; ·875. 8. $\cdot 67\dot{8}5714\dot{2}$; $\cdot 0000254\dot{6}29\dot{6}$.

9. $\cdot 8\dot{2}28571\dot{4}$; ·000015. . . . 10. ·0334821 . . . .; 82·5.

11. $1\cdot 91\dot{6}$; 14·24. 12. $114\cdot\dot{5}\dot{4}$; ·00061 . . . .

13. 75·789. . . .; $5212\cdot\dot{3}0769\dot{2}$. 14. ·01875; ·805; ·7317.

15. ·13125; $\cdot\dot{3}$. 16. $\cdot 3061\dot{2}$; ·013671875. 17. ·225; ·511.

18. ·00243 . . .; ·000080. . . 19. ·000304. . .; ·065625.

20. ·288; ·546875. 21. $\cdot 11\dot{8}2539\dot{6}$. 22. 1·59. . . . 23. ·3140625.

24. (1) 2 c. 5 m. (2) 4 c. 1⅔ m. (3) 1 c. 8¾ m. (4) 2 fl. 5 c.
(5) 5 fl. 2 c. 5 m. (6) 8 fl. (7) £5. 6 fl. 2 c. 5 m.
(8) £54. 3⅗ fl. (9) £20. 9 fl. 8 c. 1¼ m. (10) 7 fl. 6 c. $9\cdot 791\dot{6}$ m.
(11) 7 fl. 3 c. 4 m. (12) £2. 7 fl. 9 c. 6⅕ m. (13) £3. 4 c. 9 m.

ANSWERS (pp. 141–143.)

EX. XLV. (p. 141.)

I.

1. $\dfrac{1}{16}$; $\dfrac{314159}{100000}$; $3\cdot 273809523$.   2. $13\cdot 0125$.

3. $573\cdot 005754$; $573\cdot 004246$; $\cdot 43204577$; $759953\cdot 5\ldots$; $1\cdot 02515$; $1\cdot 00485$; $\cdot 01030225$; 100.

4. Yes.   5. (1) 394.   (2) $\cdot 009072$.   (3) 1.   (4) $11\cdot 8125$.

6. $1\cdot 05$ nearly.

II.

1. $\cdot 000700409$; $\dfrac{24269}{2000}$; $\cdot 0032546$.

2. Three hundred and ninety-seven thousand and eight, and four hundred and five thousand and nine millionths; $397008405\cdot 009$; $397\cdot 008405009$. Three hundred and ninety-seven millions, eight thousand four hundred and five, and nine thousandths. Three hundred and ninety-seven, and eight millions four hundred and five thousand and nine thousand-millionths.

3. $\cdot 03493$.   4. $11\cdot 025$; $\dfrac{441}{40}$; $\cdot 00053874$; $\cdot 0002$; $\cdot 0642$.

5. (1) $\cdot 000091304\ldots$ (2) $2\cdot 518$.   (3) $\cdot 625$.   (4) $10\cdot 0045$.

6. $2\cdot 4976096088$.

III.

1. $\cdot 57$ and 57000; $12644\cdot 042\ldots$ 2. (1) $\dfrac{9}{10}$ and $\cdot 9$.

(2) $\dfrac{1384}{1975}$ and $\cdot 7007\ldots$ (3) $\dfrac{968}{625}$ and $1\cdot 5488$.   (4) $\dfrac{7}{24}$ and $\cdot 2916$.

3. $2\cdot 6$; 8585; no.   5. $15\cdot 35$ miles.   6. $\dfrac{37}{240}$.

IV.

1. $124\cdot 36653$. $31\frac{457}{999}$; $1\frac{2377}{4995}$.

2. 3006005; three hundred thousand, six hundred and five-tenths.

3. In order of magnitude they stand thus $1\cdot 5 \times \cdot 75$; $2\cdot 625 \div 5$; $5 \times \cdot 05$.

4. $\cdot 0049$; $\cdot 12693$.   *Ans.* $\cdot 006545$; 542000; $\cdot 0046 20020$; $\cdot 02002$.

5. $3\cdot 14159$.   6. $\dfrac{3}{5}$.

V.

1. $3\cdot08\dot{3}$; $2\cdot87296$; $8\cdot02\dot{6}7857\dot{1}42$.

2. (1) 851.   (2) $1\cdot60546875$.   (3) $35\cdot4875$.   (4) $1\cdot68350\dot{1}$.

3. $\dfrac{333}{106}$ is the nearer.   4. $2\cdot7182818$; $\cdot00097061$; $\dfrac{97061}{100000000}$.

5. $7925\cdot7$ miles nearly.   6. $13\cdot74696$.

VI.

1. $12\cdot24362412$; $\cdot0089147\ldots$; $\cdot0730091$; $7\cdot30091$.   2. $\dfrac{451}{369}$.

3. $\dfrac{49}{410}$ $6\tfrac{139}{448}$ 4. $\dfrac{6401}{49500}$.   5. $\cdot7\dot{2}$.   6. $\cdot4375$.

VII.

1. A ought to receive 60 cents; B, 36 cents; C, 12 cents.
2. $\$655.67\tfrac{1}{97}$; $\$786.80\tfrac{49}{97}$; $\$1101.52\tfrac{56}{97}$; each person ought to pay $26\tfrac{33}{97}$ cts.
3. 45 boys.   4. $1\tfrac{1}{2}$ miles per hour; 1 hr.
5. A should have \$3.60; B, 2.88; and the boy, 72 cents.   6. $\$1.25\tfrac{1}{2}$.

Ex. XLVI. (p. 148.)

1. \$68.75; \$37.50.   2. \$41.40; \$40.   3. \$90; \$182.75.
4. \$247.50; \$437.50.   5. \$67.84; \$82.50.   6. \$228.96; \$30.24.
7. $\$71.42\tfrac{1}{2}$.   8. \$701.10.   9. \$88.20; \$81.20.
10. \$191006.73; \$184.08.   11. \$27082.20.   12. \$350.
13. £12. 15s. $11\tfrac{1}{4}d$.   14. £215. 16s. $8\tfrac{3}{4}d$.   15. £89. 6s. $1\tfrac{1}{4}d$.
16. £467. 1s. $6\tfrac{3}{4}d$. $\tfrac{3}{7}q$.   17. £12. 5s. $10\tfrac{1}{2}d$. $\tfrac{6}{7}q$.   18. £2. 15s. $11\tfrac{1}{4}d$.
19. £147. 16s. $11\tfrac{1}{4}d$. $1\tfrac{1}{50}q$.   20. £230. 16s. $8\tfrac{1}{2}d$. $\tfrac{1}{7}q$.   21. £4. 9s. $5\tfrac{1}{4}d$. $1\tfrac{1}{12}q$.
22. $\$237.06\tfrac{1}{2}$.   23. \$1350.   24. $\$16.12\tfrac{1}{2}$.

Ex. XLVII. (p. 157.)

I.

1. $10\cdot51\dot{3}8461\dot{5}$.   2. £64. 4s.   3. $\cdot625$; $\cdot32$.
4. \$2698.36.   5. \$3001.86.   6. £236. 8s. 6d.
7. \$13.50.   8. 2 francs, 13 centimes.   9. 10 wks.

II.

1. $\dfrac{10}{13}$.   2. $1668\tfrac{20}{27}$ metres.

ANSWERS (pp. 158-164.)

3. A's share=54.85\frac{5}{7}$, B's share=61.71\frac{3}{7}$, C's share=75.42\frac{6}{7}$.
4. ·0025.   5. $4\frac{1}{8}$ cts.   6. 3361·3 ... rev.
7. £4. 17s. $1\frac{17}{30}d$.   8. $5\frac{5}{11}$ hrs.   9. $74\frac{2}{27}$ hrs.

III.

1. 15 cwt., 2 qrs., 21 lbs.; $567.26.
2. 8·495; 8 chains, 4 chainlets, 9 links, 5 linklets.
4. 8d.; £3. 17s. 1d.   5. 5 mo.   6. 365·2425 days.
7. £50. 10s.   9. 666.66\frac{2}{3}$; 1133.33\frac{1}{3}$.

IV.

1. 26973.   2. ·0219238095.   3. £36. 10s. $8\frac{1}{2}d$. $\frac{2}{7}q$.
4. 9 days.   5. £4. 4s. $2\frac{3}{4}d$.; 6s. $0\frac{11}{50}d$.
6. £395. 19s. 2d.; £11614s. 16s. 8d.   7. 195 sq. yds.
8. A's share=$600, B's share=$480, C's share=$320.   9. 96 cts.

V.

1. 374 quotient, and 446 remainder.   2. $\frac{42}{315}$, $\frac{225}{315}$, $\frac{81}{315}$; $1\frac{11}{15}$.
3. $3\frac{3373}{9144}$ sov.   4. £15468750.   5. $3\frac{3}{4}$ hrs.
6. 10 hrs., 12 m.   7. $7\frac{7}{10}$ days.
8. $75\frac{925}{48620}$ cts.; 1856.72\frac{3053}{48620}$; 1465.57\frac{3050}{24313}$; 1055.15\frac{38755}{48620}$.
9. 1512.

VI.

1. £95335. 17s. 9d.   2. 4732.72\frac{8}{11}$.   3. 10 yds., 11 in.
4. 12 cents.   5. 4 cents.   6. 6 cents.
7. $635\frac{19}{31}$ hrs.   8. £1000   9. 600 trees.

VII.

1. $1; $\frac{2}{175}$; ·0042....   2. 16; £4. 19s.; £9. 18s.; £19. 16s.
3. $129\frac{57}{687}$ yrs.   4. $5\frac{118}{513}$ days.   5. 1200 men.   6. $\frac{1}{30}$; $1000
7. $309.   8. 9s.   9. 10234 fr., $66\frac{2}{3}$ cent.

VIII.

1. $2576.34.   2. $15\frac{73}{80}$.   3. $\frac{1}{12}$.   4. $978.28.
5. $\frac{3}{1280}$; ·00234375.   6. 8 cwt., $37\frac{1}{2}$lbs. (cwt. = 112 lbs.)
7. $57\frac{3}{5}$ hrs.   8. 6s. 8d.   9. $383784.80.

IX.

1. 2143.   2. ·4; ·04.   3. Yes; 70 cents.   4. 80 days.
5. £19. 8 fl. 7 c. 7 m.; $\frac{1}{13}$ m.   6. 4s.   7. The English hen.
8. 45 men.   9. 160 boys.

ANSWERS (p.p. 164-182.)

X.

1. ·661; ·017; 11·2. 2. $\frac{5}{2164}$. 4. $235. 5. $664\frac{8}{13}$ yds. 6. 6d.
7. $83.70. 8. 24 days. 9. 800000 sheets.

XI.

1. ·08270676691729323$\dot{3}$; 24.81\frac{27}{133}$. 2. $61.40.
4. $449.36. 4. 36·96 cents. 5. $2\frac{1}{4}$ in.
6. 25; 12·5; 5; 2·5; 1·$\dot{6}$; 1·25; 2·08$\dot{3}$. 7. $25\frac{9}{43}$ fr.; 215; $39\frac{181}{271}$.
8. $5\frac{1}{4}$ days. 9. 240 sov.; 720s.; 960d.

Ex. XLVIII. (p. 169.)

1. 40. 2. $\frac{16}{27}$. 3. $\frac{1}{8}$. 4. 3·9.
5. $\frac{1}{8}$. 6. ·79936. 7. 3. 8. 60.

Ex. XLIX. (p. 177.)

1. $69.12. 2. 72 yds. 3. $3.78. 4. 82 acs.
5. $324. 6. 48 wks. 7. 12. 8. 14 cwt., 1 qr., 3 lbs.
9. $22.50. 10. $86.40. 11. 3.86\frac{1}{4}$. 12. $2.28.
13. $92\frac{4}{7}$ bush. 14. 364.10\frac{4}{25}$. 15. £690. 16. $118.25.
17. $87\frac{1}{2}$ cents. 18. 8047\frac{1}{17}$. 19. $3. 20. 50.
21. 8 days. 22. 3·8709 ... days. 23. 4676 yards.
24. £5012. 2 fl., 3 c., $7\frac{1}{2}$ m. 25. $503.12. 26. 18 days.
27. $230.04. 28. 8 men. 29. £5. 6s. 8d. 30. £6. 4s. $8\frac{3}{4}d$.
31. 24 yds. 32. $2\frac{6}{7}$ mos. 33. $172. 34. 4.68\frac{3}{4}$.
35. 443.57\frac{1}{2}$. 36. 11 yds. 37. $2041.20. 38. $1\frac{11}{25}$ days.
39. 7·7 cents. 40. £8. 14s. 11$\frac{1}{2}d$. $\frac{46}{0239}$. 41. 840.92\frac{508}{751}$.
42. $89.60. 43. $13200. 44. $2.10. 45. 8080.53\frac{59}{37}$.
46. 10 hrs., 40', $36\frac{7}{48}$". 47. 70 ft., 8·232 in.
48. 4 cwt., 2 qrs., $16\frac{2}{103}$ lbs. 49. 7213\frac{3}{7}$. 50. £4. 10s.
51. $1.32. 52. 15 hrs. 53. $193.98. 54. $693.
55. 208 days. 56. $1.95. 57. $10\frac{1}{2}$ days. 58. $65\frac{5}{27}$ yds.
59. He loses $3\frac{1}{3}d$. 60. $5\frac{715}{1201}$' before 4 o'clock. 61. $7500.
62. 70 days. 63. $4166\frac{2}{3}$ yds. 64. 240000 lbs.
65. 22403\frac{529}{3927}$. 66. $157.50. 67. $168\frac{59}{1533}$ lbs.
68. Monday fortnight, at 6 h. 36 m., p.m. 69. 72 yds.
70. $3.20. 71. 466650 lbs. 72. 100 days.
73. $121800. 74. £18. 10s. 3d. nearly. 75. 80 days.
76. £211. 19s. 3d. 77. $20.16. 78. $9\frac{1}{2}$ yds.
79. $39·46125. 80. 207:82. 81. 3 days.

ANSWERS (pp. 188-199.) 321

Ex. L. (p. 188.)

1. 15 men. 2. 7 men. 3. 66 days. 4. 7200 soldiers.
5. $19\frac{7}{35}$ bus. 6. 600 ac. 7. $12\frac{27}{35}\frac{24}{23}$ mi. 8. $9\frac{19}{35}$ cwt.
9. $784. 10. 11 mo. 11. $88.
12. 59 cwt., $22\frac{8}{77}$ lbs. 13. 12 hrs. 14. $8\frac{1}{4}$ wks. 15. £20.
16. 3d., 6 hrs. 17. 10 hrs. 18. £50. 8s. 9d. 19. 500 reams.
20. 9 days. 21. 8 wks. 22. 64 days. 23. 350 men.
24. 2400 men. 25. 47 tons, 17 cwt., 66 lbs. 26. 324 men.
27. $59\frac{17}{25}$ days 28. £332. 5s. $2\frac{4}{17}d.$ 29. 2268 cub. ft.
30. 1.59\frac{779}{3816}$. 31. 8 ft. 32. $10.35 33. 1320 yds.
34. $18\frac{13}{25}$ ft. 35. $465.69\frac{3}{5}$. 36. 35 days. 37. 19·36 days.
38. 49·3 lbs. 39. $2446.08.

Ex. LI. (p. 194.)

1. $6.80. 2. $21.70. 3. $65. 4. 106.87\frac{1}{2}$.
5. $112.38. 6. $267.24. 7. $132·356. 8. £6. 18s. 10d.
9. £17. 14s. 6d. 10. £1. 4s. $8\frac{1}{2}d. \frac{1}{4}q.$ 11. $1140.
12. 3598.22\frac{3}{4}$. 13. £1554. 9 fl. 2 c. 5 m. 14. 176.49\frac{3}{4}$.
15. 2513.83\frac{19}{32}$. 16. £1733. 15s. $5\frac{11}{20}d.$
17. $96.25+ ; $471.25+ 18. $116.09; $562.59.
19. 9.62\frac{1}{2}$; 220.62\frac{1}{2}$. 20. 47.13\frac{1}{2}$; 290.93\frac{1}{2}$.
21. 2.74\frac{30}{73}$; 42.74\frac{30}{73}$. 22. 47.05\frac{232}{292}$; 367.80\frac{232}{292}$.
23. 8s. $17\frac{202}{7300}d.$; £31. 18s. $17\frac{202}{7300}d.$

Ex. LII. (p. 195.)

1. $113·63. 2. 8 per cent. 3. 4 years. 4. 6 per cent.
5. 10 years. 6. 6 per cent. 7. $600. 8. $815.
9. £345. 17s. 6d. 10. 5$\frac{1}{2}$ years. 11. 4$\frac{8}{9}$ per cent. 12. 20 years.
13. £315. 10s. 8d.; 2$\frac{1}{2}$ years. 14. $225.

Ex. LIII. (p. 198.)

1. $247·20. 2. $980·0344. 3. $44·928.
4. $845·2796 5. $274.83 nearly. 6. $12.79+.
7. $34·49+. 8. $2135.58+. 9. $739·427.
10. $1·4976. 11. $350.41+. 12. $228.28+.
13. $87·3456. 14. $226.33.... 15. £1. 11s. 7d. nearly.

Ex. LIV. (p. 202.)

1. $300.
2. 237.32\frac{4}{33}$
3. $657.
4. 912.96\frac{128}{207}$.
5. $450.
6. £375. 15s. 0½d. nearly.
7. $824.
8. $1240.
9. $1228.50 nearly.
10. £2000.
11. £262. 4s. 5¼d. ½q.
12. £765.
13. $462.47+.
14. £1953. 2s. 6d.
15. 1.25\frac{25}{57}$.
16. 78.11\frac{17}{53}$.
17. 2.61\frac{24}{41}$.
18. $2.46.
19. 13.80\frac{40}{247}$.
20. 8.29\frac{11}{41}$.
21. £140.
22. £48. 9s.
23. 25.52\frac{93}{167}$.
24. $2.
25. 2s. 11d. nearly.
26. $1·06+
28. 16⅔ per cent.
29. £2. 6s. 8d.

Ex. LV. (p. 207.)

1. $3800.
2. $800.
3. $525.
4. $4300.
5. 597\frac{1}{87}$.
6. 1059.60\frac{40}{131}$.
7. £5050.
8. $2600.
9. £3091. 10s. 2$\frac{14}{53}$d.
10. $2418.
11. $1488.
12. $2775.
13. $1834.
14. £972. 10s.
15. £4048. 11s. 7½d
16. $72.
17. $480.
18. $122.50.
19. 135.59\frac{57}{177}$.
20. £112. 0s. 3$\frac{123}{307}$d.
21. £159. 12s.
22. £111. 8s. 1$\frac{151}{377}$d
23. $1700
24. $3216.
25. $2185.71⅔.
26. £2164. 2s. 6d.
27. £875. 7s. 9¾d.
28. 6.59\frac{31}{81}$.
29. 7.44\frac{32}{47}$.
30. 8.27\frac{29}{73}$.
31. 6.79\frac{63}{103}$.
32. $594.94.
33. $7.77⅞; $8.75; 97⅜ cents.
34. 87½; $1371⅔.
35. 5297.54\frac{98}{163}$.
36. $320.
37. 331.28\frac{136}{163}$.
38. £104. 8s. 4d.
39. 15.21\frac{17}{23}$.
40. $\frac{100}{107}$, half-yearly.
41. $200; 141.50\frac{60}{53}$, half-yearly.
42. $280 half-yearly; $7350.
43. £2729$\frac{1}{131}$; £5. 10s. 3$\frac{87}{131}$d.
44. The Canadian Bank of Commerce Stock.
45. 10612.24\frac{24}{49}$.
46. £6. 5s.
47. £1666. 13s. 4d.
48. £7900·; £6310. 2s. 6d.
49. £86.
50. £25; £22. 11s. 7$\frac{11}{51}$d
51. £20000; £225000

Ex. LVI. (p. 216)

1. 187·98; 352·4625; 2255·76; 4285·944; 5639·4; 84872·97.
2. 15·625; 23·456...; 8·984375; 2·530...; ·048...
3. ·005; ·0275; ·013; ·05625; ·26$\dot{3}$; 2·3005; 5·000138.
4. $350;
5. 8s. 8d.
6. $763.60; $268.08 gain.
7. 1463·65 gals.
8. 8137·5 bus.
9. 1$\frac{37}{63}$.
10. 15⅝ gain
11. 9 cents.
12. $9.13½ +, (cwt. = 112 lbs.)
13. 66$\frac{2}{13}$ gain.

ANSWERS (pp. 216-227.)

14. 10·72...; **8·71**...; 10·44....   15. 5·24...; 20·29...; 16·11....
16. 1045678·375 persons.
17. 3 of the age of 18 years; 19 between 15 yrs. and 18 yrs.; 38 between 12 yrs. and 15 yrs.; 153 between 10 yrs. and 15 yrs., and 190 under that age.
18. $10\frac{5}{7}$.   19. $33\frac{1}{4}$.   20. $168.75.
21. Weight of oxygen = 997·12 lbs.; weight of carbon = 865·3 lbs.; weight of hydrogen = 137·58 lbs.
22. $204\frac{2}{7}$; $153\frac{211}{212}$; for the whole time, $672\frac{6}{7}$ per cent. or $42\frac{3}{56}$ per cent. per annum.
23. He loses $91.80.
24. No. of male criminals : No. of female criminals :: 5 : 4.
25. 5s. $3\frac{3}{19}d$.; 5s. 6d.   26. $2.19.   27. £37. 1s. $11\frac{1}{4}$. $\frac{2}{17}q$.
28. $1375; 37\frac{27}{29}$.   29. £27.   30. $3·456.
31. £3. 18s.   32. 47·15 cents.   33. £96. 7s. $3\frac{3}{11}d$.
34. 1s. $10\frac{1}{2}d$.; 2s. $1\frac{3}{5}d$. $\frac{4}{17}q$.; 3s. 9d.   35. 30 quarters.   36. $40.

Ex. LVII. (p. 220.)

1. (1) $895, (2) $173.50, (3) $534.25.   2. 2851507\frac{7}{11}$ qrs.   3. 1.08\frac{1}{4}$.
4. Average age of boys = $9\frac{2}{3}$ yrs. Average age of girls = $10\frac{9}{20}$ yrs. Average age of whole class = $10\frac{9}{22}$ yrs.
5. 4·447 days.   6. 3·9428571.   7. $1979.   8. £372. 18s. $1\frac{1}{4}d$. $\frac{4}{77}q$.

Ex. LVIII. (p. 226.)

1. 6, 28, 38.   2. $20.10; $64.32.
3. 516, 860, 1204, 1892; 717\frac{37}{80}$; 861\frac{7}{15}$, 820\frac{2}{33}$.
4. £179. 8s. 8d.; £142. 9s.; £99. 3s. 4d.
5. 12 cwt., 30$\frac{6}{11}$ lbs.; 3 cwt., 30$\frac{6}{11}$ lbs; 2 cwt., 50$\frac{10}{11}$ lbs.
6. $1900.80; $1555.20;   7. 2 cwt., 1 qr., 12 lbs., 13$\frac{1}{5}$ oz.
8. £3250; £2166. 13s. 4d.; £1083. 6s. 8d.   9. $1680. $2160.
10. $60, $60, $120, $240.
11. A's share = £5000, B's share = £3750; C's share = £3125.
12. $1350.   13. $5\frac{11}{13}$ months.   14. $4\frac{5}{8}$ months.
15. 12. months.   16. £3. 10s.   17. 1265.62\frac{1}{2}$.
18. A ought to have £80, B £90, and C £84.

Ex. LIX. (p. 235.)

1. $447·148.... 2. £1271. 13s. $9\frac{49}{100}d.$; $6104.11.
3. 1246 pias., $6\frac{36}{35}$ reals. 4. $4.87. 5. $4.67. 6. $5.02.
7. The direct way. 8. £19. 10s. $7\frac{1}{4}d.$; 25 francs. 9. £11. 5s. gain
10. $180·08$\frac{34}{74}$ gain; the income in England is taken at par.

Ex. LX. (p. 241.)

1. 17; 24; 38; 64. 2. 81; 145; 416. 3. 314; 193; 108.
    5. 999; 989; 908.   5. 5432; 3789; 2312.
6. 15367; 531441; 16807. 7. 543200; 2039750.
8. 12·96; 5·37; 240·1. 9. ·59049; 6·2573.
10. ·207; ·0374; ·0451. 11. 2·403; 2·403.
12. 347·6905; 490·304. 13. 4; 1·2649...; ·4; ·1264.....
14. 15·3492...; ·3162...; ·1; 2·2360...; ·7071......
15. ·02; ·0284...; 19·4901...... 16. $4\frac{1}{2}$; 12·4007...; ·5773...; $\frac{47}{99}$.

Ex. LXI. (p. 248.)

1. 12; 15; 31. 2. 38; 48; 67. 3. 88; 93; 98.
4. 134; 411; 203. 5. 631; 305; 364. 6. 258; 638; 975.
7. 3002; 6031. 8. ·73; 3·19; 45·7; ·097; ·124; ·029.
9. 1·442...; ·669...; ·310...... 10. $\frac{2}{3}$; $\frac{5}{7}$; 3·546......
11. $7\frac{2}{3}$; 1·930...; 1·442...... 12. ·046...; ·425.
13. 6·16; 1·232. 14. Each edge = 27·2 in. 15. 1369 sq. ft.
16. 3 ft., 10 in.

Ex. LXII. (251.)

1. 103544, 82321; 114111; 609, 5e8. 2. 5t594, 5e6t4; 10334.
3. 2615; 650410; 51117314; 2t3568; 574097; 2704054; 475t968; 29t96580; 14332216; 23033210.
4. 1456; e7t8; t4te; 10232; 3402. 5. 2504; 62te; 6543.
6. 5221; 1110111001111; 35, 61; 82t; 33233344; 3t4e2.
7. 10787; 418; 2t43; 16130335. 8. 9294; 344; 1465; 27t.
9. 475t968.

ANSWERS (pp. 255-265.)

Ex. LXIII. (p. 255.)

1. 11·41, 7·84; 101·332, 31·714; 4204·3514..., 664·7921... ·023843..., ·0/5; 132·14.

2. 102·501/..., 146·408373..., 173·35136...; 21·1̇4̇, 25·1̇; 27·1̇; 1259·57/13, 2218·4̇6̇1̇0̇6̇4̇2̇7̇8̇2̇, 3145·412172...; 312·5̇/381̇, 457·4̇7̇6̇5̇8̇4̇1̇2̇3̇0̇, 570·424365...

3. 22111·210111..., 3211·302323; 12012000·12211002..., 323010·22112; 102·120211...; 23·2112.

4. 572640; 26 583/6480.   5. 12 ft. 9 in.   6. £11. 16s.

Ex. LXIV. (p. 264.)

1. 4 yds. 2 ft. 8 in.; 3690 revolutions.
2. 19 sq. yds. 5 sq. ft. 113½ sq. in.
3. (1) 2 yds. 0 ft. 2 2/143 in.   (2) 124 096 ... yds.
4. (1) 733 yds. 1 ft.   (2) 27 sq. yds. 8 sq. ft. 61⅝ sq. in.
5. (1) $92.40   (2) 462 sq. yds.
6. (1) 36 ft. 6 in.   (2) 3 chs. 40 lks.
7. (1) 16·229 yds.   (2) 36 min.
8. (1) 52 ft. 7 in.   (2) 60·13... ft.   9. 156·52... mi.
10. 42 sq. yds. 4 sq. ft. 72 sq. in.   11. 322 yds. 2 ft.
12. (1) 52 sq. yds.   (2) 12 sq. yds. 4 sq. ft. 18 sq. in.
    (3) 3 sq. yds. 6 sq. ft. 108 sq. in.   (4) 2 sq. yds. 114 sq. in.
    (5) 3 sq. yds. 5 sq. ft. 12 sq. in.   (6) 15 sq. yds. 6 sq. ft. 112 sq. in.
13. (1) 85 sq. ft. 10".   (2) 76 sq. ft. 7'. 5".
    (3) 20 sq. ft. 9'. 10". 1'''. 10''''.   (4) 58 sq. ft. 8'. 8'''.
    (5) 70 sq. ft. 3'. 8". 10'''. 4''''.   (6) 37 sq. ft. 5". 4'''.
14. (1) 383⅓ c. ft.   (2) 35 c. ft.   (3) 285 c. ft. 559 cu. in.
15. (1) 6 yds.   (2) 6 yds. 9 in.   (3) 2 yds. 2 ft.   (4) 18 yds. 1 ft.
16. (1) 122 sq. yds. 2 sq. ft.   (2) 107 sq. yds. 3 sq. ft. 16 sq. in.
    (3) 131 sq. yds. 2 sq. ft 36 sq. in.   (4) 145 sq. yds. 18 sq. in.
    (5) 121 sq. yds.
17. (1) $650.   (2) $279.44 7/12.
18. (1) 26 yds. 4 in.   (2) 4 ft. 9¾ in.
19. (2) 2 ac. 4 po.   (3) 3 chs. 50 lks. off the length.
20. (1) 17 yds. 1 ft. 9⅜ in.   (2) 69½ sq. yds.
21. (1) £21.   (2) £13.   22. 250 yds.   23. $13.07½.

ANSWERS (pp. 265-267.)

24. £5 10s.   25. (1) 1968 lbs., 12 oz.   (2) 11 ft. $6\frac{85}{152}$ in.
26. (1) $365\frac{113}{125}$ times.   (2) 251·049... bus.   (3) 7·987.
27. (1) 300.   (2) 3155002 nearly.
28. (1) 2 ac., 1 ro., 39 po., 13 yds., 5 ft., 36 in.   (2) 1000 yds.
29. (1) £1046. 8s.   (2) 20.   30. (1) 18 sq. ft., 54 sq. in.
    (2) 3601 sq. ft. 72 sq. in.   31. (1) 512.   (2) 3 ft., $1\frac{1}{3}$ in.
32. $301\frac{1}{2}$ c. yds.; $165\frac{65}{339}$ lbs. (cwt. = 112 lbs.,
33. (1) $21\frac{8869}{27218}$.   (2) $487\frac{1}{2}$ nearly.   (3) 139.28...

In compliance with the request of several masters, the answers to the *EXAMINATION QUESTIONS* are not given. They appear in the Key.

www.ingramcontent.com/pod-product-compliance
Lightning Source LLC
Chambersburg PA
CBHW030014240426
43672CB00007B/941